Advances in Amorphous Semiconductors

Advances in Condensed Matter Science
Edited by D.D. Sarma, G. Kotliar and Y. Tokura

Volume 1
Strong Coulomb Correlations in Electronic Structure Calculations: Beyond the Local Density Approximation
Edited by Vladimir I. Anisimov

Volume 2
Colossal Magnetoresistive Oxides
Edited by Yoshinori Tokura

Volume 3
Spin Dependent Transport in Magnetic Nanostructures
Edited by Sadamichi Maekawa and Teruya Shinjo

Volume 4
Electronic Structure of Alloys, Surfaces and Clusters
Edited by Abhijit Mookerjee and D.D. Sarma

Volume 5
Advances in Amorphous Semiconductors
Jai Singh and Koichi Shimakawa

Advances in Amorphous Semiconductors

Jai Singh and Koichi Shimakawa

CRC PRESS

Boca Raton London New York Washington, D.C.

Library of Congress Cataloging-in-Publication Data

Catalog record is available from the Library of Congress

Direct all inquiries to CRC Press LLC, 2000 N.W. Corporate Blvd., Boca Raton, Florida 33431.

Visit the CRC Press Web site at www.crcpress.com

International Standard Book Number 0-415-28770-7
Printed in the United States of America 2 3 4 5 6 7 8 9 0
Printed on acid-free paper

Contents

Preface

The book presents some of the recent advances made in the field of amorphous semiconductors. We have tried to provide reasonably detailed derivations of most quantities described in the book to allow the reader to obtain a good grasp of the basic concepts in the field of amorphous solids. We believe that this approach would be particularly helpful to students who are beginners in the field of amorphous semiconductors.

The concept of effective mass (Chapter 3) and its applications in studying the properties of optical absorption (Chapter 4) and photoluminescence (Chapter 5) in amorphous solids represent recently developed topics. Structural properties are reviewed in Chapter 2. Chapter 6 presents the properties associated with the charge carrier–phonon interaction. Chapter 7 deals with defects, Chapter 8 with electronic transport, Chapter 9 with photoconductivity and Chapter 10 with photoinduced reversible effects in amorphous semiconductors. Amorphous semiconductors have been used in fabricating several optoelectronic devices in the last two decades. Therefore, the final chapter reviews some of the applications of amorphous semiconductors. Thus, the book covers a wide range of studies on hydrogenated amorphous silicon (a-Si : H), amorphous chalcogenides and some oxide glasses. It is expected that the book will be useful to all students, researchers and teachers in the field of amorphous solids.

We have benefited very much from consultations, help and discussions from many friends and colleagues. In particular, we would like to mention T. Aoki, D.P. Craig, E.A. Davis, S.R. Elliott, A. Ganjoo, T. Goto, H. Hosono, Noriaki Itoh, Safa Kasap, A.V. Kolobov, S. Kugler, C.H. Lai, A.R. Long, A.H. Matsui, K. Morigaki, I.-K. Oh, Ke. Tanaka and K.L. Tan. Finally, we gratefully acknowledge the enormous support that we received from our family members during the course of this task.

J. Singh, Darwin
K. Shimakawa, Gifu

1 Introduction

Amorphous materials have attracted much attention in the last two decades. The first reason for this is their potential industrial applications as suitable materials for fabricating devices, and the second reason is the lack of understanding of many properties of these materials, which are very different from those of crystalline materials. Some of their properties are different even from one sample to another of the same material. An ideal crystal is defined as an atomic arrangement that has infinite translational symmetry in all the three dimensions, whereas such a definite definition is not possible for an ideal amorphous solid (a-solid). Although an a-solid is usually defined as one that does not maintain long-range translational symmetry or has only short-range order, it does not have the same precision in its definition, because long- or short-range order is not precisely defined. A real crystal does not have infinitely long translational symmetry, because of its finite size, but that does not make it amorphous. A finite size means that a crystal has surface atoms that break the translational symmetry. In addition to surface atoms in amorphous materials, however, there are also present other structural disorders due to different bond lengths, bond angles and coordination numbers at individual atomic sites. Nevertheless, despite the vast differences, there are many properties of amorphous materials, which are found to be similar to those of crystalline solids (c-solids).

Amorphous semiconductors (a-semiconductors) and insulators are used for fabricating many opto-electronic devices. Amorphous silicon (a-Si) and its alloys are probably the most widely used a-semiconductors for fabricating thin film solar cells, thin film transistors and other opto-electronic devices. Amorphous chalcogenides (a-chalcogenides) are used in fabricating memory storage discs, etc. Crystalline semiconductors (c-semiconductors) are also used for fabricating these devices, but usually such devices are more efficient, stable and expensive. Although one of the amorphous forms of solid is glass, well known and well used by human beings for many centuries, the use of a-semiconductors for fabricating devices started only in the early part of the 1960s, well after the technological developments with crystalline materials. As a result, there is a general tendency to apply the theory developed for crystalline materials to understand many properties of a-solids and often that works at least qualitatively. One of the reasons for this is that the theory of crystalline structure is very well advanced; every textbook in solid state or condensed matter physics deals mainly with the theory of crystalline materials. On the other hand, the theory of amorphous systems is relatively difficult, because some of the techniques of simplification applicable for deriving analytical results in crystals cannot be applied to amorphous structures. One has to depend very heavily on numerical simulations using computers, which itself is a relatively new field.

There exists a kind of hierarchy in every field, but more so in physics. As the theory of crystalline materials was developed first, people try to understand the physics of a-semiconductors in terms of that of c-solids. As stated above, it is commonly established that a-solids have mainly three types of structural disorders, which do not exist in c-solids. These are: (1) different bond lengths; (2) different bond angles; and (3) under- and over-coordinated sites, although varying bond lengths and bond angles are not usually regarded as defects in a-solids. However, this does not mean that every individual atom is randomly distributed in an amorphous material. For example, in silicon (Si), as each Si atom has four valence electrons to contribute for the covalent bonding, whether it is amorphous or crystalline silicon, these electrons per Si atom must be covalently bonded and shared with the neighboring Si atoms. Thus each Si atom forms four covalent bonds with its neighbors, so the coordination number in crystalline Si is 4 at every site. In a-Si also all Si atoms are bonded covalently but it is not necessary that all atomic sites have the same coordination number 4. Some are under coordinated, which means that one or more covalent electrons on a Si atom cannot form covalent bonds with the neighboring atoms, as shown in Fig. 1.1. These uncoordinated bonds are called dangling bonds. The density of dangling bonds in a-Si is very high, which reduces the photoconductivity of the material, and also prevents it from doping. Therefore, a-Si cannot be used for the fabrication of devices. In addition to dangling bonds, a-Si network also has many weak and strained bonds, which are usually longer than a fully coordinated Si–Si bond. A fully coordinated Si–Si bond has a bond length of 2.5 Å, but a strained or weak bond can have a bond length between 2.5 and 3 Å. A dangling bond is regarded as longer than 3 Å (Fedders *et al.*, 1992).

In order to reduce the density of dangling bonds in a-Si, the technique used these days is to hydrogenate it to produce hydrogenated a-Si, denoted usually by a-Si : H. The hydrogenation of a-Si saturates many of the dangling bonds and makes it a semiconductor more suitable for fabricating devices. Even after hydrogenation the typical dangling bond density in a-S : H is $\leq 10^{16}\,\text{cm}^{-3}$. Another effect of hydrogenation of a-Si is that it softens the a-Si network.

The presence of strained and weak bonds gives rise in a-Si : H to what are called band tail states, which are also found in other a-semiconductors and insulators (Street, 1991). A c-semiconductor has quite well-defined valence and conduction band edges, and hence a very well defined electron energy gap between the top of the valence band and bottom of the conduction band, as shown in Fig.1.2(a). In a-semiconductors however, the neutral dangling bond states (D^0) lie in the middle of the energy band gap, and bonding and anti-bonding orbital of weak bonds lie above the valence band and below the conduction band edges respectively, as

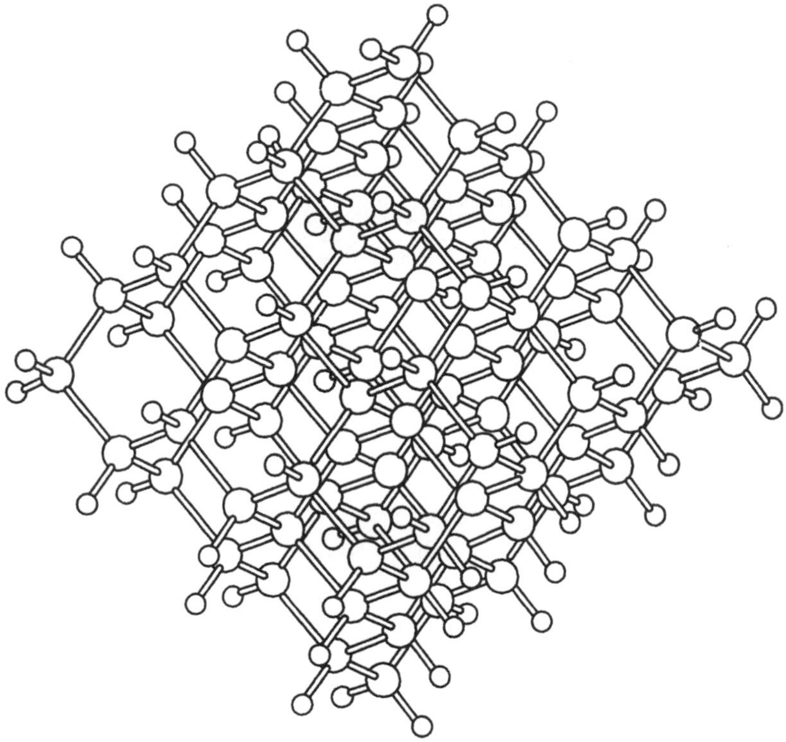

Figure 1.1 Structure of atomic network of a-Si showing dangling bonds.

shown in Fig.1.2(b). In addition to the neutral dangling bond states, there may also exist charged dangling bond states. If the charge carrier–phonon interaction is very strong in an a-solid, due to the negative U effect, then the positive charged dangling bond states (D^+) would lie above the neutral dangling bond state, but below the conduction band edge. The negative charged dangling bond states (D^-) would lie below the neutral dangling bond state, but above the valence band edge. In the case of weak charge carrier–phonon interaction, the positions of D^+ and D^- get reversed on the energy scale. These energy states found within the energy gap in a-solids are localized states and any charge carrier created in these states will be localized on some weak, strained or dangling bonds. As removal of these localized states from the band gap means the removal of amorphousness from an amorphous solid, these states are inevitable in amorphous materials. The interesting point is that, as the band tail states lie above the valence and below the conduction band edges, these states are usually the highest occupied and lowest empty energy states in any amorphous semiconductor. Therefore, band tail states play the dominant role in most optical and electronic properties of a-semiconductors, particularly in the low temperature region.

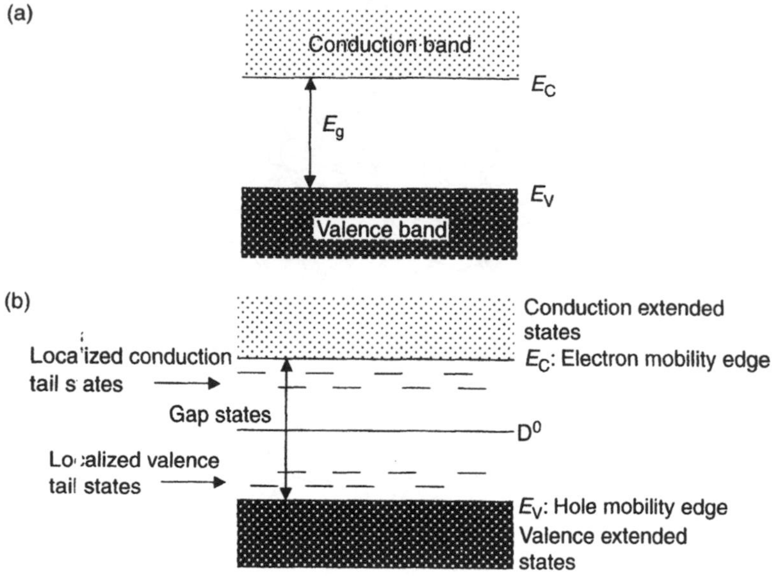

Figure 1.2 (a) Schematic illustration of energy bands in c-semiconductors. (Note the forbidden energy gap contains no free electron energy states.) (b) Energy states in a-solids.

In view of the above description of electronic energy states of a-semiconductors/insulators, it is obvious that these consist of delocalized states like valence and conduction bands, commonly known as the extended states and the localized states like band tail and dangling bond states. The extended states arise due to short-range order, and tail states due to disorder. It is therefore important to present here a concise review of the theoretical aspects of both the extended and tail states.

The theory of extended states in a-semiconductors is similar to that of the valence and conduction bands in c-semiconductors. In a-semiconductors, these are formed through the covalent bonding of atoms. Only the top filled electronic states of individual atoms take part in the bonding. Let us first describe very briefly the nature of covalent bonding, and then extend that concept to develop the theory of extended states.

1.1 Covalent bonding

In covalent bonding, the interaction between the nearest neighbor atoms plays the most important role (Ibach and Lüth, 1991). The essential features of covalent bonding can therefore be obtained from the basic quantum theory of chemical bonding between two atoms, a diatomic molecule, with a single bonding electron as shown in Fig. 1.3(a). Two homonuclear atoms located at A and B at a distance

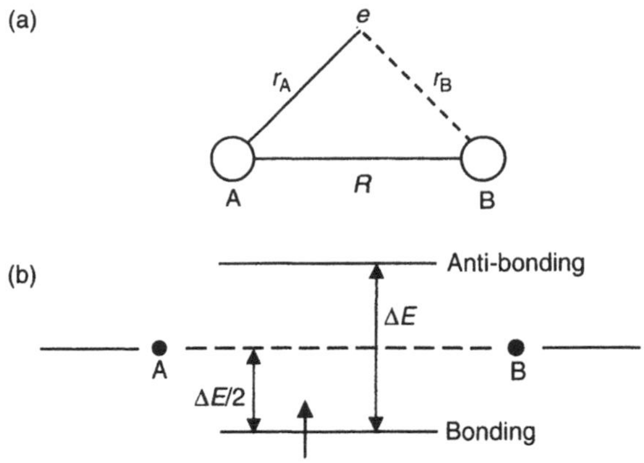

Figure 1.3 (a) Structure of a diatomic molecule with a single covalently bonding electron between two atoms A and B. (b) The bonding and anti-bonding orbitals of a diatomic molecule.

R apart, and their electronic wavefunctions, representing one electron in the top filled state and with all core states filled, are denoted by ϕ_A and ϕ_B, respectively. The Hamiltonian of the diatomic molecule can be written as

$$H = -\frac{\hbar^2}{2m_e}\nabla^2 - \kappa\left[\frac{Ze^2}{r_A} - \frac{Ze^2}{r_B} + \frac{Z^2e^2}{R}\right], \tag{1.1}$$

where $\kappa = 1/4\pi\varepsilon_0 = 9.0 \times 10^9\,\mathrm{m\,F^{-1}}$, $\varepsilon_0 = 8.8542 \times 10^{-12}\,\mathrm{F\,m^{-1}}$ is the permittivity of vacuum, Z is the atomic number of the atoms and m_e is the electronic mass. Defining a molecular orbital ψ as the linear combination of atomic wavefunctions, we can write

$$\psi = c_A\phi_A + c_B\phi_B. \tag{1.2}$$

The Schrödinger equation:

$$H\psi = E\psi, \tag{1.3}$$

can then be solved to calculate E as

$$E = \frac{\int \psi^* H\psi \, d^3\mathbf{r}}{\int \psi^*\psi \, d^3\mathbf{r}}. \tag{1.4}$$

We thus obtain E as

$$E = \frac{c_A^2 H_{AA} + c_B^2 H_{BB} + 2c_A c_B H_{AB}}{c_A^2 + c_B^2 + 2c_A c_B S}, \tag{1.5}$$

where

$$H_{AA} = H_{BB} = \int \phi_A H \phi_A \, d^3\mathbf{r} = \int \phi_B H \phi_B \, d^3\mathbf{r}, \tag{1.6}$$

$$H_{AB} = H_{BA} = \int \phi_A^* H \phi_B \, d^3\mathbf{r} = \int \phi_B^* H \phi_A \, d^3\mathbf{r}, \tag{1.7}$$

and the overlap integral

$$S = \int \phi_A^* \phi_B \, d^3\mathbf{r}. \tag{1.8}$$

Minimizing E with respect to c_A and c_B we get two solutions, E_+ and E_-, for the energy eigenvalues of the molecular orbital as

$$E_\pm = \frac{H_{AA} \pm H_{AB}}{1 \pm S}, \tag{1.9}$$

separated by an energy gap, $\Delta E = E_- - E_+$.

The result derived in Eq. (1.9) is very well known in quantum chemistry. It shows that the degenerate energy state of two isolated atoms splits into two energy levels when the two atoms are brought together to form a bond. The lower energy state is called the bonding orbital or bonding state and its energy is lower than the energy of the state of isolated individual atoms, and the upper energy state is called the anti-bonding orbital or anti-bonding state. The single electron occupies the bonding orbital after the bonding, as shown in Fig. 1.3(b). As an electronic state can have up to two electrons, if both atoms contribute one electron each for covalent bonding, both the electrons will occupy the bonding state. However, it is important to note the following two points. (1) It is not possible to identify which one of the electrons, occupying the bonding orbital, comes from which atom. The bonding orbital is an energy state of the bonded diatomic molecule not of individual atoms. (2) Both the interaction energy, H_{AB}, and overlap integral S depend on the inter-atomic distance R. If $R \to \infty$, both H_{AB} and S are zero. This means that the energy separation between bonding and anti-bonding states reduces as the bond length or inter-atomic distance R increases.

Another important aspect of chemical bonding must be noted here. For a diatomic molecule the probability amplitude coefficients c_A and c_B are obtained from the fact that the electron has an equal probability of occupying each of the two atoms. This gives, $c_A = \pm c_B = 1/\sqrt{2}$, which implies that after the diatomic molecule is formed, the electron spends equal time on each of the two atoms.

The above theory can be extended to determine the electronic energy states in solids as well, which consist typically of a very large number ($\approx 10^{22}\,\mathrm{cm}^{-3}$) of closely packed atoms. In this case the gap between the bonding and anti-bonding orbitals gets filled by very closely packed, $\approx 10^{22}\,\mathrm{cm}^{-3}$ energy states; one from every atom in the solid. Thus, each atomic energy state of an isolated atom gets grouped into an electronic energy band in solids as described below.

1.2 Electronic energy bands

For the time being let us not be concerned with whether we are dealing with an a- or a c-solid. We consider a solid consisting of N atoms whose top occupied shells consist of s and p atomic states, for example, carbon with $2s^2 2p^2$, silicon with $3s^2 3p^2$, and germanium with $4s^2 4p^2$. In these cases when the atoms are brought close together, the overlap of electronic wavefunctions first causes the s and p bands to overlap and become one band with a capacity of $2N(s) + 6N(p) = 8N$ occupying electrons. The band is not fully occupied as each atom contributes only 4 electrons to a total of $4N$ electrons, and therefore a solid thus formed should be a good conductor, which these solids are not. However, in this situation the atoms in these solids are not as closely packed as they are in tetrahedrally bonded systems. They move still closer due to sp^3 hybridization of the top occupied orbitals of individual atoms with covalent bonds directed towards the vertices of a tetrahedron for optimal directional covalent bonding, as shown in Fig. 1.4. Thus, the coordination number in these solids becomes 4 and each atom can bond with 4 other atoms in a fully coordinated structure. Then the single overlapping energy band gets divided into two separate bands, each with the capacity of $4N$ occupying electrons (Krane, 1996). The lower band gets completely filled with the available 4 electrons per atom from the sp^3 hybridization in these solids and the upper energy band remains completely empty. The two bands get separated by an energy gap, which depends on the inter-atomic separations as shown in Fig. 1.5. Depending on the size of the energy gap a solid thus formed will behave either like an insulator, as carbon (diamond) is, or a semiconductor, as Si and Ge are. The inter-atomic separation in Si and Ge is larger than that in carbon at equilibrium, which results in a smaller energy gap in these solids than that in diamond. In this regard, the energy gap in these solids has the same characteristic as the energy gap between the bonding and anti-bonding orbitals ΔE obtained from Eq. (1.9).

The above description of the formation of energy bands is the same for both c- and a-solids, and therefore the existence of energy bands is not a consequence of the translational symmetry. The electronic wavefunction of a solid thus formed

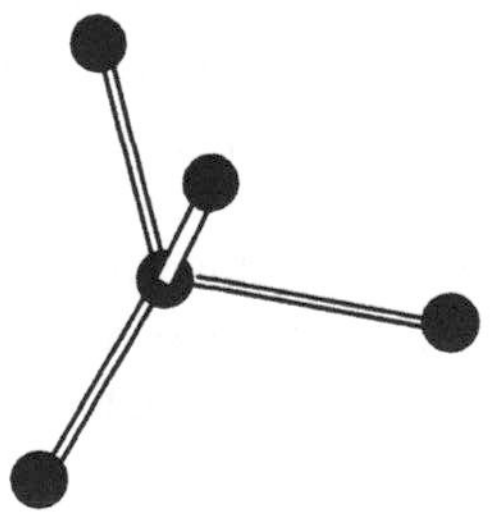

Figure 1.4 Tetrahedral optimal directional covalent bonding in sp^3-hybridized systems.

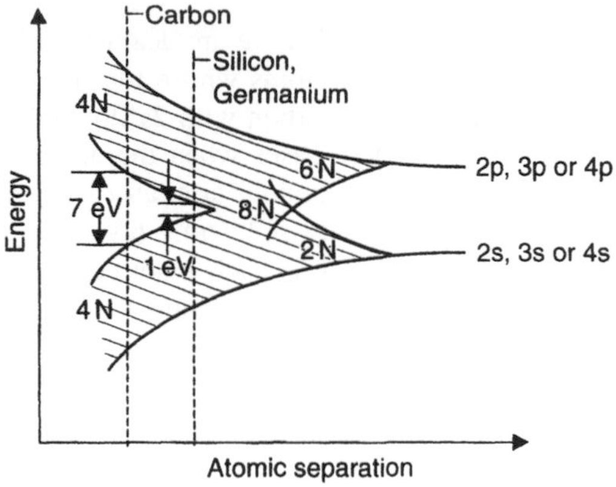

Figure 1.5 Band structure for carbon, silicon and germanium. The combined ns + np band splits into two bands due to sp^3 hybridization (tetrahedral bonding). Atomic separation in carbon corresponds to a very large band gap so it is an insulator; Ge and Si are semiconductors because their band gaps are small.

can be written according to Eq. (1.2) as

$$\psi = \sum_n C_n \phi_i(\mathbf{r} - \mathbf{r}_n), \tag{1.10}$$

where $\phi_i(\mathbf{r} - \mathbf{r}_n)$ is the electronic wavefunction of the ith state of an atom located at $\mathbf{r}_n$ and C_n is the coefficient of linear expansion. In principle, the wavefunction in Eq. (1.10) is applicable to both c- and a-solids. However, because of the translational symmetry in c-solids, the coefficient of expansion C_n has an analytical expression given by

$$C_n = N^{-1/2} \exp(i\mathbf{k} \cdot \mathbf{r}_n), \tag{1.11}$$

where $\mathbf{k}$ is the reciprocal lattice vector, which is well defined and regarded as a good quantum number only in c-solids. In a-solids, the absence of long-range translational symmetry does not allow the coefficient of linear expansion to be written as in Eq. (1.11). It may be noted that Eq. (1.11) also gives equal probability for a charge carrier to be localized on every atomic site, as in the case of the diatomic molecule described above. Thus even if one considers systems with short range order, a charge carrier in a covalent bond is going to be delocalized between nearest neighbors, that is, it will spend equal time on each of the neighboring bonded atoms.

The interesting point is that using Eq (1.11) in (1.10), we get the electronic wavefunction for a c-solid in the reciprocal lattice vector $\mathbf{k}$-space in the form of Bloch functions, and then the electronic energy bands are also obtained as a function of **k**. The top filled energy band is called the valence band and the lowest empty band is the conduction band, separated by an energy gap, which also thus depends on **k**. This gives rise to two types of c-solids. The one in which the maximum of the valence band and minimum of the conduction band is found to be at the same **k**, is called a direct band gap solid, and the other with no such occurrence is an indirect band gap solid. For semiconductors we usually use the terms direct and indirect semiconductors. Any transition from the valence to conduction band must obey the momentum $\hbar\mathbf{k}$ conservation in a c-solid, usually referred to as **k**-conservation.

In a-solids, on the other hand, we cannot have such a distinction between semiconductors on the basis of **k**. There is no **k** associated with the energy states in a-solids.

Thus, the interaction with so many other atoms in solids is the origin of formation of the quasi-continuous states, whether we call them bands or not. The energy gap arises because of the discrete energy levels of atoms, which are the constituents of every solid. However, due to the dependence of energy bands and energy gap on the reciprocal lattice vector in c-solids being known for so long, it was initially very puzzling when the occurrence of an energy gap was found in a-solids that have no well-defined reciprocal lattice vector **k**. This remained so until Weaire (1971) showed, using a tight-binding approach similar to chemical bonding and considering only the nearest neighbor interaction, that there exists a discontinuity in the density of states in tetrahedrally bonded solids, separating the valence and conduction bands. Nevertheless, as these quasi-continuous states in a-solids do not have any **k** dependence, it is a good reason to distinguish them from the crystalline energy bands. For this reason, in this book, the terms valence and conduction extended states will be used for a-solids instead of valence and conduction bands, which are used for c-solids. The valence extended states in a-semiconductors are the conducting states for holes, and conduction extended states are the conducting states for electrons, analogous to valence and conduction bands, respectively, in c-semiconductors. In any solid, as individual atoms have a fixed number of bonding electrons, in the amorphous form it cannot be expected to be very different in terms of inter-atomic distances from those in the crystalline form. Hence it is only natural to have similar electronic states in both. However, the ideal translational symmetry does not exist in the former. The width of extended states in a-semiconductors depends dominantly on the nearest neighbor interaction energy, which is nearly the same as in c-solids. For these reasons the optical and electronic properties of a-semiconductors originating from their extended states can be expected to be very closely related to those of c-semiconductors originating from their energy bands.

1.3 Band tail, mobility edge and dangling bond states

In addition to the fully coordinated covalent bonds, as stated above, a-semiconductors also have many weak, strained and even uncoordinated (dangling) bonds. The bonding and anti-bonding states of weak and strained bonds lie close to the valence and conduction extended state edges, respectively, because these bonds have relatively larger bond lengths than those giving rise to the extended states. This is obvious from Eq. (1.9), which shows that the energy gap between bonding and anti-bonding states reduces if the distance between the bonding atoms increases. It is also consistent with Fig. 1.5, which shows that for larger separation between atoms the energy band gap, which is similar to the separation between bonding and anti-bonding states, will be smaller. Thus, the presence of the weak and strained bond states give rise to what are commonly known as band tail states or tail states within the energy gap and near the extended state edges in a-semiconductors. These states are localized states.

It is important to realize that in a-solids, in principle, all states are localized, including the extended states, because their eigenfunctions cannot be written using the probability amplitude coefficient given in Eq. (1.11). However, the interstate energy separation is so negligible that the extended states form a quasi-continuum, as stated above. In that sense the extended states can be regarded as delocalized states, because transitions from one to another are easy, although may be not as easy as in crystalline solids. This is another reason why a-semiconductors, like a-Si and a-Ge, usually do not exhibit as good carrier mobility as their crystalline forms. On the other hand, as the concentration of weak or strained bonds is usually relatively lower, the resulting tail states may not form quasi-continuum, remaining localized.

The edge separating the conduction extended states and tail states is called the mobility edge. As the tail states are localized energy states, no conduction is expected to occur when excited electrons occupy these states. Therefore, at 0 K only conduction can occur when excited electrons are in the extended states above the conduction tail states, and that defines the mobility edge, that is, the energy edge above which the electronic conduction can occur at 0 K (Mott and Davis, 1979). Likewise, one can define a similar edge separating the valence extended states from valence tail states. Thus there are two mobility edges, electron mobility edge at the bottom of the conduction extended states and hole mobility edge at the top of valence extended states.

Dangling bonds contribute to non-bonded states, and therefore they can be regarded as equivalent to the states of isolated atoms, which lie in the middle of bonding and anti-bonding energy states as is obvious from Figs 1.3(b) and (1.5). Accordingly, in a-solids of atoms with ns^2np^2 outer electrons, the dangling bond states lie in the middle of the energy gap between the edges of the valence and conduction extended states. The typical density of states of a-semiconductors is shown in Fig. 1.6, which clearly indicates that the energy gap in a-semiconductors is not as well defined as that in c-semiconductors due to the presence of tail states. The dangling bond states are also localized energy states.

The influence of the presence of dangling bonds on the electronic and optical properties of a-semiconductors can be very significant, depending on their numbers, because these bonds do not facilitate electronic conduction except through hopping or tunneling. For example, pure a-Si inevitably has very high density of dangling bonds, which prevents it from doping and exhibiting photoconductivity (Street, 1991). Therefore, a-Si cannot not be used as a semiconductor

Figure 1.6 Schematic illustration of density of states in an a-semiconductor.

for fabricating devices. These dangling bonds have to be reduced in number or pacified, as it is commonly called, by hydrogenation during the deposition, before the material can exhibit characteristics suitable for device fabrication (Lewis *et al.*, 1974). Following the work of Chittick and coworkers (Chittick *et al.*, 1969; Chittick and Sterling, 1985) and later that of Spear and LeComber (1975), the a-Si thus produced is the a-Si : H, which is used for fabricating devices.

There are three kinds of possible dangling bonds: neutral, positively charged and negatively charged, denoted usually by D^0, D^+ and D^-, respectively. The energy state of D^0 dangling bond lies in the middle of the energy gap and that of D^- and D^+ depends on whether a material has negative U or positive U. In materials with negative U, resulting from the strong charge carrier–phonon interaction, the energy states of D^+ and D^- lie above and below D^0 states, respectively. This is reversed in materials with positive U or weak carrier–phonon interaction. The negative and positive U are discussed further in Chapter 6. Free or excited charge carriers (e: electrons and h: holes) can be captured by these dangling bonds according to the following processes:

$$\mathrm{e} + \mathrm{D}^0 \rightarrow \mathrm{D}^-,$$

$$\mathrm{e} + \mathrm{D}^+ \rightarrow \mathrm{D}^0,$$

$$\mathrm{h} + \mathrm{D}^0 \rightarrow \mathrm{D}^+,$$

and

$$\mathrm{h} + \mathrm{D}^- \rightarrow \mathrm{D}^0.$$

Accordingly, as both electrons and holes can be captured by dangling bonds, these bonds act as trapping and recombination centers for charge carriers, and reduce the photoconductivity of materials they are present in. The above processes also suggest that for any material containing D^0, D^+ and D^-, when exposed to photons of energy equal or greater than the energy gap, the number of these pre-existing dangling bonds will be altered. The excited e–h pairs will be captured, changing one type of dangling bonds into another. However, as the number of photo-excited electrons and holes is the same and both have equal probability of being captured, the number of total pre-existing dangling bonds may not alter due to photo-excitations. This is the current status of understanding of dangling bonds in a-semiconductors.

1.4 Preparation

Amorphous materials are prepared through non-equilibrium processes. The physical properties of these materials therefore greatly depend on how samples are prepared. The preparation methods used at present can be classified, in principle, into the following two categories; (1) quenching from the liquid state (melt) and (2) condensation from the gas phase. There are of course other methods, for example, irradiation by ionizing particles (ion bombardment), sol–gel process, etc. Ion bombardment of c-solids produces structural damage and amorphizes the material, in particular surface layers, and is of great technological importance in the current crystalline Si industry. Through implanting the dopant atoms (e.g. P or B) the electronic properties are controlled. Amorphization with implantation is removed easily by laser annealing. The technique of the sol–gel method has great technological advantages in silica glass technology (glassy SiO_2-related materials) (see, e.g. Mukherjee, 1980).

We will predominantly discuss covalently bonded materials. Some of these fall into the category of tetrahedrally bonded materials (e.g. such as Si and Ge, etc.) and others into that of amorphous chalcogenides (e.g. Se, $As_2Se(S)_3$, $GeSe(S)_2$, etc.). These materials are not prepared by ion implantation or sol–gel technique; they are usually prepared using techniques (1) and (2) and hence we will briefly describe these below.

1. *Quenching from liquid*: Some materials do not crystallize below their melting temperature and they become a supercooled liquid. A schematic illustration in Fig. 1.7 shows the volume versus temperature curve in glassy (amorphous), crystalline and liquid states. With decreasing temperature from liquid states, the materials undergo the so-called *glass transition*, which occurs at a certain temperature, T_g, below which they become glasses. Most chalcogenide and oxide glasses can be prepared by the melt quenching (MQ) method. It is of interest to note that $T_g \approx 2T_m/3$ is empirically obtained for most glasses, where T_m is the melting temperature (Mott and Davis, 1979). The nature of the glass transition is poorly understood while there are numerous studies (see, e.g. Elliott, 1990) devoted to this topic. Factors that determine the glass-transition temperature are still not known, while the correlation between T_g and the average coordination number Z has been found to obey $\ln T_g \approx 1.6Z + 2.3$ (Tanaka, 1985).

Amorphous materials produced by the MQ technique are usually called *glasses*. The term, glasses, may have some historical reasons, since so-called glasses are obtained from the MQ technique. It is not clear why certain materials readily form glasses on cooling a melt (glass forming ability depending on composition). This is one of the fundamental questions, together with the glass transition, that still remains unanswered in glass science. As will be discussed in Chapter 2 (structure), the constraint theory developed by Phillips (1979) may help understanding some

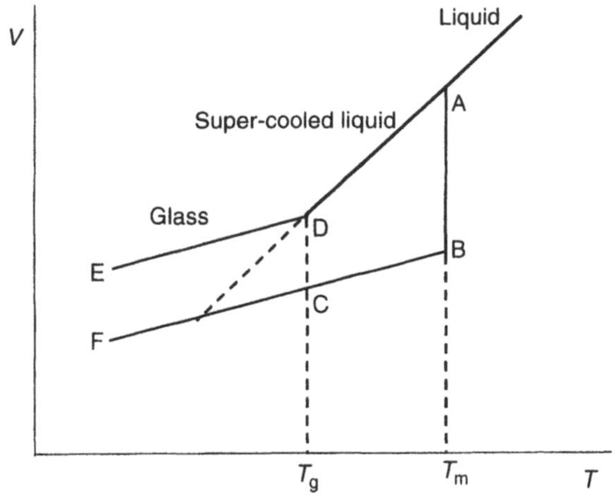

Figure 1.7 Schematic illustration of the volume versus temperature curve. T_g and T_m denote the glass transition and melting temperatures, respectively.

of these mysteries. Note that tetrahedrally bonded a-Si, a-Ge, etc. are not prepared by the MQ method.

2. *Condensation from the gas phase*: Amorphous materials prepared by this technique are not called "glasses." Amorphous Si, Ge and C are only prepared from the gas phase and therefore these materials are in the form of films. a-chalcogenides are also prepared by this technique. The types of condensation from gas phase can be classified into physical vapor deposition (PVD) and chemical vapor deposition (CVD). Traditionally a well-known method for PVD is *vapor deposition* in which ingots (or powders) of materials are melted in a chamber in vacuum around 10^{-6} torr. The evaporated gas is condensed on a substrate and the thickness is usually controlled by the deposition time. For materials that have higher melting temperatures (e.g. Si and C), techniques of electron-beam or arc vaporization can be employed for melting. The other PVD technique is *sputtering deposition* in which ions (e.g. Ar) accelerated by DC or AC (rf frequency) high voltage in a chamber in vacuum $\sim 10^{-6}$ torr supply kinetic energy to atoms on the surface of a target. Atoms dissociated from the target are deposited on the substrate in the form of films. When a magnetic field is applied perpendicular to the electrodes, the deposition is enhanced, because the ionized gas gets confined in a spiral motion by the magnetic field. This method is called *magnetron sputtering*. Reactive gases such as H_2 and/or N_2 are also used for sputtering, and the technique is known as *reactive sputtering*, in which such gases are incorporated into the depositing films (a-Si : H, a-$Si_{1-x}N_x$, a-$Si_{1-x}N_x$: H, a-C : H, a-$C_{1-x}N_x$, a-$C_{1-x}N_x$: H, etc.).

The CVD technique uses gases such as silane (SiH_4), germane (GeH_4), etc., which are dissociated due to energies supplied either by thermal (thermal CVD),

Table 1.1 Optical band gap E_o, activation energy for electrical conductivity ΔE, dielectric constant ε_∞, room-temperature dc conductivity σ_{RT}, and glass-transition temperature T_g are listed for some representative a-semiconductors

Materials	E_o (eV)	ΔE (eV)	ε_∞	σ_{RT} (S cm^{-1})	T_g (K)
a-Si : H	1.80	0.70	12	$\sim 10^{-9}$	
a-Ge : H	1.15	0.54	16	$\sim 10^{-3}$	
a-Se	2.05	1.10	6.3	$\sim 10^{-16}$	320
a-As_2Se_3	1.76	0.91	9.9	$\sim 10^{-12}$	470
a-As_2S_3	2.40	1.14	7.8	$\sim 10^{-17}$	480
a-$GeSe_2$	2.20	0.95	7.2	$\sim 10^{-11}$	700
a-GeS_2	3.07	0.72	7.0	$\sim 10^{-14}$	770

plasma (plasma CVD) or optical (photo CVD). These techniques are nowadays considered to be modern semiconductor technologies and are widely used in a-Si industries (see, e.g. Hirose, 1984; Street, 1991). Using plasma CVD and mixing phosphene (PH_3) or diborane (B_2H_6) into silane makes possible p- or n-type doping in a-Si (Spear and LeComber, 1975), which is a well known story in the development of a-semiconductor technology. A recent method, which principally is classified into thermal CVD, called the catalytic CVD or hot-wire-assisted CVD, has received much attention, since it produces high-quality a-Si : H (Matsumura, 1986, 1998; Crandall *et al.*, 1998).

Diamond-like amorphous carbon (DLAC) or tetrahedrally bonded amorphous carbon (ta-C) is prepared by a filtered cathodic vacuum arc (FCVA), in which n-type doping is possible by the addition of nitrogen or phosphorus. This material may also have technological importance in preparing devices such as field emission devices (Milne, 1996, 1997).

Finally, some important physical parameters of representative a-semiconductors are listed in Table 1.1.

1.5 Electronic excitations in a-semiconductors

First of all, let us review the case of c-semiconductors for a comparison. In c-semiconductors, an electronic excitation can occur from the top of the valence band to the bottom of the conduction band creating a hole in the valence band and an electron in the conduction band. Such an excitation is caused by a photon of energy $\hbar\omega = E_g$, E_g being the minimum energy gap. There is no absorption for $\hbar\omega < E_g$ in c-semiconductors, and for $\hbar\omega > E_g$ higher energy states are excited and the excited charge carriers relax rapidly to their band edges. Then they either recombine radiatively to give photoluminescence or form an exciton state due to their Coulomb interaction and recombine radiatively, later producing excitonic photoluminescence. If we measure energy from the top of the valence band, the exciton states lie below the conduction band edge by an energy equivalent to the binding energy of excitons (Knox, 1965; Singh, 1994). The purpose of describing these well known results here is to illustrate the point that the valence and conduction band edges, being the lowest energy states for exciting holes and electrons optically, play very significant roles in the electronic excitations of any intrinsic c-semiconductors.

The lowest energy states for exciting electron–hole pairs in a-semiconductors are not the valence and conduction extended state edges (mobility edges). This is due to the presence of tail and dangling bond states within the energy gap, as described in Section 1.3. The lowest possible energy states become the tail states, and therefore the photoluminescence of a-semiconductors has to be dominantly influenced by the tail states. Thus, all those optical and electronic properties that originate from the lowest energy states may be expected to be different in a-semiconductors in comparison with those in c-semiconductors.

One of the typical characteristics of c-semiconductors is to exhibit excitonic photoluminescence in the emission spectra, particularly at low temperatures as shown in Fig. 1.8. A detailed description of excitonic states will be presented in Chapter 5, however, here we would like to address some general issues for comparison of excitonic states in c- and a-semiconductors. Any photo-excited pair of e and h is subjected to their attractive Coulomb interaction, regardless of whether they are far apart or close to each other. The Coulomb interaction between the excited e and h in an excited pair enables them to form a bound state similar to that of a hydrogen atom, which is a bound state between an electron and a proton. The bound state thus formed between a pair of excited e and h, is called an *exciton*, and the excitonic energy states lie below the edge of the conduction band by an energy that is the binding energy of an exciton. It is important to realize that

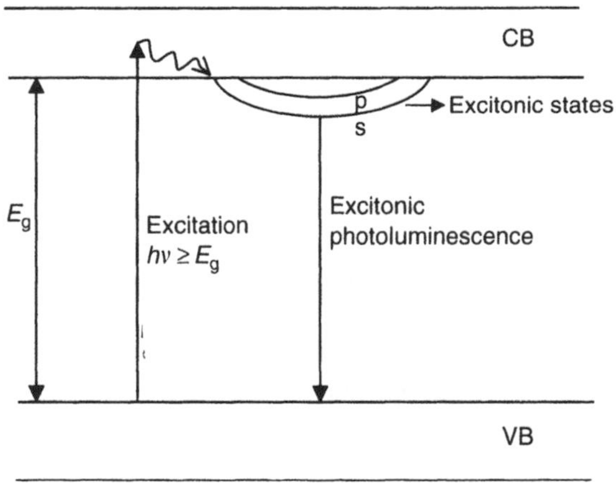

Figure 1.8 Schematic illustration of excitonic photoluminescence in c-solids at low temperature.

although the exciton state lies within the energy gap and near the conduction band edge, it is a delocalized state. An exciton has a translational momentum $\hbar\mathbf{k}$, with which its center of mass moves throughout the crystal. Therefore, in this view, the theory of excitons can be regarded as applicable only to c-semiconductors and insulators. In inorganic c-semiconductors, like Si and Ge, it is considered that the excited e and h are delocalized and far apart from each other. Therefore, the name of large radii orbital excitons is used for Wannier or Wannier–Mott excitons in inorganic materials (Elliot, 1962; Knox, 1965; Singh, 1994). The concept of large radii orbital has another significance. In developing the theory of Wannier–Mott excitons, it is noted that the interaction between the excited e and h consists of two parts: (1) Coulomb interaction and (2) exchange interaction. The exchange interaction is assumed to be of a short range (Elliot, 1962; Singh, 1994), and therefore its contribution is neglected in calculating the binding energy of Wannier–Mott excitons. However, it is only the exchange interaction that is spin dependent, therefore, its omission leads to a binding energy independent of the spin. Consequently, one theoretically obtains the same binding energy for both singlet and triplet Wannier–Mott excitons (Singh, 1994).

As the amorphous materials lack translational symmetry, the formal theory of Wannier–Mott excitons cannot be applied to describe such excited states in these materials. An effective mass approach, as described in Chapter 3, can be applied to develop the theory of excitonic states in a-semiconductors. Nevertheless, the following points need to be noted:

1 Although the concept of large radii orbital excitons is well established for crystalline inorganic semiconductors due to large overlap of the interatomic electronic wavefunctions, it may not be strictly applicable in a-semiconductors. The overlap of electronic wavefunctions will still be significant, but the lack of translational symmetry will act against creating a large separation between an excited pair of e and h.
2 If free e and h are excited in the extended states initially, they will tend to move closer due to their Coulomb interaction as they relax down to their edges and then to tail states. In this case there will be a continuous loss of energy from the excited pair, but any peaked photoluminescence structure cannot be expected. For photoluminescence a quantum transition must occur, and therefore it is necessary for the excited pair to form an exciton-like quantized energy state.
3 The excited e and h in a-semiconductors can form an exciton-like bound state between them in one of the following four possibilities: The excited electron and hole can form an excitonic state when (a) electron is in the conduction extended state and hole is in the valence extended state, (b) electron is in the conduction extended state and hole is in the valence tail state, (c) electron is in the conduction tail states and hole is in the valence extended state, and (d) electron is in the conduction and hole is in the valence tail states. The influence of these four possibilities on the photoluminescence from a-semiconductors will be presented in detail in Chapter 3.

1.6 Carrier transport

It should be quite clear now that the transport of charge carriers in a-semiconductors becomes quite complicated due to the presence of tail states. It was suggested by Mott and Davis (1979) that delocalized extended and localized tail states cannot coexist in a-solids. Extending the idea to transport of charge carriers, it may be considered that a charge carrier created in the extended states can move freely, but in the tail states its motion will be restricted. This can be adopted in a quantum mechanical model by considering the height of the electron mobility edge in a-solids as a potential barrier that exists between different sites. If the energy of an electron is higher than the barrier height its motion remains wavelike, because the nature of the wavefunction does not change in the barrier region. However, if the electron energy is less than that of the barrier height, the electron wavefunction decays exponentially in the barrier region, and the electron has to tunnel through the barrier to move to the other side. A similar model can also be applied for the motion of holes in the valence states. The theory associated with the model can be developed for the transport of excited charge carriers in a-solids at low temperatures as follows.

At very low temperatures, it can be expected that any optical excitation will excite an electron from states near the hole mobility edge at the top of the valence extended states. Thus the hole will be created near its mobility edge, but the state of the electron will depend on the energy of the exciting photon. Let us first focus on an excited electron in the conduction states; similar results can then also be derived for holes. Consider only the nearest neighbor interaction for an excited electron–hole pair and assume that the electron mobility edge is at an energy E_c above the hole mobility edge. Assume that the electron is excited with energy E, and at the same time it is interacting with the excited hole. Before the electron can move to another site it has to cross a barrier of height E_c, as shown in Fig. 1.9. Figure 1.9 shows three regions. In region I, the electron is only interacting with the excited hole. In region II, it is under the influence of both Coulomb potential and the barrier potential E_c, and in region III, it has moved to an energy state of another site but it is still under the influence of the Coulomb potential due to the

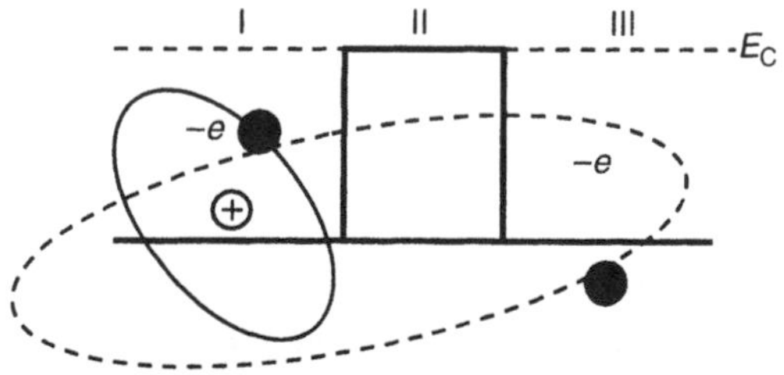

Figure 1.9 Schematic illustration of excitation of an e–h pair under the influence of Coulomb interaction in a-solids. The excited electron is in the conduction tail states.

same hole. The Schrödinger equation describing the motion of an excited electron in regions I and III can be written as

$$-\frac{\hbar^2\nabla^2\psi(\mathbf{r})}{2\mu_e} - k\frac{e^2}{\varepsilon r}\psi(\mathbf{r}) = E\psi(\mathbf{r}), \tag{1.12}$$

where μ_e is the reduced effective mass of electron and hole, and $\psi(\mathbf{r})$ is the wavefunction of electron. ε is the static dielectric constant of the material. The repulsive interaction with other excited electrons is obviously neglected here for simplification. The Schrödinger equation (1.12) is the same as that of the relative motion between electron and hole in a exciton or electron in an hydrogenic state. Its energy eigenvalues will be discussed in detail in Chapter 5.

In region II (Fig. 1.9), the Schrödinger equation can be written as

$$-\frac{\hbar^2\nabla^2\psi(\mathbf{r})}{2\mu_e} + \left(E_c - k\frac{e^2}{\varepsilon r}\right)\psi(\mathbf{r}) = E\psi(\mathbf{r}), \tag{1.13}$$

where E_c is the energy of the electron mobility edge measured from the hole mobility edge, see Fig. 1.9. Equation (1.13) represents the motion of an electron subjected to a Coulomb potential due to the excited hole, and a constant potential of height E_c acting as a barrier that the electron must overcome before it can move to the neighboring atom. The Coulomb potential is reduced by the dielectric constant, and as most semiconductors have a dielectric constant of more than 10, the effect of Coulomb interaction on the motion of the electron can be neglected in comparison with E_c. This reduces Eq. (1.13) to

$$\nabla^2\psi(\mathbf{r}) + \frac{2\mu_e\cdot(E - E_c)}{\hbar^2}\psi(\mathbf{r}) = 0. \tag{1.14}$$

The solution of Eq. (1.14) is well known (e.g. Schiff, 1968). Defining $k_1^2 = 2\mu_e(E - E_c)/\hbar^2$, and $k_2^2 = 2\mu_e(E_c - E)/\hbar^2$, we get two different solutions. One for $E < E_c$ that the excited electron is in the tail states and the other for $E > E_c$ that the electron is in the extended states, as given below

$$\psi(\mathbf{r}) = Ae^{i\mathbf{k}_1\cdot\mathbf{r}} \quad \text{for } E > E_c \tag{1.15}$$

and

$$\psi(\mathbf{r}) = Be^{-\mathbf{k}_2\cdot\mathbf{r}} \quad \text{for } E < E_c. \tag{1.16}$$

In deriving Eqs (1.15) and (1.16) it is considered that the electron does not get reflected back from the barrier E_c. If reflections are considered then we will get another term in both the Eqs (1.15) and (1.16) obtained by replacing $\mathbf{k}_1$ and $\mathbf{k}_2$ by their negative vectors. Nevertheless, it is obvious from Eq. (1.15) that if the energy of the excited electron is above the mobility edge ($E > E_c$), the electron will behave like a free wave. It is important to realize that $\mathbf{k}_1$ and $\mathbf{k}_2$ are not any reciprocal lattice vectors, although depending on E_c, they can have three components (k_x, k_y, k_z) in the three-dimensional systems. For $E < E_c$, the electronic eigenfunction decreases exponentially within the barrier. This is the case when an excited electron moves from one tail state to another. The process is called non-radiative tunneling. The probability p_t, of tunneling is obtained as (e.g. Eisberg and Resnick, 1974)

$$p_t \approx \exp(-2k_2 R). \tag{1.17}$$

The rate of non-radiative tunneling, R_t, will depend on the rate of excitation, G as

$$R_t = Gp_t = G\exp(-2k_2 R). \tag{1.18}$$

It is to be noted that the non-radiative tunneling rate does not involve the assistance of phonons. The transfer is from one localized state to another having the same energy. A transition from a localized state of lower energy to higher energy has to be phonon assisted. The rate of such phonon assisted transition, R_{tp}, is given by (Mott and Davis, 1979)

$$R_{tp} = \omega_0 \exp(-R/R_0)\exp(-\Delta E/kT), \tag{1.19}$$

where $\omega_0 = 10^{13}\,\mathrm{s}^{-1}$ is approximately a phonon frequency. $R_0 = [\hbar^2/m_e^* (E_c - E)]^{1/2}$, with $E = E_1$ or E_2, and $\Delta E = E_1 - E_2$ is the difference between the initial state energy E_1 and final state energy E_2 such that $E_2 > E_1$. For $E_1 > E_2$, the rate of transition is given by

$$R_{tp} = \omega_0 \exp(-R/R_0), \tag{1.20}$$

which is similar to that in Eq. (1.18) obtained from the quantum tunneling through a barrier. In fact, for $m_e^* = m_h^*$, both R_t and R_{tp} have the same exponential dependence on the inter-defect separation R. More on carrier transport will be described in Chapter 9.

The effective mass of an electron or hole is not very well understood in a-solids. In physical quantities derived for a-solids, usually the free electron mass is used, but is not correct. Calculation of the effective mass of charge carriers in a-solids is described in Chapter 3.

References

Chittick, R.C., Alexander, J.H. and Stirling, H.F. (1969). *J. Electrochem. Soc.* **116**, 77.

Chittick, R.C. and Stirling, H.F. (1985). In: Adler, D. and Schwartz, B.B. (eds), *Tetrahedrally Bonded Amorphous Semiconductors.* Plenum, New York.

Crandall, R.S., Liu, X. and Iwaniczko, E. (1998). *J. Non-Cryst. Solids* **227–230**, 23.

Eisberg, R. and Resnick, R (1974). *Quantum Physics of Atoms, Molecules, Solids, Nuclei and Particles.* Wiley, New York.

Elliot, R.J. (1962). In: Kuper, K.G. and Whitfield, G.D. (eds), *Polarons and Excitons.* Oliver and Boyd, Edinburgh and London, p. 269.

Elliott, S.R. (1990). *Physics of Amorphous Materials*, 2nd edn. Longman, London.

Fedders, P.A., Fu, Y. and Drabold, D.A. (1992). *Phys. Rev. Lett.* **68**, 1888.

Hirose, M. (1984). In: Pankove, J.I. (ed.), *Semiconductors and Semimetals*, Part A, Vol. 21. Academic Press, New York, p. 109.

Ibach, H. and Lüth, H. (1990). *Solid State Physics*. Springer-Verlag, Heidelberg.

Knox, R.S. (1965). In: Seitz, F., Turnbull, D. and Ehrenreich, H. (eds), *Solid State Physics*. Academic Press, New York.

Krane, K. (1996). *Modern Physics*. Wiley, New York.

Lewis, A.J., Connell, G.A.N., Paul, W., Pawlik, J. and Tenkin, R. (1974). *AIP Conf. Proc.* **20**, 27.

Matsumura, H. (1986). *Jpn. J. Appl. Phys.* **25**, L949.

Matsumura, H. (1998). *Jpn. J. Appl. Phys.* **37**, 3175.

Milne, W.I. (1996). *J. Non-Cryst. Solids* **198–200**, 605.

Milne, W.I. (1997). In: Marshall, J.M., Kirov, N., Vavrek, A. and Maud, J.M. (eds), *Future Directions in Thin Film Science and Technology*. World Scientific, Singapore, p. 160.

Mott, N.F. and Davis, E.A. (1979). *Electronic Processes in Non-Crystalline Materials*. Oxford University Press, Oxford.

Mukherjee, S.P. (1980). *J. Non-Cryst. Solids* **42**, 477.

Phillips, J.C. (1979). *J. Non-Cryst. Solids* **34**, 153.

Schiff, L.I. (1968). *Quantum Mechanics*, 3rd edn. McGraw-Hill, Singapore.

Singh, J. (1994). *Excitation Energy Transfer Processes in Condensed Matter*. Plenum, New York

Spear, W.E. and LeComber, P.G. (1975). *Solid State Commun.* **17**, 1193.

Street, R.A. (1991). *Hydrogenated Amorphous Silicon*. Cambridge University Press, Cambridge.

Tanaka, Ke. (1985) *Solid State Commun.* **54**, 867.

Weaire, D. (1971). *Phys. Rev. Lett.* **26**, 1541.

2 Structure

A fundamental understanding of the properties of condensed matter, whether electronic, optical, chemical, or mechanical, requires a detailed knowledge of its microscopic structure (atomic arrangement). The structure of a crystalline solid (c-solid) is determined by studying its structure within the unit cell. The structure of the crystal as a whole is then determined by stacking unit cells. Such a procedure is impossible for determining the structure of amorphous solids (a-solids). Due to the lack of long range periodicity in a-solids, unlike c-solids, determination of structure is very difficult. There is no technique to provide atomic resolution in a-solids comparable with that in crystals.

The first well-known work on structure of amorphous semiconductors (a-semiconductors) were the electron diffraction studies on amorphous carbons (a-C) (Kakinoki *et al.*, 1960). They proposed a microcrystalline model in which graphitic (sp^2) and diamond-like (sp^3) domains coexist. Practically, since the end of 1960, a-semiconductors have become popular materials. Different diffraction techniques, using electrons, X-rays or neutrons, can be useful to obtain structural information. Among these the neutron diffraction measurement can be the best technique, although a large volume of sample is required (see, e.g. Elliott, 1990).

The diffraction measurements give the structure factor $S(\mathbf{Q})$ with scattering vector $\mathbf{Q}$. The Fourier transform of $S(\mathbf{Q})$ produces the radial distribution function (RDF). The RDF studies show that the structure of many a-solids is non-random and there is a considerable degree of local ordering (short-range order; SRO) despite the lack of long-range order (LRO). There are comprehensive reviews on structural studies of disordered solids (Ziman, 1979; Zallen, 1983; Elliott, 1990) and hence here we will review these only briefly in typical a-semiconductors.

The RDF, $J(r)$, is defined as the number of atoms lying at distances between r and $r + \mathrm{d}r$ and is given by

$$J(r)\,\mathrm{d}r = 4\pi r^2 \rho(r)\,\mathrm{d}r, \tag{2.1}$$

where the density function $\rho(r)$ is an atomic pair correlation function. As shown in Fig. 2.1, $\rho(r)$ exhibits an oscillatory behavior, because peaks of the probability function represent average interatomic separations. The RDF hence oscillates about the average density parabola given by the curve $4\pi r^2 \rho_0$. As shown in Fig. 2.2,

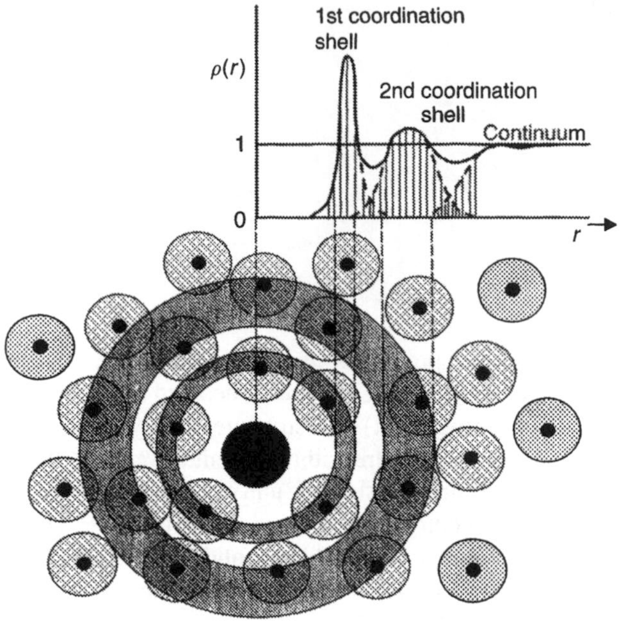

Figure 2.1 Schematic illustration of the structural origin of certain features in the radial distribution function (RDF). The density function $\rho(r)$ plotted as a function of r for amorphous solid exhibits oscillatory behavior and the shaded area under a given peak gives the effective coordination number.

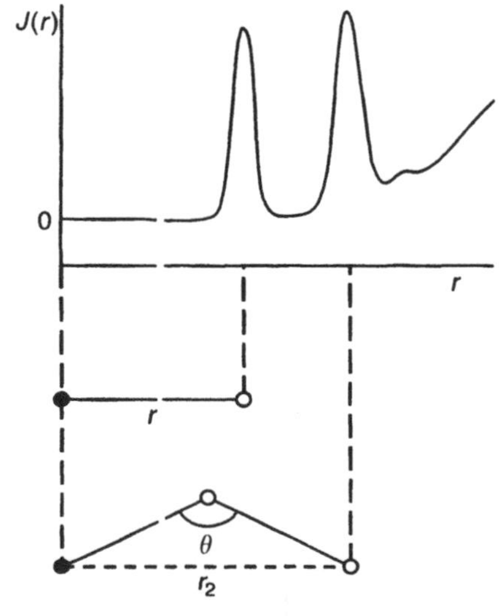

Figure 2.2 The relationship between short-range structural parameters: first and second nearest-neighbor lengths, r_1 and r_2, and bond angle θ for first and second shell, corresponding to the first two peaks in RDF.

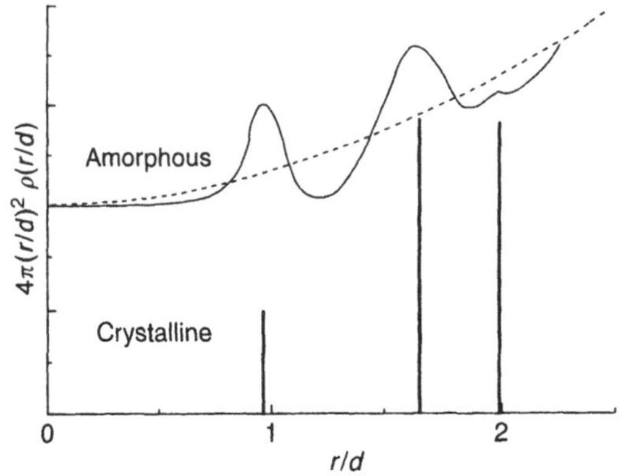

Figure 2.3 RDF of a- and c-Si, both of which take sp^3 configuration, plotted as functions of r/d, where d is the bond length. The first and second coordination shells in a-Si are almost identical in spacing as those in the perfect diamond lattice (c-Si). The third peak that appears in c-Si disappears in a-Si.

the position of the first peak in the RDF produces the average nearest-neighbor bond length r_1 and the position of second peak gives the next-nearest-neighbor distance r_2. The knowledge of r_1 and r_2 yields the bond angle θ as

$$\theta = 2\sin^{-1}\frac{r_2}{2r_1}.$$

The area under a peak gives the coordination number of the structure. The second peak is generally wider than the first for covalent a-solids, which can be attributed to a static variation in the bond angles θ. If no bond-angle variation exists, then the width of the first two peaks should be equal.

Figure 2.3 shows, for example, the RDF for crystalline and amorphous Si, both in sp^3 configuration. The first- and second-coordination shells are almost identical in spacing and numbers with those in the perfect diamond lattice. However, as the rotation of tetrahedra about their common bond, shown in Fig. 2.4, changes the distance to third neighbors, the third peak disappears in the RDF when one approaches the amorphous structure from crystalline structure of Si (Fig. 2.3); from the staggered configuration to an eclipsed configuration. Note that the dihedral angle ϕ is 60° and 0° for the staggered and the eclipsed configurations, respectively.

Note that the RDF is a one-dimensional (1D) representation of a three-dimensional (3D) structure, and hence carries only a limited amount of structural information. This is why modeling studies are required. In the next section, we will briefly review the theory (modeling studies) applied in typical a-semiconductors.

2.1 Theory (modeling)

In the last two decades, structural study has focused on developing realistic model structures by computer simulation. There are two principal ways for constructing disordered structures: one is Monte Carlo (MC) type simulation and other is

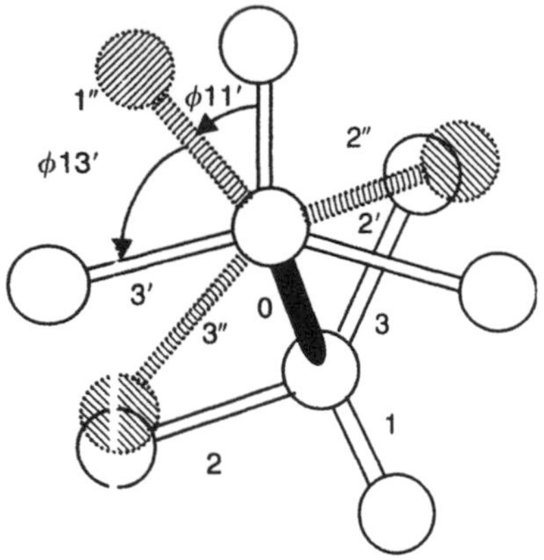

Figure 2.4 Rotation of bond tetrahedra from the staggered configuration, 1″, 2″ and 3″, towards an eclipsed configuration, resulting in disappearance of the third peak.

Molecular Dynamics (MD) simulation. In the following a brief review on MC and MD will be given.

2.1.1 Monte Carlo simulation

Principally, MC simulation looks for a network with local minimum in energy on the hypersurface by changing bond lengths and bond angles. We should know the potential energy of interatomic interaction to perform traditional MC simulation. To do this an empirical potential where the free parameters are fitted to the experimental data or a potential is fitted to quantum mechanical calculations (e.g. tight-binding, Hartree–Fock, density functional, or *ab initio* type, etc.).

The Keating potential V (Keating, 1966) as described below has been used in MC for constructing a-Ge and a-Si model structures (Wooten *et al.*, 1985). This is known as the best computer model for constructing continuous random network (CRN) and is often called the WWW algorithm. Starting from an f.c.c. crystalline lattice, bonds are switched randomly, hence breaking and reforming between different pairs of atoms occur repeatedly. After a sufficient number of such operations and relaxations (Monte-Carlo procedures), an amorphous structure is produced. The Keating potential provides a good empirical description of the bonding forces with only two measurable parameters and has been found to be quite adequate for model building (Wooten and Weaire, 1994). It is given by

$$V = \frac{3}{16}\frac{\alpha}{d^2}\sum_{l\{ij\}}\left(r_{li}\cdot r_{lj} - d^2\right)^2 + \frac{3}{8}\frac{\beta}{d^2}\sum_{l\{i,i'\}}\left(r_{li}\cdot r_{li'} + \frac{1}{3}d^2\right)^2, \qquad (2.2)$$

where α and β are the bond-stretching and bond-bending force constants, respectively, and d is the strain-free equilibrium bond length in the crystal, which is 2.35 Å for c-Si, for example. The first sum is over l denoting the atomic sites and their four neighbors designated by i and j, in the second term the sum is over distinct pairs of neighbors i and i', and r_{li} represents the vector position of the ith neighbor from atom l.

The correlation function $\rho(r)$ obtained by the WWW model agrees remarkably well with experiment for tetrahedrally bonded a-Ge and a-Si (Wooten *et al.*, 1985; Kugler *et al.*, 1989), which will be discussed in Section 2.2. The WWW method has been successively applied to the modeling of diamond-like amorphous carbon (ta-C) (Kugler and Naray-Szabo, 1991), SiO_2 glass and hydrogenated amorphous silicon (a-Si : H) (Mousseau and Lewis, 1990; Holender *et al.*, 1993). The WWW method for a-Si has also been used for calculations of the density of electronic states (DOS) (Hickey and Morgan, 1986; Bose *et al.*, 1988), thermal conductivity (Feldman *et al.*, 1993), optical properties (Weaire *et al.*, 1993), substitutional doping (Kadas *et al.*, 1998) and charge fluctuations (Kugler *et al.*, 1988). Using the WWW algorithm (Drabold *et al.*, 1994), the electronic structure has also been calculated for tetrahedral carbon (ta-C).

2.1.2 Reverse Monte Carlo simulation

A new technique, so-called Reverse Monte Carlo (RMC) simulation has been developed (McGreevy and Pusztai, 1988). In RMC, the analysis of the experimental RDF and modeling of construction are performed at the same time. The basic algorithm is described as follows: (1) Start with an initial set of Cartesian coordinates, calculate its $S(\mathbf{Q})$ or RDF and find the discrepancy from the experimental $S(\mathbf{Q})$ or RDF. (2) Generate a new configuration by random motion of a particle (MC step), calculate the corresponding $S(\mathbf{Q})$ or RDF and again find the discrepancy from the experimental $S(\mathbf{Q})$ or RDF. (3) If the new discrepancy is smaller than the previous one, the new configuration becomes the starting configuration, that is, the move is accepted, otherwise it is accepted with a probability according to the MC strategy. (4) Steps (1)–(3) are then repeated until the discrepancy in the calculated and experimental $S(\mathbf{Q})$ or RDF converges to the zero value. To speed up the simulation and achieve more realistic configurations, some conditional constraints are usually applied; for example, coordination number, the lowest limit of bond length, bond angles, etc.

RMC has several advantages over other modeling methods, including the system size, which is much larger than is possible with other methods. No interatomic potential is needed to perform RMC. There are no specific needs for high speed computers. This new technique has been applied to a-Si (Kugler *et al.*, 1993a), a-Ge and a-C (Gereven and Pusztai, 1994), which will be discussed in Section 2.2.

2.1.3 Molecular Dynamics simulation

The MD method begins with the Newton equations of motion of N atoms. To solve Newton equations, the Verlet algorithm (Verlet, 1967) is usually used with time steps of the order of a femtosecond. The new position $r(t + \Delta t)$ of a given particle is written as

$$r(t+\Delta t) = 2r(t) - r(t-\Delta t) + r''(t)(\Delta t)^2 + O[(\Delta t)^4], \qquad (2.3)$$

where $r''(t)$ is the second derivative with respect to t.

The motion of a particle is determined by repeating this procedure. It may, however, be worth noting that MD is based on classical mechanics and hence it is applicable only under the condition that the atomic deBroglie wavelength, $\lambda = h/\sqrt{2ME}$, where E is the kinetic energy and M is the mass of atom, is sufficiently smaller than the interatomic separation.

Similar to the MC simulations, MD also needs local potentials to calculate $r''(t)$. For covalently bonded systems, the empirical potentials like Morse or Lenard–Jones (pair potentials) is not applicable, hence a three-body potential is taken into consideration. Some of the examples of empirical potentials used in the calculations are Keating potential as described in the previous section, force field potential developed by Warshel and Lifson (1970), Tersoff (1988, 1989), Oligschleger *et al.* (1996); Vink *et al.* (2000), etc.

Although simulations based on empirical potentials have achieved remarkable successes, an extension of the simulation method to quantum mechanics can be important. The density-functional theory (DFT) may be used for a proper choice of potential in a quantum mechanical way. However, DFT consumes computing time. The first-principles MD simulation (or *ab initio* MD simulation), overcoming the difficulty of large computing time, treats both atomic and electronic states of the system self-consistently, which has been developed by combining MD and DFT (Car and Parrinello, 1985). The interatomic forces are directly derived from the instantaneous electronic ground state. The application of these for a-Si : H will be discussed in Section 2.2.

To understand the complex dynamics of mechanisms, such as growth, larger systems are required, and tight-binding MD (TBMD), which will be discussed in the following section may be useful.

2.2 Current understanding of structures

Let us first discuss tetrahedrally bonded (sp^3) a-semiconductors. As already stated in Section 2.1.1, the correlation function $\rho(r)$ obtained by the WWW model agrees remarkably well with the experimentally obtained $\rho(r)$ for a-Ge and a-Si. Figure 2.5 shows a comparison between the correlation function for a-Ge obtained experimentally and that obtained from the original 216-atom model of a-Si, scaled to a-Ge (Wooten *et al.*, 1985). A computer-generated picture clearly shows that the sp^3 configuration piles up in a random manner in both a-Ge and a-Si when the rms angular deviation of 10.9° and rms bond-length deviation of 2.7% are introduced in the crystalline values. The extension of the above MC method to a-Si : H shows that the theoretical correlation functions thus obtained produce bond lengths of Si–Si and Si–H bonds in good agreement with those obtained experimentally (Mousseau and Lewis, 1989).

The *ab initio* MD calculation provides structural and electronic configurations in a-Si : H (Buda *et al.*, 1991; Yonezawa *et al.*, 1991), in which the structure around

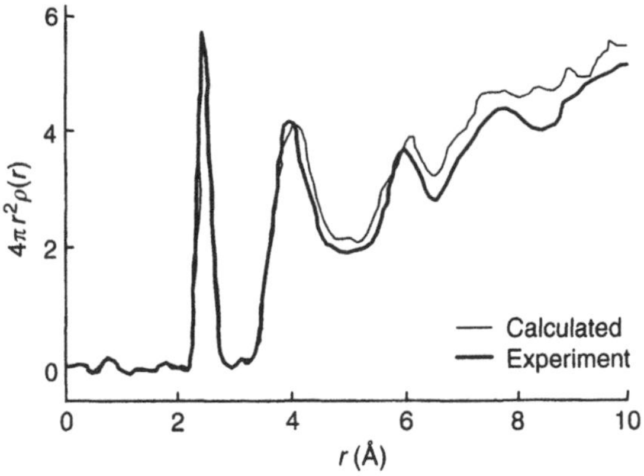

Figure 2.5 Comparison of correlation functions for a-Ge (experimental) and the original 216 atom model of a-Si (scaled to a-Ge) (Wooten *et al.*, 1985).

Si–H bonds and the hydrogen motion have been elucidated applying this method. Buda *et al.* (1991) concluded that monohydride complexes are prevalent and there exists a strong tendency for hydrogen atoms to form small clusters. The average structural, dynamical and electronic properties are calculated by this method and a detailed analysis of structural relaxation processes from liquid to amorphous Si has been also given in a-Si (Stich *et al.*, 1991). Their *ab initio* results also suggest that the main coordination defects are five-fold coordinated (floating bonds) or weak bonds, while the possibility of existing usual dangling bonds due to three-fold coordinated atoms is not excluded. The analysis of the structural and dynamical (phonon DOS) properties of a-Si has been presented by applying the MD with the Tersoff empirical potential (Tersoff, 1988, 1989). This has produced excellent agreement between theoretical and experimental results (Ishimaru *et al.*, 1997). The a-Si structure generated by the Tersoff potential contains defects consisting of five-fold coordinated atoms, whose concentration decreases with the cooling rate from liquid state.

As described above the *ab initio* MD simulations using the local density approximation have been successful in dealing with the melted Si and amorphous structures. However, due to the requirement of huge computing time, the *ab initio* MD has only been performed on small systems (~100 atoms) and short time scales (~10 ps). Both larger systems and longer time scales are required to apply it to more practical cases such as growth and epitaxy. Applying tight-binding MD, instead of the first-principles, may improve upon the computing difficulty and it can handle up to a few hundred atoms. In accuracy it is also known to be comparable with the *ab initio* calculations (Xu *et al.*, 1992; Wang *et al.*, 1993; Kwon *et al.*, 1994; Kugler *et al.*, 1999).

The MD models described above are applied to amorphous states quenched from liquid states. Practically, however, most a-semiconductors are obtained from the vapor phase and hence the tight-binding MD has been applied to a-Si produced from the vapor phase (Kohary and Kugler, 2000, 2001). Interestingly, a significant number of three membered rings (i.e. triangles) and near planar squared rings exist in the simulated network. This is consistent with the neutron diffraction measurements for a-Si analyzed by the RMC technique (Kugler *et al.*, 1993a) and systematic analysis of the Cambridge Structural Database (Kugler and Varallyay, 2001).

Yang and Singh (1998) have applied a TBMD model for simulating Si–H bonds in a-Si : H. They have generated a-Si : H structure models with different hydrogen concentrations. The calculated RDF agree very well with the experimental results. They have also calculated the total average binding energy per silicon atom as a function of the hydrogen concentration in their a-Si : H models, as shown in Fig. 2.6, and found it to be minimum when the hydrogen concentration is within the range 8–14%, implying that the a-Si : H models are most stable in this hydrogen concentration range.

Although a-C is not the main objective of this monograph, in comparison with other elements, carbon is unique in its ability to form strong chemical bonds with varying coordination numbers, for example, two (linear chain), three (graphite) and four (diamond). Therefore, there are numerous structural studies done on a-C (see, e.g. Robertson, 1986), some of which will be discussed here briefly. The most important parameter in a-C is the ratio of the number of atoms with sp^2 coordination to that with sp^3 coordination, which dominates its electronic and optical properties. The neutron diffraction measurements confirm that a vapor-deposited a-C is constructed mostly from sp^2 configuration (Kugler *et al.*, 1993b), while a diamond-like carbon (ta-C) is constructed mostly from sp^3 (Gilkes *et al.*, 1995). The structures of a-C are also constructed by the RMC technique (Gereben and

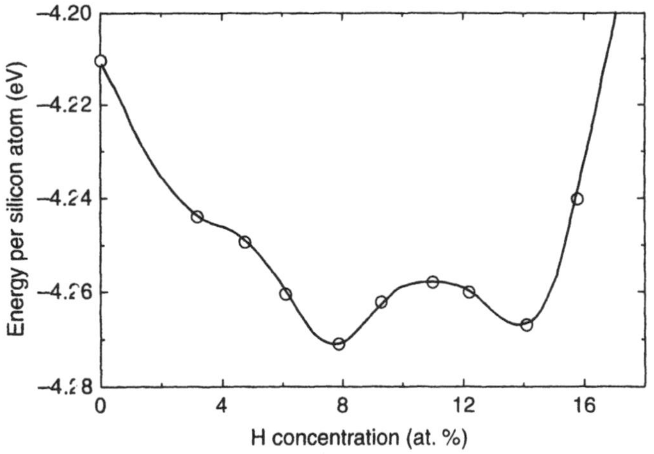

Figure 2.6 The average total binding energy per Si atom, E_{S_i}, in eV calculated as a function of the hydrogen concentration. The region of minimum energy indicates the range of H concentration in which a-Si : H can be expected to be most stable.

Pusztai, 1994). The study of electronic and structural properties of both a-C and ta-C have been very much advanced with the *ab initio* MD (Galli *et al.*, 1989; Marks *et al.*, 1996) and tight-binding MD techniques (Xu *et al.*, 1992; Wang *et al.*, 1993; Kwon *et al.*, 1994). Such studies can explain the structural, dynamic and electronic properties of a-C.

Note again that in these MD simulations, solid materials are quenched from their melts and not from their vapor phase. However, for more realistic simulation one should start from the vapor phase, for example, starting with 120 atoms in *vapor* phase the structure of a-C has been simulated applying the tight-binding MD (Kohary and Kugler, 2000, 2001). The RDF thus calculated agrees well with the experimentally obtained RDF. It has also been confirmed that more sp^3 configuration is achieved when the kinetic energy of carbon atoms is increased.

Finally, we must discuss the structure of amorphous chalcogenides (a-Ch). Typical a-Ch are Se, $As_2Se(S)_3$ and $GeSe(S)_2$, whose structures cannot be as simple as tetrahedrally bonded a-Ge and a-Si, because these systems contain two-fold (Se or S), three-fold (As), and four-fold (Ge) coordinated materials. In fact, crystalline Se (c-Se) takes the hexagonal structure being composed of chains of two-fold coordinated Se, while crystalline As_2S_3 (c-As_2S_3) takes the layered structure whose basic configuration is AsS_3 pyramidal unit. Schematic structures of c-Se and c-As_2S_3 are shown in Figs 2.7 and 2.8, respectively. $GeSe(S)_2$ in the crystalline form, on the other hand, takes corner- and edge-sharing of $GeSe(S)_4$ tetrahedra, in which both types of connections are found in varying proportions. A proportion of edge-sharing connection between neighboring tetrahedra leads to the layered structure as shown in Fig. 2.9. As will be discussed later, these crystalline structures are retained locally in amorphous states. It is known that SiO_2 in both glassy and crystalline forms is characterized by corner-sharing connections, leading to a 3D-CRN in which only covalent bonding is dominant (Elliott, 1990).

As far as the SRO in a-Ch is concerned, there is no controversial argument; that is, the SRO is characterized by the unit of $AsSe(S)_3$ pyramidal or $GeSe(S)_4$ tetrahedra. Note that the SRO is characterized by SiO_4 tetrahedra. We may expect that the structure of a-$GeSe(S)_2$ is the same as that of a-SiO_2. It is known, however, that the structure over medium-range order (MRO) of a-$GeSe(S)_2$ is different from that of a-SiO_2 (Elliott, 1990). This suggests that the type of connection between polyhedra can differ in each case.

The most prominent feature in the diffraction study of a-Ch is the presence of the "pre-peak" in the structure factor $S(\mathbf{Q})$ (see, e.g. Elliott, 1990), which is now called the first sharp diffraction peak (FSDP). The pre-peak in a-Ch was first discovered by Vaipolin and Porai-Koshits (1963) in $As_2S(Se)_3$ glasses. It is observed in a wide variety of compounds of a-Ch. The pre-peak occurs at $\mathbf{Q} \approx 1\,\text{Å}^{-1}$ with the half width $\Delta\mathbf{Q}$ around $0.2\,\text{Å}^{-1}$, implying the existence of ordered structures with a periodic distance of $\approx 5\,\text{Å}$ and a correlation length in the range 20–30 Å. The Fourier transformation of $S(\mathbf{Q})$, both including and omitting this peak, produces an indistinguishable real-space correlation function. This indicates that the pre-peak does not contain structural information about the SRO. The existence of a pre-peak is therefore indicative of the existence of MRO.

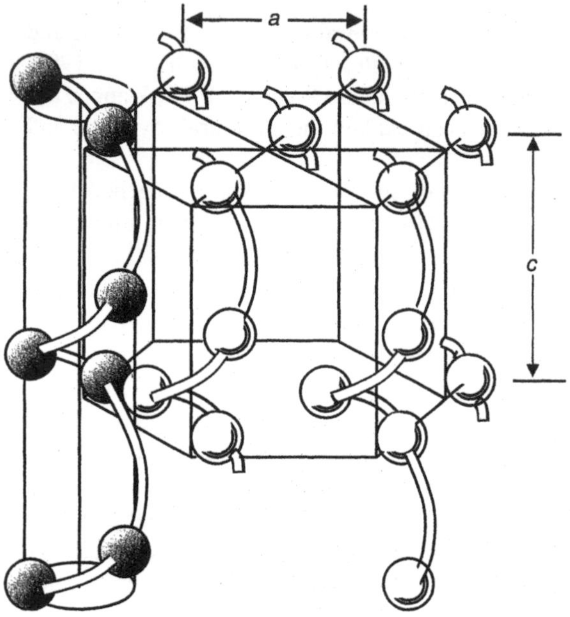

Figure 2.7 Structure of crystalline Se, with hexagonal lattice structure and composed of helical chains of two-fold coordinated Se. *a* and *c* denote the size of unit cell.

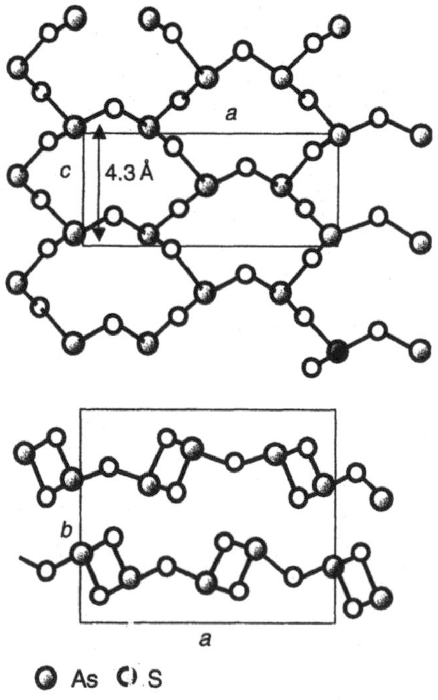

Figure 2.8 The crystalline structure of As_2S_3. The upper panel shows the sheet (layered) structure in *c*-*a* plane (along *b* axis) and the lower panel shows the view of *a*-*b* plane.

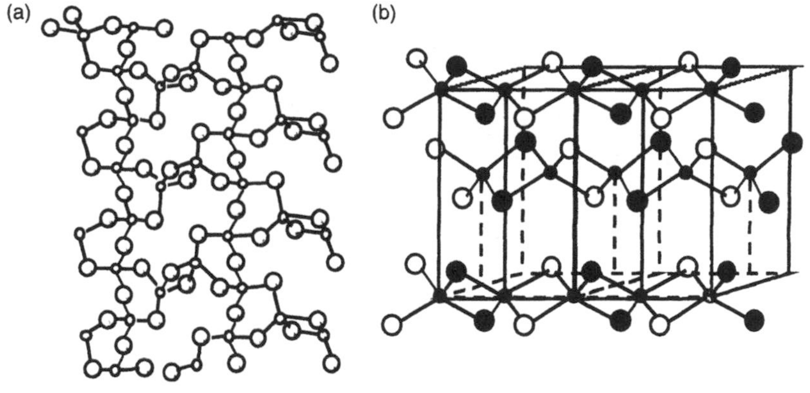

Figure 2.9 (a) Structure of crystalline $GeSe_2$. It shows a layer-structure is formed from edge- and corner-sharing tetrahedral, called a *raft*. (b) Crystal structure of $SiSe_2$: one-dimensional chains formed from edge-sharing tetrahedra.

The pre-peak is also observed in a-SiO_2, a-GeO_2, a-As and a-P, but it is not present in a-Se, a-Ge and a-Si. Its position in a-SiO_2 and a-GeO_2 is at $\mathbf{Q} \approx 1.5\,\text{Å}^{-1}$, which is relatively larger than the $1.0\,\text{Å}^{-1}$ found in other systems. Furthermore, $\Delta\mathbf{Q}$ for a-SiO_2 and a-GeO_2 is 0.3 and $0.5\,\text{Å}^{-1}$, respectively, which is broader than $0.2\,\text{Å}^{-1}$ found in other systems. This suggests that the pre-peak in a-SiO_2 and a-GeO_2 may not have the same meaning as discussed for a-Ch. Nevertheless, if the structure factor $S(\mathbf{Q})$ is plotted against the reduced variable $\mathbf{Q}r_1$, where r_1 is the nearest-neighbor bond length, $\mathbf{Q}r_1 \approx 2.5$ is obtained for all the systems examined (Write *et al.*, 1985). The pre-peak is regarded to have a common origin (universality) (Elliott, 1991). Some examples of FSDP are shown in Fig. 2.10.

The FSDP shows anomalous behavior as a function of temperature and pressure. The intensity of the FSDP increases with increasing temperature, while that of all other peaks in $S(\mathbf{Q})$ decreases, in accordance with the normal behavior of the Debye-Waller factor. The FSDP is still pronounced in the liquid state of $GeSe_2$ (Uemura *et al.*, 1978), but its intensity decreases with a shift to higher values of $\mathbf{Q}$ with increasing pressure (Tanaka, 1987).

The structural origin of the FSDP is still not clear and there has been much controversy about it (Elliott, 1990; Tanaka, 1998). This means that the structure of a-Ch over the MRO scale is not clear and hence, unlike a-Si or a-Ge, understanding the form of the overall structure for a-Ch is still in conjecture. In what follows we will review the current understanding of structures of a-Ch with the help of its SRO and the MRO properties.

Let us start with a-Se. As Se is normally two-fold coordinated, the structural configurations can be like chains or Se_8-rings. Using the quasi-random coil model for a-Se and considering 60% of trans-like and 40% of *cis*-like chain configurations,

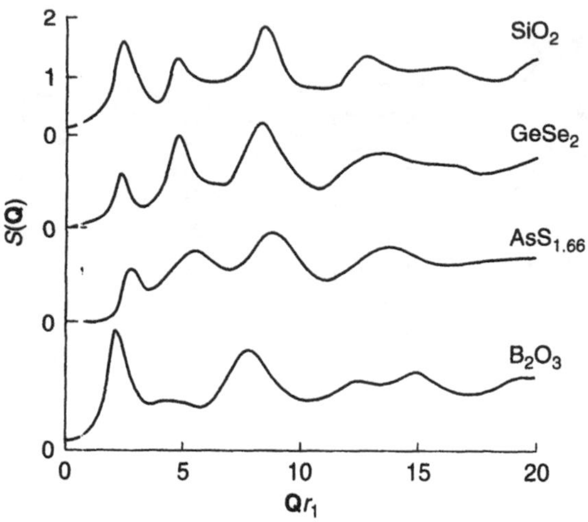

Figure 2.10 Structure factor for chalcogenide and oxide glasses plotted as a function of the reduced variable $\mathbf{Q}r_1$, where r_1 is the nearest-neighbor bond length in each case (Wright *et al.*, 1985).

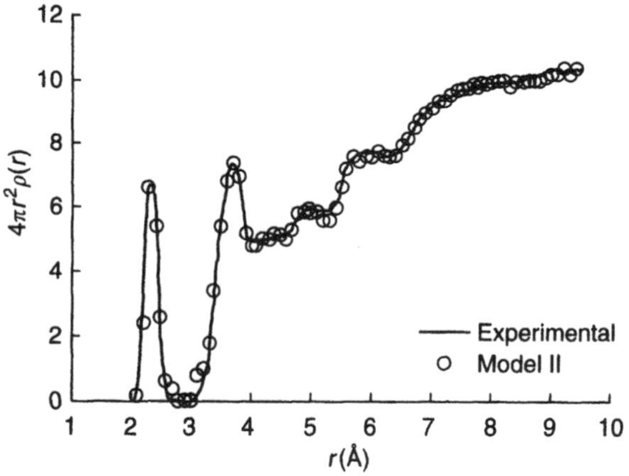

Figure 2.11 RDF for a-Se obtained experimentally and predicted by quasi-random coil model (Corb *et al.*, 1982).

we find that $J(r)$ in good agreement with the X-ray diffraction results, as shown in Fig. 2.11 (Corb *et al.*, 1982).

Next, we discuss a-$As_2Se(S)_3$ and a-$GeSe(S)_2$. As already stated, the pyramidal unit of $AsSe(S)_3$ and tetrahedron unit of $GeSe(S)_4$ determine the SRO in a-$As_2Se(S)_3$ and a-$GeSe(S)_2$, respectively. The problem is how these units are piled up in the network. Is the whole structure 2D- or 3D-like in these systems? This question is directly related to the origin of the FSDP.

A number of explanations has been proposed for the FSDP. Depending on how the peak in $S(\mathbf{Q})$ is assumed to originate (Elliott, 1990, 1991; Tanaka, 1998), these can principally be classified into two categories. First is the "crystalline" model, originally proposed as "distorted layer" model by Vaipolin and Porai-Koshits (1963). The model is based on the fact that the layered structure can hold to some extent even in amorphous states because the related crystals have layered structures. This model provides plausible explanations for some experimental observations (Mori *et al.*, 1983; Busse, 1984; Lin *et al.*, 1984; Tanaka, 1989a,b; Matsuda *et al.*, 1992). In the crystalline model, the FSDP arises from the interlayer correlation around 5 Å. Thus we expect that a restricted local layer-like structure can hold in both a-$As_2Se(S)_3$ and a-$GeSe(S)_2$. Note also that the MRO in a-$GeSe_2$ is also explained by 2D character of the "raft" bordered Se–Se wrong bonds (Phillips, 1981). The above ideas have led to a concept of network dimensionality (Zallen, 1983).

However, the crystalline model is not generally accepted, because the FSDP has also been discovered in molten states (Uemura *et al.*, 1978; Penfold and Salmon, 1991). Furthermore, similar FSDPs are observed in many disordered matters including a-SiO_2 which is known to have 3D-CRN structure. This suggests that layer-like structure is not a prerequisite to the existence of FSDP in a-Ch. Thus a universality for the FSDP to be present in disordered materials has been suggested. This is the second model, which contradicts the crystalline model (Moss and Price, 1985; Wright *et al.*, 1985; Elliott, 1991, 1994; Pfeiffer *et al.*, 1991), that has been put forward.

More recent work, however, seems to support the crystalline model (Tanaka, 1998), although the nature of the FSDP for a-Ch is qualitatively similar to that for a-SiO_2. Quantitatively, however, unlike a-SiO_2, the FSDP position in a-Ch exhibits dramatic changes with temperature and pressure, suggesting that the FSDP is related to the layer structure, since the interlayer spacing is greatly changed by application of pressure (Tanaka, 1989). We thus suggest that the FSDP does not have universally the same origin in all systems. As far as a-Ch is concerned the crystalline model, that is, distorted layer model, may be valid. Although the precise structure for a-$As_2Se(S)_3$ and a-$GeSe(S)_2$ cannot be conclusively determined, layer-like structure should be locally retained.

Finally, it is of interest to discuss the compositional dependence of structural and electronic properties, which will be scaled with the average coordination number Z. For A_xB_{1-x} composition Z is defined as

$$Z = xn_c(A) + (1-x)n_c(B), \tag{2.4}$$

where n_c is the coordination number of each atom obeying the $8-N$ rule (Mott, 1967) and is given by $n_c = 8 - N$, where N is the number of valence electrons. The values of N for S (Se), As (P) and Ge (Si) are therefore 2, 3 and 4, respectively. Then, for $As_2S(Se)_3$ and $GeS(Se)_2$ stoichiometric compositions,

$Z = 2.40$ and 2.67, respectively. Following the argument of "network constraint" originally proposed by Phillips (1979), the number of topological constraints Z_c for an atom in 3D space is defined as

$$Z_c = \frac{Z}{2} + (2Z - 3), \tag{2.5}$$

where the first term on the right-hand side represents a constraint from a covalent bond- length. The second term corresponds to the degrees of freedom due to bond angles. If $Z = 2$, then there is only one angle $(2Z - 3 = 1)$. Every additional atom bonded to another atom located at the origin adds two more degrees of freedom (Döhler *et al.*, 1980; Thorpe, 1983; Phillips and Thorpe, 1985).

A hypothetical atom having $Z_c = 3$ is regarded as "just rigid" in 3D space, which from Eq. (2.5) gives $Z = 2.40$, at which a glass becomes stable. The percolative arguments also support this conclusion and the way of counting the number of zero-frequency modes (Thorpe, 1983; He and Thorpe, 1985). A network with $Z \geq 2.40$ is called "over-constraints" (non-flexible) and that with $Z < 2.40$ is called "under-constraints" (flexible).

The Phillips transition at $Z = 2.40$, however, does not always occur. Other transitions in electronic and structural parameters are found to occur at $Z = 2.67$ through the detailed compositional dependence of structural, elastic and electronic properties (Tanaka, 1989). One of the examples is shown in Fig. 2.12: the atomic volume is plotted as a function Z for various a-Ch at $Z = 2.67$. In order to understand the transition at $Z = 2.67$, the constraint for an atom in 2D plane is defined as

$$Z_c = \frac{Z}{2} + (Z - 1), \tag{2.6}$$

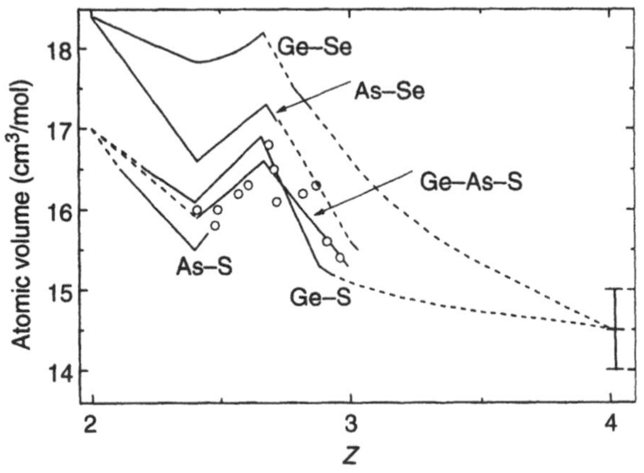

Figure 2.12 The atomic volumes for various chalcogenide glasses as a function of the average coordination number Z (Tanaka, 1989).

where only one angular term is taken into consideration and the rotational degrees of freedom in this case is only one. Again by $Z_c = 3$ we get $Z = 2.67$; that is, a 2D glass will be stabilized in 3D space at $Z = 2.67$, which can be called the Tanaka transition.

On a length scale of a few angstrom (<5 Å), 3D behavior may occur in the physical parameter related to this length scale, while in a medium-range scale (>10 Å), 2D behavior may appear in the properties of interest. Thus, the Tanaka transition may take place in the physical conditions which can be dominated by a medium-range scale of structure. A maximum value in the FSDP intensity, in fact, occurs at $Z = 2.67$ (Tanaka, 1989).

References

Bose, S.K., Winer, K. and Andersen, O.K. (1988). *Phys. Rev. B* **37**, 6262.

Buda, F., Chiarotti, G.L., Car, R. and Parrinello, M. (1991). *Phys. Rev. B* **44**, 5908.

Busse, L.E. (1984). *Phys. Rev. B* **39**, 3639.

Car, R. and Parrinello, M. (1985). *Phys. Rev. Lett.* **55**, 2471.

Corb, B.W., Wei, W.D. and Averbach, B.L. (1982). *J. Non-Cryst. Solids* **53**, 29.

Döhler, G.H., Dandoloff, R. and Bilz, H. (1980). *J. Non-Cryst. Solids* **42**, 87.

Drabold, D.A., Fedders, P.A. and Stumm, P. (1994). *Phys. Rev. B* **49**, 16415.

Elliott, S.R. (1990). *Physics of Amorphous Materials*, 2nd edn. Longman Scientific & Technical, London.

Elliott, S.R. (1991). *Nature* **354**, 445.

Elliott, S.R. (1994). *J. Non-Cryst. Solids* **182**, 40.

Feldman, J.L., Kluge, M.D., Allen, P.B. and Wooten, F. (1993). *Phys. Rev. B* **48**, 12589.

Galli, G., Martin, R.M., Car, R. and Parrinello, M. (1989). *Phys. Rev. Lett.* **62**, 555.

Gereven, O. and Pusztai, L. (1994). *Phys. Rev. B* **50**, 14136.

Gilkes, K.W.R., Gaskell, P.H. and Robertson, J. (1995). *Phys. Rev. B* **51**, 12303.

He, H. and Thorpe, M.F. (1985). *Phys. Rev. Lett.* **54**, 2107.

Hickley, B.J. and Morgan, G.J. (1986). *J. Phys. C* **19**, 6195.

Holender, J., Morgan, J. and Jones, R. (1993). *Phys. Rev. B* **47**, 3991.

Ishimaru, M., Munetoh, S. and Motooka, T. (1997). *Phys. Rev. B* **56**, 15133.

Kadas, K., Ferenczy, G.G. and Kugler, S. (1998). *J. Non-Cryst. Solids* **227–230**, 367.

Kakinoki, J., Katada, K. and Hanawa, T. (1960). *Acta. Cryst.* **13**, 171.

Keating, P.N. (1966). *Phys. Rev.* **145**, 637.

Kohary, K. and Kugler, S. (2000). *J. Non-Cryst. Solids* **266–268**, 746.

Kohary, K. and Kugler, S. (2001). *Phys. Rev. B* **63**, 193404.

Kugler, S. and Naray-Szabo, N. (1991). *Jpn. J. Appl. Phys.* **30**, L1149.

Kugler, S. and Varallyay, Z. (2001). *Philos. Mag. Lett.* **81**, 569.

Kugler, S., Surjan, P.R. and Naray-Szabo, N. (1988). *Phys. Rev. B* **37**, 9069.

Kugler, S., Moluar, G., Peto, G., Zsoldos, E., Rostoc, L., Meuelle, A. and Bellissent, R. (1989). *Phys. Rev. B* **40**, 8030.

Kugler, S., Pusztai, L., Rosta, L., Chieux, P. and Bellissent, R. (1993a). *Phys. Rev. B* **48**, 7685.

Kugler, S., Shimakawa, K., Watanabe, T., Hayashi, K., Laszlo, I. and Bellissent, J. (1993b). *J. Non-Cryst. Solids* **164–166**, 831.

Kugler, S., Laszlo, I., Kohary, K. and Shimakawa, K. (1999). *Func. Mater.* **6**, 459.

Kwon, I., Biswas, R., Wang, C.Z., Ho, K.M. and Soukoulis, C.M. (1994). *Phys. Rev. B* **49**, 7242.

Lin, C., Busse, L.E., Nagel, S.R. and Faber, J. (1984). *Phys. Rev. B* **29**, 5060.

Marks, N.A., McKenzie, D.R., Pailthorpe, B.A., Bernasconi, M. and Parrinello, M. (1996). *Phys. Rev. Lett.* **76**, 768.

McGreevy, R.L. and Pustai, L. (1988). *Molec. Sim.* **1**, 369.

Matsuda, O., Inoue, K., Nakane, T. and Murase, K. (1992). *J. Non-Cryst. Solids* **150**, 202.

Mori, T., Yasuoka, H., Saegusa, H., Okawa, K., Kato, M., Arai, T., Fukunaga, T. and Watanabe, W. (1983). *Jpn. J. Appl. Phys.* **22**, 1784.

Moss, S.C. and Price, D.L. (1985). In: Adler, D. Fritzshe, H. and Ovshinsky, S.R. (eds), *Physics of Disordered Matter*. Plenum, New York, p. 77.

Mott, N.F. (1967). *Adv. Phys.* **16**, 49.

Mousseau, N. and Lewis, L.J. (1989). *J. Non-Cryst. Solids* **114**, 202.

Mousseau, N. and Lewis, L.J. (1990). *Phys. Rev. B* **41**, 3702.

Oligschleger, C., Jones, R.O., Reimann, S.M. and Schober (1996). *Phys. Rev. B* **53**, 6165.

Penfold, I.T. and Salmon, P.S. (1991). *Phys. Rev. Lett.* **67**, 973.

Pfiffer, G., Paesler, M.A. and Agarwal, S.C. (1991). *J. Non-Cryst. Solids* **130**, 111.

Phillips, J.C. (1979). *J. Non-Cryst. Solids* **34**, 153.

Philips, J.C. (1981). *J. Non-Cryst. Solids* **43**, 37.

Phillips, J.C. and Thorpe, M.F. (1985). *Solid State Commun.* **53**, 699.

Robertson, J (1986). *Adv. Phys.* **35**, 317.

Stich, I., Car, A. and Parrinello, M. (1991). *Phys. Rev. B* **44**, 11092.

Tanaka, Ke. (1987). *J. Non-Cryst. Solids* **90**, 363.

Tanaka, Ke. (1989a). *Phys. Rev. B* **39**, 1270.

Tanaka, Ke. (1989b). In: Brrisov, M., Kirov, N. and Vavrek, A. (eds), *Disordered Systems and New Materials*. World Scientific, Singapore, p. 290.

Tanaka, Ke. (1998). *Jpn. J. Appl. Phys.* **37**, 1747.

Tersoff, J. (1989). *Jpn. J. Appl. Phys.* **39**, 5566.

Tersoff, J. (1998). *Phys. Rev. B* **38**, 9902.

Thorpe, M.F (1983). *J. Non-Cryst. Solids* **57**, 355.

Uemura, O., Sagara, Y., Muno, D. and Satow, T. (1978). *J. Non-Cryst. Solids* **30**, 155.

Vaipolin, A.A. and Porai-Koshits, E.A. (1963). *Sov. Phys.-Solid State* **5**, 497.

Verlet, L. (1967). *Phys. Rev.* **159**, 98.

Vink, R.L.C , Barkema, G.T., van der Weg, W.F. and Mousseau, N. (2001). *J. Non-Cryst. Solids* **282**, 248.

Wang, C.Z., Ho, K.M. and Chan, C.T. (1993). *Phys. Rev. Lett.* **70**, 611.

Warshel, A. and Lifson, S. (1970). *J. Chem. Phys.* **53**, 582.

Weaire, D., Hobbs, D., Morgan, G.J., Holender, J.M. and Wooten, F. (1993). *J. Non-Cryst. Solids* **164–166**, 877.

Wooten, F. and Weaire, D. (1994). In: Marshall J.M., Kirov, N. and Vavrek, A. (eds), *Electronic, Optoelectronic and Magnetic Thin Films*. John Wiley & Sons Inc., New York, p. 197.

Wooten, F., Winer, K. and Weaire, D. (1985). *Phys. Rev. Lett.* **54**, 1392.

Write, A.C., Sinclair, R.N. and Leadbetter, A.J. (1985). *J. Non-Cryst. Solids* **71**, 295.

Xu, C.H., Wang, C.Z., Chan, C.T. and Ho, K.M. (1992). *J. Phys.: Condens. Matter* **4**, 6047.

Yang, R. and Singh, J. (1998). *J. Non-Cryst. Solids* **240**, 29.

Yonezawa, F., Sakamoto, S. and Hori, M. (1991). *J. Non-Cryst. Solids* **137–138**, 135.

Zallen, R. (1983). *The Physics of Amorphous Solids*. John Wiley & Sons, New York.

Ziman, J.M. (1979). *Models of Disorder*. Cambridge University Press, Cambridge.

3 Theory of effective mass

In amorphous solids (a-solids) when determining various electronic transport related quantities that require electron or hole mass, the free electron mass is usually used. However, a charge carrier in any solid including a-solid is not free. It can only be made free by associating it with an effective mass through the effective mass approximation. Thus the use of free electron mass is not only inappropriate, but it may lead to wrong results and conclusions for a-solids. For instance, it is well established that the charge carrier states in a-solids are of two types: (1) extended states, which are delocalized states and (2) tail states, which are localized. Should one use the same free electron mass for a charge carrier in both types of state? The answer is obviously no, because delocalized states correspond to a lower effective mass than localized states. Therefore, the use of the free electron mass for both is unjustified, and there is a need to determine the effective mass of charge carriers in both the states of a-solids. One of the problems in calculating the effective mass of charge carriers in a-solids is the absence of translational symmetry, resulting into the wavevector **k** not being a good quantum number. Although Kivelson and Gelatt (1979) have illustrated that the effective mass of charge carriers can also be derived in the real coordinate space, the concept has not gone very far in determining and using the effective mass of charge carriers in a-solids. In this chapter, first a general concept of the effective mass of charge carriers will be described, then the theory of effective mass in crystalline solids (c-solids) will be reviewed for developing an understanding, and finally a theory for determining the effective mass of charge carriers in a-solids will be presented.

3.1 Concept of effective mass in solids

The definition of a free particle is that it is not subjected to any external force, which means it is either subjected to no potential or a constant potential of interaction. This comes from the basic relation between a conservative force $F(x)$ and potential $V(x)$, given by

$$F(x) = -\frac{\partial V(x)}{\partial x}, \tag{3.1}$$

yielding zero force for any constant potential in space. The energy E of such a particle can then be written as

$$E = \frac{p^2}{2m}, \tag{3.2}$$

where p is the linear momentum and m is its mass. In the case of an electron, m is the free electron mass. A particle has only kinetic energy in this situation.

Let us now consider a charge carrier moving in a solid. Regardless of whether the solid is crystalline or not, the charge carrier will be subjected to the ionic Coulomb potential from all atoms, and carrier–carrier Coulomb interaction potential from other charge carriers present in the solid, neither of which is constant. In this case the energy E of the particle will consist of both kinetic energy and potential energy, which may be written as

$$E = \frac{p^2}{2m} + V, \tag{3.3}$$

where V is the potential energy, which will depend on the kind of solid, say crystalline or amorphous, and it may as well depend on the location of the particle within the solid. However, by redefining the mass of the particle such that the energy, E, can still be written as in Eq. (3.2), we can write Eq. (3.3) as

$$E = \frac{p^2}{2m^*}, \tag{3.4}$$

where

$$m^* = \frac{p^2}{[p^2 + 2mV]}m. \tag{3.5}$$

The modified mass m^* is called the effective mass of the charge carrier. Equation (3.5) is a very simplified form of energy of a charge carrier in a solid. The expression of the effective mass given in Eq. (3.5), although accurate, is not very suitable for solids. One often comes across a situation when the energy, E, is such that it consists of a constant term, independent of the linear momentum, and the rest of it depends on the linear momentum. In that case it is more convenient to determine the effective mass by writing the energy as

$$E = E_0 + \frac{p^2}{2m^*}, \tag{3.6}$$

where E_0 is the constant energy term independent of p. The form of Eq. (3.6) is very useful in separating the energy of the charge carriers in different energy states. For example, for a hole in the valence band of a crystal, E_0 can be set to zero, but for an electron in the conduction band it will be the minimum energy gap E_g, that is, $E_0 = E_g$. An approximation used to write the energy of a charge carrier in the form of Eq. (3.6) is called the effective mass approximation. Accordingly,

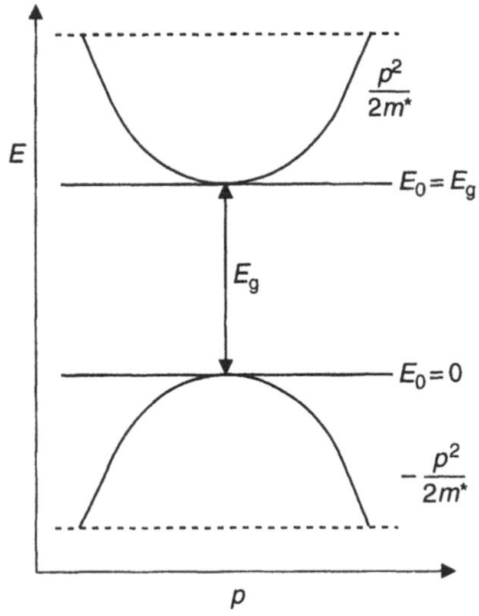

Figure 3.1 Use of effective mass approximation in expressing the energy of a charge carrier as a parabolic function of linear momentum according to Eq. (3.6).

if the energy is plotted as a function of the linear momentum, one gets a parabola. The vertex of the parabola, for the valence band, would be at the top edge of the band and that for the conduction band at the bottom edge of the band, as shown in Fig. 3.1. Thus, the form of Eq. (3.6) separates parabolas of different energy bands, whereas if one uses Eq. (3.5) all parabolas will have the same vertices, which will not be of much use. (Note in Fig. 3.1 that the parabola of a hole in the valence band is inverted.) Having described the concept of effective mass, let us now look at the theory of deriving the effective mass in c-solids, which is presented in the next section.

3.2 Effective mass of an electron in c-solids

For determining the effective mass, as described in Section 3.1, we need to determine the energy of the particle as a function of the linear momentum by solving the corresponding Schrödinger equation. As solids are many particle systems, it is convenient to develop the theory in the second quantisation. The electronic Hamiltonian $\hat{H}$ for a c-solid can be expressed in the reciprocal lattice vector space as (Singh, 1994)

$$\hat{H} = \hat{H}_1 + \hat{H}_2, \tag{3.7}$$

where

$$\hat{H}_1 = \sum_{t,\mathbf{k},\sigma} \left\langle t, \mathbf{k}, \sigma \left| -\frac{\hbar^2 \nabla^2}{2m_e} + V(\mathbf{r}) \right| t, \mathbf{k}, \sigma \right\rangle a^+_{t\mathbf{k}}(\sigma) a_{t\mathbf{k}}(\sigma), \tag{3.8a}$$

$$\begin{aligned} \hat{H}_2 = \frac{1}{2} & \sum_{t_1,\mathbf{k}_1,\sigma_1} \sum_{t_2,\mathbf{k}_2,\sigma_2} \sum_{t_3,\mathbf{k}_3,\sigma_3} \sum_{t_4,\mathbf{k}_4,\sigma_4} \\ & \times \langle t_1, \mathbf{k}_1, \sigma_1; t_2, \mathbf{k}_2, \sigma_2 | U(|\mathbf{r}_1 - \mathbf{r}_2|) t_3, \mathbf{k}_3, \sigma_3, t_4, \mathbf{k}_4, \sigma_4 \rangle \\ & \times a^+_{t_1\mathbf{k}_1}(\sigma_1) a^+_{t_2\mathbf{k}_2}(\sigma_2) a_{t_3\mathbf{k}_3}(\sigma_3) a_{t_4\mathbf{k}_4}(\sigma_4), \end{aligned} \tag{3.8b}$$

where $\hat{H}_1$ and $\hat{H}_2$ are the single and two particle interaction operators, respectively. $V(\mathbf{r})$ is the ionic periodic potential that an electron is subjected to, and $U(|\mathbf{r}_1 - \mathbf{r}_2|)$ is the repulsive potential between two electrons. These are given as

$$V(\mathbf{r}) = -\sum_n \frac{\kappa Z e^2}{|\mathbf{r} - \mathbf{R}_n|} = \sum_n V_n(\mathbf{r}), \tag{3.9}$$

and

$$U(|\mathbf{r}_1 - \mathbf{r}_2|) = \frac{\kappa e^2}{|\mathbf{r}_1 - \mathbf{r}_2|}, \quad \kappa = (4\pi\varepsilon_0)^{-1}. \tag{3.10}$$

$|t, \mathbf{k}, \sigma\rangle = |t, \mathbf{k}\rangle|\sigma\rangle$ represents the eigenvector of an electron in the band $t = 0$ for the valence band and $t = 1$ for the conduction band, with wavevector $\mathbf{k}$ and spin σ. $|t, \mathbf{k}\rangle = |\phi_{t\mathbf{k}}(\mathbf{r})\rangle$ is written in terms of Bloch functions as

$$\phi_{t\mathbf{k}}(\mathbf{r}) = V_0^{-1/2} u_{t\mathbf{k}}(\mathbf{r}) \exp(\mathrm{i}\mathbf{k} \cdot \mathbf{r}), \tag{3.11}$$

where V_0 is the crystal volume. $a^+_{t\mathbf{k}}(\sigma)[a_{t\mathbf{k}}(\sigma)]$ is the creation (annihilation) operator of an electron in band t, with wavevector $\mathbf{k}$ and spin σ.

Let us consider creating an electron in the conduction band ($t = 1$), with wavevector $\mathbf{k}_e$ and spin σ_e, and we want to calculate the energy of this electron. We consider a vacuum state $|0\rangle$ such that the valence band is completely filled and conduction band is completely empty, then the eigenvector of the electron in the conduction band can be written as

$$|1, \mathbf{k}_e, \sigma_e\rangle = a^+_{1\mathbf{k}_e}(\sigma_e)\,|0\rangle. \tag{3.12}$$

It is important to note that the same eigenvector can be written in the real crystal space as

$$|1, \mathbf{k}_e, \sigma_e\rangle = \sum_{\mathbf{l}} C_{\mathbf{l}}(\mathbf{k}_e) a^+_{1\mathbf{l}}(\sigma_e)\,|0\rangle, \tag{3.13}$$

where the probability amplitude coefficient $C_{\mathbf{l}}(\mathbf{k})$ is given by

$$C_{\mathbf{l}}(\mathbf{k}) = N^{-1/2} \exp(\mathrm{i}\mathbf{k} \cdot \mathbf{l}). \tag{3.14}$$

Here N is the number of atoms in the crystal and $\mathbf{l}$ position coordinate of an atom. From Eq. (3.14) one gets

$$C_{\mathbf{l}}^{*}(\mathbf{k})C_{\mathbf{l}}(\mathbf{k}) = 1/N, \quad \text{and} \quad \sum_{\mathbf{l}} C_{\mathbf{l}}^{*}(\mathbf{k})C_{\mathbf{l}}(\mathbf{k}) = 1. \tag{3.15}$$

The first part of Eq. (3.15) suggests that the probability of finding the particle at every site is the same, and the second part ensures that the created particle is within the crystal.

For determining the energy, $W_1(\mathbf{k}_e)$, of the electron in the conduction band, we need to solve the following Schrödinger equation:

$$\hat{H}\,|1, \mathbf{k}_e, \sigma_e\rangle = W_1(\mathbf{k}_e)\,|1, \mathbf{k}_e, \sigma_e\rangle. \tag{3.16}$$

Operating by the complex conjugate of the eigenvector in Eq. (3.12) from the left-hand side of Eq. (3.16), and then applying the anticommutation rules of fermion operators given by

$$\begin{aligned} \{a_{t\mathbf{k}}(\sigma), a_{t'\mathbf{k}'}^{+}(\sigma')\}_{+} &= \delta_{t,t'}\delta_{\mathbf{k},\mathbf{k}'}\delta_{\sigma,\sigma'}, \\ \{a_{t\mathbf{k}}^{+}(\sigma), a_{t'\mathbf{k}'}^{+}(\sigma')\}_{+} &= \{a_{t\mathbf{k}}(\sigma), a_{t'\mathbf{k}'}(\sigma')\}_{+} = 0, \end{aligned} \tag{3.17}$$

the energy eigenvalue $W_1(\mathbf{k}_e)$ is obtained as

$$W_1(\mathbf{k}_e) = W_0 + E_1(\mathbf{k}_e), \tag{3.18}$$

$$\begin{aligned} W_0 = {} & 2\sum_{\mathbf{k}}\left\langle 0, \mathbf{k}\left| -\frac{\hbar^2\nabla^2}{2m_e} + V(\mathbf{r})\right| 0, \mathbf{k}\right\rangle \\ & + \sum_{\mathbf{k}_1,\mathbf{k}_2}\big[2\langle 0, \mathbf{k}_1; 0, \mathbf{k}_2|U(|\mathbf{r}_1 - \mathbf{r}_2|)|0, \mathbf{k}_2; 0, \mathbf{k}_1\rangle \\ & - \langle 0, \mathbf{k}_1; 0, \mathbf{k}_2|U(|\mathbf{r}_1 - \mathbf{r}_2|)|0, \mathbf{k}_1; 0, \mathbf{k}_2\rangle\big], \end{aligned} \tag{3.19}$$

and

$$\begin{aligned} E_1(\mathbf{k}_e) = {} & \left\langle 1, \mathbf{k}_e\left| -\frac{\hbar^2\nabla^2}{2m_e} + V(\mathbf{r})\right| 1, \mathbf{k}_e\right\rangle \\ & + \sum_{\mathbf{k}_2}\big[2\langle 1, \mathbf{k}_e; 0, \mathbf{k}_2|U(|\mathbf{r}_1 - \mathbf{r}_2|)|0, \mathbf{k}_2; 1, \mathbf{k}_e\rangle \\ & - \langle 1, \mathbf{k}_e; 0, \mathbf{k}_2|U(|\mathbf{r}_1 - \mathbf{r}_2|)|1, \mathbf{k}_e; 0, \mathbf{k}_2\rangle\big], \end{aligned} \tag{3.20}$$

where W_0 is the total electronic energy of all electrons in the valence band before an electron was created in the conduction band, $E_1(\mathbf{k}_e)$ is the total energy of the electron in the conduction band including its interaction energy with all the electrons in the valence band. The factor of 2 in Eqs (3.19) and (3.20) comes from the summation over the spin quantum number. It is to be noted that in obtaining

Eq. (3.18) the definition of the vacuum state is used, that is, $a^{+}_{0\mathbf{k}}(\sigma)|0\rangle = 0$, and $a_{1\mathbf{k}}(\sigma)|0\rangle = 0$.

It is obvious from Eq. (3.20) that the energy of the electron in the conduction band is a function of its wavevector $\mathbf{k}_e$, but the exact nature of this function is not known. As Bloch functions are usually not known, it is not possible to derive $E_1(\mathbf{k}_e)$ analytically. It can only be evaluated numerically, which gives the band structure of the conduction band for any crystal. However, for calculating the effective mass of an electron at any particular value of the wavevector, say $\mathbf{k}_0$, one can expand $E_1(\mathbf{k}_e)$ in Taylor's series about that point and terminate the series at the second order term as

$$E_1(\mathbf{k}_e) \approx E_1(\mathbf{k}_0) + \sum_i (\mathbf{k}_e - \mathbf{k}_0)_i \left(\frac{\partial E_1(\mathbf{k}_e)}{\partial k_{ei}}\right)_{k_{ei}=k_{0i}} + \frac{1}{2}\sum_{i,j} (\mathbf{k}_e - \mathbf{k}_0)_i (\mathbf{k}_e - \mathbf{k}_0)_j \left(\frac{\partial^2 E_1(\mathbf{k}_e)}{\partial k_{ei}\partial k_{ej}}\right)_{k_{ei}=k_{0i}, k_{ej}=k_{0j}}, \tag{3.21}$$

where $i, j = x, y, z$ denote the components of the wavevector. Two points are very important to note in this expansion. (i) The difference $(\mathbf{k}_e - \mathbf{k}_0)$ is assumed to be small, so that the neglect of higher order terms in the expansion is justified, and (ii) the second term, which is the first-order term in the expansion, is close to zero. If these two conditions are satisfied, the second-order term can be used to determine the effective mass tensor of electron, m^*_{eij}, in the conduction band, and then the energy of the electron in Eq. (3.21) can be written as

$$E_1(\mathbf{k}_e) \approx E_1(\mathbf{k}_0) + \frac{\hbar^2}{2}\sum_{i,j} (\mathbf{k}_e - \mathbf{k}_0)_i (\mathbf{k}_e - \mathbf{k}_0)_j [m^*_{eij}]^{-1}, \tag{3.22}$$

where

$$[m^*_{eij}]^{-1} = \hbar^{-2} \sum_{i,j} \left.\frac{\partial^2 E_1(\mathbf{k}_e)}{\partial k_{ei}\partial k_{ej}}\right|_{k_{ei}=k_{0i}, k_{ej}=k_{0j}}. \tag{3.23}$$

The approximation used for writing the electron energy in the form of Eq. (3.22) is called the *effective mass approximation*. Using Eq. (3.22) in Eq. (3.18), the electronic energy of a crystal with one electron in the conduction band can be written as

$$W_1(\mathbf{k}_e) = W_0 + E_1(\mathbf{k}_0) + \frac{\hbar^2}{2}\sum_{i,j} (\mathbf{k}_e - \mathbf{k}_0)_i (\mathbf{k}_e - \mathbf{k}_0)_j [m^*_{eij}]^{-1}. \tag{3.24}$$

For a band with isotropic effective mass, Eq. (3.24) reduces to

$$W_1(\mathbf{k}_e) = W_0 + E_1(\mathbf{k}_0) + \frac{\hbar^2(\mathbf{k}_e - \mathbf{k}_0)^2}{2m^*_e}. \tag{3.25}$$

In Eq. (3.25), W_0, being the total energy of all valence electrons before the excitation, is essentially the Fermi energy, and $E_1(\mathbf{k}_0)$ is the energy of the bottom

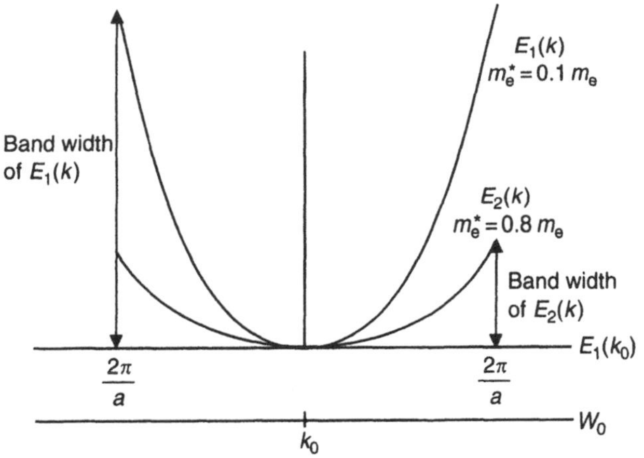

Figure 3.2 Applying effective mass approximation, two energy bands with $m_e^* = 0.1m_e$ and $m_e^* = 0.8m_e$ are plotted. The width of the lighter electron is 8 times that of the heavier electron.

of the conduction band at $\mathbf{k}_0$. Using $\mathbf{p} = \hbar(\mathbf{k}_e - \mathbf{k}_0)$ in Eq. (3.25) we get an expression similar to that obtained in Eq. (3.6) in the preceding section. For two effective masses of electron, $m_e^* = 0.1m_e$ and $m_e^* = 0.8m_e$, Eq. (3.25) is plotted in Fig. 3.2, which illustrates very clearly that an energy band associated with a heavier effective mass has a smaller band width and hence flatter energy band. The objective of describing the effective mass approximation in crystalline solid in this book is to be able to compare its results with those in a-solids obtained in the next section.

3.3 Effective mass of charge carriers in a-solids

As $\mathbf{k}$ is not a good quantum number in a-solids, the derivation of the effective mass of charge carriers has to be done in the real co-ordinate space. In studying the problem of electron self-trapping, it was first shown by Toyozawa (1959, 1963) that the effective mass of an electron interacting with acoustic phonons can be obtained in the real crystal space. He illustrated the point that under a certain condition of coupling between the electron and the lattice, the electron effective mass increases by a few orders of magnitude and the electron becomes localized. Kivelson and Gelatt (1979) have developed a theory for calculating the effective mass of charge carriers in a-solids in the real co-ordinate space. In developing the theory of Frenkel excitons, Singh (1994) has also derived an expression for the effective mass of a Frenkel exciton in the real crystal space. In this section, we will follow an approach similar to that applied for calculating the effective mass of Frenkel excitons. For this we need to express the electronic Hamiltonian, $\hat{H} = \hat{H}_1 + \hat{H}_2$, of Eq. (3.7) for a-solids in the real coordinate space with $\hat{H}_1$ and

$\hat{H}_2$ given by

$$\hat{H}_1 = \sum_{\mathbf{l},\mathbf{m},j,\sigma} E_{\mathbf{lm}} a^{+}_{j,\mathbf{l}}(\sigma) a_{j,\mathbf{m}}(\sigma), \tag{3.26}$$

$$\hat{H}_2 = \frac{1}{2} \sum_{\mathbf{l}_1,j_1,\sigma_1} \sum_{\mathbf{l}_2,j_2,\sigma_2} \sum_{\mathbf{l}_3,j_3,\sigma_3} \sum_{\mathbf{l}_4,j_4,\sigma_4} \langle j_1\mathbf{l}_1,\sigma_1;\, j_2,\mathbf{l}_2,\sigma_2 | U(|\mathbf{r}_1 - \mathbf{r}_2|)$$

$$\times\, | j_3,\mathbf{l}_3,\sigma_3;\, j_4,\mathbf{l}_4,\sigma_4\rangle a^{+}_{j_1\mathbf{l}_1}(\sigma_1) a^{+}_{j_2\mathbf{l}_2}(\sigma_2) a_{j_3\mathbf{l}_3}(\sigma_3) a_{j_4\mathbf{l}_4}(\sigma_4), \tag{3.27}$$

where

$$E_{lmj} = \left\langle j,\mathbf{l},\sigma \left| -\frac{\hbar^2\nabla^2}{2m} + V(\mathbf{r}) \right| j,\mathbf{m},\sigma \right\rangle, \tag{3.28}$$

and $|j,\mathbf{l},\sigma\rangle = |\phi_{j,\mathbf{l},\sigma}(\mathbf{r})\rangle$ is the ket vector of an electron in the electronic state j of an atom localized at $\mathbf{l}$ with spin $\sigma = +\frac{1}{2}$ or $-\frac{1}{2}$, and $\phi_{j,\mathbf{l},\sigma}(\mathbf{r})$ is the corresponding electronic wavefunction. A brief description of the form of $\phi_{j,\mathbf{l},\sigma}(\mathbf{r})$ is given at the end of this chapter in Section 3.4. The operators $a^{+}_{j,\mathbf{l}}(\sigma)$ and $a_{j,\mathbf{l}}(\sigma)$ are, respectively, the creation and annihilation operators of an electron with spin σ in the electronic state j of an atom localized at $\mathbf{l}$.

3.3.1 Effective mass of an electron in conduction states

The electronic states of interest to us in a-solids are the valence and conduction states. Valence states consist of the extended states below the hole mobility edge, E_v, and tail states above E_v. Likewise the conduction states consist of extended states above the electron mobility edge, E_c, and tail states below E_c. We will define the vacuum state, $|0\rangle$, for an a-solid as the one with completely filled valence states and totally empty conduction states. Let us create an electron with energy E and spin σ_e in the conduction states of such an a-solid. Following Eq. (3.13), the eigenvector of such an electronic state in an a-solid can be written as (Singh, 2002, 2003)

$$|1, E, \sigma_e\rangle = \sum_{\mathbf{l}} C_{1\mathbf{l}}(E) a^{+}_{1\mathbf{l}}(\sigma_e)\,|0\rangle, \tag{3.29}$$

where $j = 1$ denotes conduction states. It is to be noted that although the eigenvector in Eq. (3.29) is similar to that of Eq. (3.13) used for c-solids, it is not the same. The difference is in the form of probability amplitude coefficients, $C_{1\mathbf{l}}(E)$. In c-solids the probability amplitude coefficient is written in the **k**-space as given in Eq. (3.14), but for a-solids it can only be written in the energy space, or the real coordinate space. Here we will assume it to be of the following form:

$$C_{1\mathbf{l}}(E) = N^{-1/2} \exp(\mathrm{i}\boldsymbol{\rho}_e \cdot \mathbf{l}), \tag{3.30}$$

where

$$\rho_e(E) = \sqrt{\frac{2m_e^*(E - E_c)}{\hbar^2}}. \tag{3.31}$$

Here m_e^* is the effective mass of electron in the conduction states, which is yet to be determined. The choice of $C_{1l}(E)$ is based on the fact that for energy $E > E_c$ the electron will move as a free wave with an effective mass m_e^*, but for $E < E_c$, when the electron will be in the tail states, $\rho_e(E)$ becomes imaginary and the probability amplitude coefficient becomes a real exponentially decreasing function. In this case, the electron will get localized at the atomic site corresponding to that energy state and the atomic site will act as the center of localization. In this way, as the electron becomes localized in the tail states, it may have a different effective mass than that in the extended states.

For calculating the energy eigenvalue of electron, W_{1e}, we need to solve the following Schrödinger equation:

$$\hat{H}\,|1, E, \sigma_e\rangle = W_{1e}|\,1, E, \sigma_e\rangle. \tag{3.32}$$

Using Eqs (3.26)–(3.29) in Eq. (3.32) and applying the anti-commutation rules [Eq. (3.17)] for fermion operators, we can evaluate

$$\langle 1, E, \sigma_e\,|\hat{H}|\,1, E, \sigma_e\rangle = \langle 1, E, \sigma_e\,|W_{1e}|\,1, E, \sigma_e\rangle, \tag{3.33}$$

to get

$$W_{1e} = W_0 + \sum_{\mathbf{l},\mathbf{m}} C_{1\mathbf{l}}^*(E)C_{1\mathbf{m}}(E)E_{1\mathbf{lm}}, \tag{3.34}$$

where

$$\begin{aligned} W_0 = {} & 2\sum_{\mathbf{l}}\left\langle 0, \mathbf{l}\left|-\frac{\hbar^2\nabla^2}{2m_e} + V(\mathbf{r})\right|0, \mathbf{l}\right\rangle \\ & + \sum_{l_1, l_2}[2\langle 0, \mathbf{l}_1; 0, \mathbf{l}_2|U(|\mathbf{r}_1 - \mathbf{r}_2|)|0, \mathbf{l}_2; 0, \mathbf{l}_1\rangle \\ & - \langle 0, \mathbf{l}_1; 0, \mathbf{l}_2|U(|\mathbf{r}_1 - \mathbf{r}_2|)|0, \mathbf{l}_1; 0, \mathbf{l}_2\rangle], \end{aligned} \tag{3.35}$$

and

$$\begin{aligned} E_{1\mathbf{lm}} = {} & \left\langle 1, \mathbf{l}\left|-\frac{\hbar^2\nabla^2}{2m_e} + V(\mathbf{r})\right|1, \mathbf{m}\right\rangle \\ & + \sum_{\mathbf{l}'}[2\langle 1, \mathbf{l}; 0, \mathbf{l}'|U(|\mathbf{r}_1 - \mathbf{r}_2|)|0, \mathbf{l}'; 1, \mathbf{m}\rangle \\ & - \langle 1, \mathbf{l}; 0, \mathbf{l}'|U(|\mathbf{r}_1 - \mathbf{r}_2|)|1, \mathbf{m}; 0, \mathbf{l}'\rangle]. \end{aligned} \tag{3.36}$$

Equation (3.34) is obtained by applying the following condition:

$$\sum_{\mathbf{l}} C_{1\mathbf{l}}^*(E)C_{1\mathbf{l}}(E) = 1, \tag{3.37}$$

which ensures that the created electron is within the solid. Here again W_0 is the total energy of all valence electrons in the solid before an electron was introduced in the extended states, that is, the Fermi energy. $E_{\mathbf{llm}}$ is the energy of the electron introduced in the conduction states, including its interaction energy with all the electrons in the valence states as it can be seen from Eq. (3.36). Using Eq. (3.30) in Eq. (3.34), we get

$$W_{1\mathrm{e}} = W_0 + N^{-1} \sum_{\mathbf{l},\mathbf{m}} \exp[i\boldsymbol{\rho}_\mathrm{e} \cdot (\mathbf{m} - \mathbf{l})] E_{\mathbf{llm}}. \tag{3.38}$$

It should be noted here that the eigenvalue W_1 thus calculated does not depend on the spin of the electron. From now on in this section, ρ_e will be regarded as a variable, otherwise by substituting the form in Eq. (3.31) one can get meaningless results. Equation (3.38) can be used to evaluate the effective mass of the electron in the conduction states. We will expand Eq. (3.38) about $\rho_\mathrm{e} = \mathbf{0}$ in Taylor's series as

$$W_{1\mathrm{e}} \approx W_0 + N^{-1} \sum_{\mathbf{l}} E_{\mathbf{lll}} + i\boldsymbol{\rho}_\mathrm{e} \cdot N^{-1} \sum_{\mathbf{l}\neq\mathbf{m}} (\mathbf{m} - \mathbf{l}) E_{\mathbf{llm}}$$
$$- \frac{\rho_\mathrm{e}^2}{2} \cdot N^{-1} \sum_{\mathbf{l}\neq\mathbf{m}} (\mathbf{m} - \mathbf{l})^2 E_{\mathbf{llm}}, \tag{3.39}$$

where the series is terminated at the second order. As stated above, the expansion is carried out about $\rho_\mathrm{e} = \mathbf{0}$, which in terms of energy means at $E = E_\mathrm{c}$, it is the electron mobility edge. The second term on the right-hand side of Eq. (3.39) gives the average energy of the excited electron in the conduction states. The third term is zero, because the summation is over both $\mathbf{l}$ and $\mathbf{m}$, such that $\mathbf{l} \neq \mathbf{m}$. The last term can be used to define the effective mass of the electron, m_e^*, in the conduction states as (Singh, 1994, 2002; Singh *et al.*, 2002)

$$[m_\mathrm{e}^*]^{-1} = -\hbar^{-2} N^{-1} \sum_{\mathbf{l}\neq\mathbf{m}} (\mathbf{m} - \mathbf{l})^2 E_{\mathbf{llm}}. \tag{3.40}$$

The main achievement of this approach is to get an expression for the electron effective mass in the real coordinate space. Although a similar expression for the effective mass of an electron in a-solids has been derived by Kivelson and Gelatt (1979), the approach followed here is relatively simpler. The advantage of the expression in Eq. (3.40) is that it can be used to derive distinct effective masses of an electron in the extended states and tail states.

For calculating m_e^* from Eq. (3.40) one requires first to evaluate all the integrals involved in Eq. (3.36). As described later in this section, for sp^3 hybridized systems, one has first to form linear combinations of the atomic orbitals and then calculate the integrals. Even then, however, for amorphous semiconductors (a-semiconductors) one has to do these calculations for every sample separately, because different samples may have different atomic configurations. This is usually going to be a very time consuming task. Alternatively, one can apply certain approximations to estimate the effective mass from Eq. (3.40). As the nearest coordinating atoms play the most significant role in a-solids, let us evaluate Eq. (3.40) by applying the nearest neighbor approximation. For this purpose, it is more convenient to write Eq. (3.40) as a sum of two terms as

$$[m_e^*]^{-1} = -\hbar^{-2}\left[aN_l^{-1}\sum_{\mathbf{l}\neq\mathbf{m}}(\mathbf{m}-\mathbf{l})^2 E_{\mathbf{llm}} + bN_2^{-1}\sum_{\mathbf{l}'\neq\mathbf{m}'}(\mathbf{m}'-\mathbf{l}')^2 E_{\mathbf{ll'm'}}\right] \tag{3.41}$$

where $a = N_l/N < 1$, N_l is the number of atoms contributing to the extended states and $b = N_2/N < 1$, N_2 is the number of atoms contributing to the tail states, such that $a + b = 1$ $(N = N_l + N_2)$. In the first term of Eq. (3.41), **l** and **m** denote positions of those atoms that contribute to the extended states and in the second term **l**′ and **m**′ represent atoms that contribute to the tail states. The first term gives the inverse of the effective mass of an electron in the extended states and second term gives that in the tail states.

For evaluating the effective mass of an electron in the conduction extended states, we denote the average bond length between nearest neighbors of atoms contributing to the extended states by L_l, and then we can write the first term of Eq. (3.41) as

$$[m_{ex}^*]^{-1} \approx -\hbar^{-2}L_l^2\, aN_l^{-1}\sum_{\mathbf{l}\neq\mathbf{m}} E_{\mathbf{llm}}, \tag{3.42}$$

where m_{ex}^* is used to represent the effective mass of the electron in the conduction extended states. The double summation over **l** and **m** ($\mathbf{l} \neq \mathbf{m}$) in Eq. (3.42) gives the total inter-atomic interaction energy, which is the full width of the conduction extended states. This can be intuitively understood from the fact that when atoms of a solid are isolated (far apart), the electronic energy levels are the same as those of atoms. However, when atoms are brought together to form a solid energy bands are formed due to inter-atomic interactions.

It should also be remembered that the energy matrix element $E_{\mathbf{llm}} < 0$, because it is the energy of an electron in a bound state of a solid. To evaluate the full width of the conduction extended states, it is easier and more convenient experimentally to determine the half width and then twice of that will give the full width. The middle of the extended states occurs at an energy where the imaginary part of the dielectric constant becomes maximum. Let us denote this energy by E_2 (Kivelson and Gelatt, 1979), and then the half width of conduction extended states becomes $= E_2 - E_c$. Using this in Eq. (3.42), we can write:

$$-N_l^{-1}\sum_{\mathbf{l}\neq\mathbf{m}} E_{\mathbf{llm}} = 2(E_2 - E_c), \tag{3.43}$$

Substituting Eq. (3.43) in Eq. (3.42), we get

$$[m_{ex}^*]^{-1} \approx \hbar^{-2}L_l^2[2(E_2 - E_c)a]. \tag{3.44}$$

For further simplification of Eq. (3.44), we denote by E_{L_1} to the constant energy given by

$$E_{L_1} = \frac{\hbar^2}{m_e L_1^2}, \tag{3.45}$$

where m_e is the free electron mass, and L_1 can be expressed as $L_1 = [3/4\pi n_1]^{1/3}$ with $n_1 = N_1/V$; V being the volume of sample. Expressing $n_1 = an$, one can write $L_1 = L/\sqrt[3]{a}$, where $n = N/V$, and $L = [3/4\pi n]^{1/3}$. Thus L becomes the average bond length in a sample, and then Eq. (3.45) becomes

$$E_{L_1} = a^{2/3} E_L, \tag{3.46}$$

where

$$E_L = \frac{\hbar^2}{m_e L^2}. \tag{3.47}$$

Using Eqs (3.46) and (3.47), we get from Eq. (3.44) the effective mass of an electron in the conduction extended states as

$$m_{ex}^* \approx \frac{E_L}{2(E_2 - E_c)a^{1/3}} m_e. \tag{3.48}$$

Likewise, for determining the electron effective mass in the conduction tail states, we consider the second term of Eq. (3.41). Denoting by L_2 the average separation between nearest atoms contributing to the tail states and applying the nearest neighbor approximation, we can write the second term of Eq. (3.41) as

$$[m_{et}^*]^{-1} \approx -\hbar^{-2} b L_2^2 N_2^{-1} \sum_{\mathbf{l}' \neq \mathbf{m}'} E_{\mathbf{l}\mathbf{l}'\mathbf{m}'}, \tag{3.49}$$

where m_{et}^* denotes the effective mass of an electron in the conduction tail states. The tail states lie below E_c, and if we denote by E_{ct} to the energy of the end of the conduction tail states, we get its width as

$$-N_2^{-1} \sum_{\mathbf{l}' \neq \mathbf{m}'} E_{\mathbf{l}\mathbf{l}'\mathbf{m}'} = (E_c - E_{ct}). \tag{3.50}$$

This is not to be confused with the effective width of the tail states obtained from the density of states used in Chapter 8 for the transport of charge carriers. Using Eq. (3.50) in (3.49), we get the effective mass of an electron in the conduction tail states as

$$m_{et}^* \approx \frac{E_{L_2}}{(E_c - E_{ct})b} m_e, \tag{3.51}$$

where E_{L_2} is obtained from Eq. (3.45) replacing L_1 by L_2, which can be estimated from the density of atoms $n_2 = N_2/V$ as $L_2 = [3/4\pi n_2]^{1/3} = L/\sqrt[3]{b}$. Using this

Table 3.1 Effective mass of an electron m^*_{ex} in the conduction extended states calculated using Eq. (3.48) with $a = 0.99$, and m^*_{ct} in the conduction tail states using Eq. (3.52) with $b = 0.01$. Energy E_L calculated from Eq. (3.47) is also given. Letters in square parentheses indicate references

	E_2 (eV) [a]	L (nm)	E_L	E_c (eV)	$E_c - E_{ct}$ (eV)	m^*_{ex} (Eq. (3.48))	m^*_{ct} (Eq. (3.52))
a-Si : H	3.6 [f]	0.235 [b]	1.23	1.80 [g]	0.8 [c]	$0.34m_e$	$7.1m_e$
a-Ge : H	3.6 [f]	0.245 [b]	1.14	1.05 [h]	0.53	$0.22m_e$	$10.0m_e$
a-As_2S_3	4.0 [d]	0.227 [e]	1.47	2.40 [i]	1.20	$0.46m_e$	$5.68m_e$

Notes
a Kivelson and Gelatt, 1979.
b Morigaki, 1999.
c Spear, 1988.
d Connell, 1979.
e Elliot, 1991.
f Ley, 1984.
g Street, 1991.
h Aoki *et al.*, 1999.
i Elliott, 1990.

in Eq. (3.51) we get:

$$m^*_{et} \approx \frac{E_L}{(E_c - E_{ct})b^{1/3}} m_e. \tag{3.52}$$

Using Eqs (3.48) and (3.52) and the values of parameters involved, the effective mass of an electron can be calculated in the extended and tail states in any a-semiconductors. Considering, for example, the density of weak bonds contributing to the tail states as 1 at.%, that is, $b = 0.01$, and $a = 0.99$, the effective mass calculated from Eqs (3.48) and (3.52) and the energy E_L calculated from Eq. (3.47) for some a-solids are given in Table 3.1. Usually one gets $m^*_{et} > m^*_e$, which is consistent with the concept that localized states have a relatively larger effective mass in comparison with delocalized states (as illustrated in Fig. 3.2).

Using Eqs (3.41), (3.48) and (3.52) in Eq. (3.39), the energy eigenvalue of an electron in the conduction states can be written within the effective mass approximation as

$$W_{1e} \approx W_0 + E_1 + \frac{p_e^2}{2m^*_e} - \frac{p_{et}^2}{2m^*_{et}}, \tag{3.53}$$

where $E_1 = N^{-1} \sum_l E_{1ll}$, is the energy of an electron in the conduction states (without the kinetic energy), and $\mathbf{p}_e = \hbar\boldsymbol{\rho}_e$ ($\mathbf{p}_{et} = \hbar\boldsymbol{\rho}_{et}$) is the linear momentum of an electron in the extended (tail) states. The negative kinetic energy in the tail states comes from the fact that according to Eq. (3.31), ρ_e becomes imaginary in the tail states. It may be pointed out here that when an electron is excited to the conduction extended states, only the first term of the kinetic energy should be considered in Eq. (3.53). When it has relaxed to the tail states, then only the second kinetic energy term corresponding to the tail states needs to be considered. However, near the mobility edge both terms together may become important. If $m_{\mathrm{et}}^{*} \gg m_{\mathrm{ex}}^{*}$, then the negative kinetic energy term can be neglected and electronic motion will be little affected by the tail states. When this is not so the influence of the tail states cannot be ignored.

3.3.2 *Effective mass of a hole in valence states*

An approach analogous to that developed in the preceding subsection for calculating the effective mass of an electron in the conduction states of a-semiconductors can be followed for calculating the effective mass of a hole in the valence states. For this purpose let us consider that a hole is created in the valence states, denoted by 0, with energy E and spin σ_{h}. Following Eq. (3.29), the eigenvector of such a state of an a-semiconductor can be written as

$$|0, E, \sigma_{\mathrm{h}}\rangle = \sum_{\mathbf{l}} C_{0\mathbf{l}}(E) d_{0\mathbf{l}}^{+}(\sigma_{\mathrm{h}})|0\rangle, \tag{3.54}$$

where $d_{0\mathbf{l}}^{+}(\sigma_{\mathrm{h}})$ $[d_{0\mathbf{l}}(\sigma_{\mathrm{h}})]$ is the creation (annihilation) operator of a hole in the valence states 0, localized at the site $\mathbf{l}$ and with spin σ_{h}. The hole operators are defined as follows

$$d_{0\mathbf{l}}^{+}(\sigma) = a_{0\mathbf{l}}(-\sigma), \quad d_{0\mathbf{l}}(\sigma) = a_{0\mathbf{l}}^{+}(-\sigma) \quad \text{and} \quad d_{0\mathbf{l}}(\sigma)|0\rangle = 0. \tag{3.55}$$

Using Eqs (3.26) and (3.53), we can calculate the energy eigenvalue, W_{h}, for an a-solid with one hole in the valence states by solving the following Schrödinger equation:

$$\hat{H}|0, E, \sigma_{\mathrm{h}}\rangle = W_{0\mathrm{h}}|0, E, \sigma_{\mathrm{h}}\rangle. \tag{3.56}$$

For calculating W_{h} from Eq. (3.56), the easiest way would be to convert hole operators into electron operators in Eq. (3.53) according to Eq. (3.55), and then follow the approach used for electron in Eq. (3.33). We thus obtain

$$W_{0\mathrm{h}} = W_0 - \sum_{\mathbf{l},\mathbf{m}} C_{0\mathbf{l}}^{*}(E) C_{0\mathbf{m}}(E) E_{0\mathbf{lm}}, \tag{3.57}$$

where

$$\begin{aligned} E_{0\mathbf{lm}} = & \left\langle 0, \mathbf{l} \left| -\frac{\hbar^2 \nabla^2}{2m_{\mathrm{e}}} + V(\mathbf{r}) \right| 0, \mathbf{m} \right\rangle \\ & + \sum_{\mathbf{l}'} [2\langle 0, \mathbf{l}; 0, \mathbf{l}' | U(|\mathbf{r}_1 - \mathbf{r}_2|) | 0, \mathbf{l}'; 0, \mathbf{m}\rangle \\ & - \langle 0, \mathbf{l}; 0, \mathbf{l}' | U(|\mathbf{r}_1 - \mathbf{r}_2|) | 0, \mathbf{m}; 0, \mathbf{l}'\rangle]. \end{aligned} \tag{3.58}$$

is the total energy of the hole in the valence states, including its interaction energy with all the electrons in the valence states. However, it is to be noted that, unlike the corresponding expression for the electronic energy in Eq. (3.36), $E_{0\mathbf{lm}} > 0$, because a hole has the positive charge. Following Eq. (3.30), the probability amplitude coefficients, $C_{0l}(E)$, for holes can be expressed as (Singh, 2002; Singh *et al.*, 2002)

$$C_{0l}(E) = N^{-1/2} \exp(i\boldsymbol{\rho}_h \cdot \mathbf{l}), \tag{3.59}$$

where

$$\rho_h(E) = \sqrt{\frac{2m_h^*(E_v - E)}{\hbar^2}}. \tag{3.60}$$

Here m_h^* is the effective mass of hole in the valence states and it will be determined in the same way as the effective mass of an electron. E_v is the energy of the hole mobility edge in the valence states; energy states below this are the extended states and above this are the tail states. Thus, here again the choice of the probability amplitude coefficient is such that it separates the valence extended states from valence tail states. Using Eq. (3.59) in Eq. (3.57), we get

$$W_{0h} = W_0 - N^{-1} \sum_{\mathbf{l},\mathbf{m}} \exp[i\boldsymbol{\rho}_h \cdot (\mathbf{m} - \mathbf{l})] E_{0\mathbf{lm}}, \tag{3.61}$$

which is similar to the energy of electron in the conduction states, as obtained in Eq. (3.38). We can calculate the effective mass of a hole from Eq. (3.61) following exactly the same steps as were used for determining the effective mass of an electron, that is, following Eqs (3.39)–(3.52). Then within the effective mass approximation one gets the energy eigenvalue of a hole as

$$W_{0h} \approx W_0 - E_h + \frac{p_h^2}{2m_{hx}^*} - \frac{p_{ht}^2}{2m_{ht}^*}, \tag{3.62}$$

where $E_h = N^{-1} \sum_l E_{0ll}$ is the interaction energy of a hole in the valence states, $\mathbf{p}_h = \hbar\boldsymbol{\rho}_h$ and $\mathbf{p}_{ht} = \hbar\boldsymbol{\rho}_{ht}$ are linear momenta, and m_{hx}^* and m_{ht}^* are effective masses of hole in the valence extended and tail states, respectively, obtained as

$$m_{hx}^* \approx \frac{E_L}{2(E_v - E_{v2})a^{1/3}} m_e, \tag{3.63}$$

and

$$m_{ht}^* \approx \frac{E_L}{(E_{vt} - E_v)b^{1/3}} m_e, \tag{3.64}$$

where E_{v2} and E_{vt} are energies corresponding to the half width of valence extended states and the end of the valence tail states, respectively.

3.3.3 Comparison with other theories

It may be considered desirable here to compare the results of the effective mass obtained above with those derived/calculated using other theories. As stated above, an earlier theory developed for calculating the effective mass of charge carriers in the real coordinate space for c-solids is by Kivelson and Gelatt (1979). The other theory developed is for Frenkel excitons by Singh (1994), who has derived an expression for the effective mass of Frenkel excitons in the real coordinate space. Applying the f-sum rule, Kivelson and Gelatt have derived an analytical expression for the average effective mass of an electron in the conduction band of c-solids as

$$\left[\frac{m_e^*}{m_e}\right]^{-1} \approx \frac{7E_2}{4E_g} - 1, \tag{3.65}$$

where E_g is the energy of the minimum energy gap. Kivelson and Gelatt have indicated that the formula given in Eq. (3.65) yields results correct to within 40% for the crystalline tetrahedral semiconductors in rows three and four of the Periodic Table. As their result is derived from the zeroth order Hamiltonian, they have argued that the same expression can be applied to a-solids as well. Their argument is based on the following two points: (1) The tight-binding Hamiltonian matrix elements, similar to $E_{\mathbf{1lm}}$ (Eq. (3.36)), are largely molecular in nature and so will be of comparable magnitude in both crystalline and a-solids. (2) The bond distances are also of comparable magnitude in both the systems. Both of their arguments are reasonable and applicable to the present theory.

As Eq. (3.65) is derived for c-solids, no tail states are considered to exist. Applying the same condition in the present theory (Section 3.3.1), we get $N_2 = 0$, $N_1 = N$ and $a = 1$. Under this condition Eq. (3.48) becomes

$$m_e^* \approx \frac{E_L}{2(E_2 - E_c)} m_e. \tag{3.66}$$

It should be noted that the form of E_L in Eq. (3.47) suggests that it is the ground state energy of a particle of effective mass $= 0.5m_e$ confined in an infinite square potential well of width L. The energy of a particle of mass m confined in a one-dimensional square well of width L is given by

$$E_n = \frac{\hbar^2}{2mL^2}\frac{1}{n^2}, \quad n = 1, 2, 3, \ldots .$$

This is the effect of the nearest neighbor approximation, implying that at every bond the electron is confined in the same way as if it is confined within the a-solid, because at any instant of time the electron cannot realize the presence of any other atoms beyond the nearest neighbors. Thus if all bond lengths are the same, which would constitute all regions contributing to extended states, E_L becomes the lowest energy state associated with the extended state. Within the nearest neighbor approximation, therefore, the energy E_L provides a good estimate for the energy of the electron mobility edge for a-solids ($E_L \approx E_c$) and that of the minimum energy gap in c-solids ($E_L \approx E_g$), that is, $E_L \approx E_c \approx E_g$. Within this approximation, the effective mass of an electron in the conduction extended states may also be approximately expressed as

$$m_e^* \approx \frac{E_c}{2(E_2 - E_c)} m_e. \tag{3.67}$$

For $E_c \approx E_g$, Eqs (3.65) and (3.67) produce similar results as described below: For a direct comparison of results obtained in Eqs (3.65) and (3.67), f rst we need to replace either E_g by E_c in Eq. (3.65) or E_c by E_g in Eq. (3.67). As we are mainly concerned here with a-solids, let us apply the former and replace E_g by E_c in Eq. (3.65), which can then be written as

$$\left[\frac{m_e^*}{m_e}\right]^{-1} \approx \frac{7E_2}{4E_c} - 1 = \frac{2(E_2 - E_c) - \left(\frac{1}{4}E_2 - E_c\right)}{E_c}. \tag{3.68}$$

Equation (3.68) demonstrates clearly that if $\frac{1}{4}E_2 \approx E_c$ (or $\frac{1}{4}E_2 \approx E_g$), then both Eqs (3.65) and (3.67) give the same result for the effective mass of electron. Table 3.2 shows the results obtained from Eq. (3.67), replacing E_c by E_g, and those obtained from Eq. (3.65) by Kivelson and Gelatt for a few c- and a-solids. It is clear from Table 3.2 that Eq. (3.67) with $E_c = E_g$ gives results identical to those obtained by Kivelson and Gelatt for all solids, with the exception of a-GaAs. In a-GaAs also the discrepancy is minimal. It is to be noted that the input data used for calculating the results given in Table 3.2 are the same as those used by Kivelson and Gelatt (1979).

The next question that arises is, if one calculates the effective mass of an electron in the conduction states and hole in the valence states of a-semiconductors using the above theory, then are they going to be the same or different? In general, like in c-solids they can be different. However, for sp^3-hybridized systems, it can be

Table 3.2 Average effective mass of electron determined using Eq. (3.65) (Kivelson and Gelatt, 1979) and Eq. (3.67) with $E_c = E_g$. Letters in square parentheses indicate references

	E_2 (eV)	E_g (eV)	m_e^*/m_e (Eq. (3.65))	m_e^*/m_e (Eq. (3.67)) with $E_c = E_g$	m_e^*/m_e (Exptl.)
c-Si	4.5	1.1	0.16	0.16	0.23 [a]
a-Si	~3	~1	~0.25	0.25	[b]
c-Ge	4.4	0.66	0.09	0.09	0.08 [c]
a-Ge	~3	~1	~0.25	0.25	[b]
c-GaAs	5.1	1.43	0.19	0.19	0.13 [a]
a-GaAs	3.5	1.17	0.24	0.25	[d]

Notes

a Moss *et al.*, 1973.

b Adler, 1971.

c Morigaki, 1999.

d Georghi and Theye, 1978.

expected that both m_e^* and m_h^* would be the same. This is because in such systems valence and conduction states are two equal halves of one band states formed from overlapping s and p bands (see Fig. 1.5). Therefore, their widths, which determine the effective mass, are equal. This is different from c-solids, where electron and hole effective masses can be different in sp^3 hybrid systems. This difference conforms to the amorphous nature in the same way as in such solids there are no direct and indirect semiconductors, which do exist in c-solids. In solids where such equal splitting due to sp^3-hybridization does not occur, the valence and conduction states may have different energy widths and hence different effective masses.

It may also be noted that expressions for the effective mass derived in Eqs (3.48), (3.52), (3.63) and (3.64) provide only an estimate of the effective masses. The effective mass of a charge carrier in the tail states is usually expected to be higher as these states are localized states and have relatively smaller widths. This is consistent with the concept of localization known in the charge carrier–lattice interaction, which increases the charge carrier's effective mass. Accordingly, for example, an electron's effective mass increases when it occupies a polaronic band (see Chapter 6), whose band width is smaller than the free electron band width. Then if the electron–lattice interaction becomes very strong, the electron reaches the state of self-trapping. In a self-trapping state, the electron's effective mass can become enormously large as first shown by Toyozawa (1959, 1962) for organic solids, and the width of such bands vanishes.

Equations (3.48), (3.52), (3.63) and (3.64) suggest that the effective mass of a charge carrier is inversely proportional to the width of the corresponding energy states at a fixed concentration of atoms contributing to the tail states. Such a dependence on the width of energy bands has also been found by Mott and Davis (1979) in discussing Anderson's localization but not found in the theory of Kivelson and Gelatt.

Present theory of the effective mass in a-solids enables one to determine effective masses in the extended and tail states separately and therefore provides a way of studying the differences in the free electron-like properties of a charge carrier in the two types of states. However, one may question the application of the effective mass concept to the tail states, which are localized states. From the theory of chemical bonding, the only difference between the extended and tail states is that the energy separation between different bonding and anti-bonding orbitals in the extended states is much smaller (quasi-continuous) than that in the tail states (discrete or localized). That means the average separation between atoms contributing to the tail states is much longer than that between atoms contributing to the extended states. In this view, it is hard to find any reason for the formalism presented in Section 2 to be applicable only for the quasi continuous but not for the discrete states. One problem may, of course, be encountered is how to determine the width of the tail states or in other words how to determine the locations of E_{ct} (Eq. (3.52)) and E_{vt} (Eq. (3.64)). According to the theoretical model, the density of states for the tail states has exponential dependence on energy and that means E_{ct} and E_{vt} cannot be defined. For calculating the effective mass it is more appropriate to use the width of the conduction tail states as the energy separation between the mobility edge and Fermi level. Accordingly, following Spear (1988), $E_c - E_{ct} = 0.8\,\text{eV}$ is used in calculating m^*_{et} for a-Si : H (see Table 3.1). This is very close to the half of the optical band gap of 0.9 eV in a-Si : H. Therefore, for an estimate of the effective mass in the tail states, one may locate E_{ct} and E_{vt} at the middle of the optical energy gap. In this way one would get the same effective mass for the hole in the valence tail states as that for the electron in conduction tail states. The effective mass thus obtained from Eq. (3.52) for $E_c - E_{ct} = 0.9\,\text{eV}$ is 6.3 m_e instead of $7.1m_e$ obtained for $E_c - E_{ct} = 0.8\text{eV}$. Using $E_c - E_{ct} = E_c/2$, the tail effective masses for a-Ge : H and a-As_2S_3 are obtained as $10.0\,m_e$ and $5.68\,m_e$, respectively, as listed in Table 3.1.

It may, therefore, be concluded that the effective mass approach can be applied to a-solids and effective masses of charge carriers can be calculated accordingly. The effective mass thus calculated should be used in calculating the properties of a-semiconductors. The notion that the effective mass concept cannot be applied to a-solids is merely an incorrect assumption.

3.4 Atomic wavefunctions

The form of the atomic wavefunctions $\varphi_{jl\sigma}(r)$ used in Section 3.3 is described here. For describing the states of excited electron and hole pairs in a solid, usually the two bands model is considered adequate. In that case $\varphi_{jl\sigma}(r)$ are used as the atomic wavefuntions associated with the valence and conduction states in writing the Hamiltonian given in Eqs (3.26)–(3.28). However, in the case sp^3-hybridized systems of semiconductors such a choice is not possible, because both valence and conduction states arise from the same atomic states of individual atoms. In this situation, each atomic wavefunction is a linear combination given by

$$\phi_{nsnpl}(r) = c_1\phi_{nsl} + c_2\phi_{np_xl} + c_3\phi_{np_yl} + c_4\phi_{np_zl},$$

where the probability amplitude coefficients are obtained as

$$c_1 = c_2 = c_3 = c_4 = \pm\tfrac{1}{2}.$$

Thus, a set of four functions is obtained to be used in forming the antisymetrized product of wavefunctions.

References

Adler, D. (1971). *CRC Crit. Rev. Solid State Sci.* **2**, 317.

Aoki, T., Shimada, H., Hirao, N., Yoshida, N., Shimakawa, K. and Elliott, S.R. (1999). *Phys. Rev. B* **59**, 1579.

Connell, G.A.N. (1979). In: Brodsky, M.H. (ed.), *Amorphous Semiconductors*. Springer-Verlag, Berlin-Heidelberg, p. 73.

Elliott, S.R. (1990). *Physics of Amorphous Materials*, 2nd edn. John-Wiley, London.

Georghi, A. and Theye, M.-L. (1978). In: Spear, W.E. (ed.), *Amorphous and Liquid Semiconductors*. University of Edinburgh, Dundee.

Kivelson, S. and Gelatt, C.D. Jr. (1979). *Phys. Rev. B* **19**, 5160.

Ley, L. (1984). In: Joanpoulos, J.D. and Lucovsky, G. (eds), *The Physics of Hydrogenated Amorphous Silicon II*. Springer-Verlag, Berlin, p. 61.

Morigaki, K. (1999). *Physics of Amorphous Semiconductors*. World-Scientific-Imperial College Press, London.

Moss, T.S., Burrell, G.J. and Ellis, B. (1973). *Semiconductors and Opto-electronics*. Butterworths, London.

Mott, N.F. and Davies, E.A. (1979). *Electronic Processes in Non-Crystalline Materials*. Clarendon Press, Oxford.

Singh, J. (1994). *Excitation Energy Transfer Processes in Condensed Matter*. Plenum, New York.

Singh, J. (2002). *J. Non-Cryst. Solids* **299-302**, 444.

Singh, J. (2003). *J. Mat. Science* (in press).

Singh, J., Aoki, T. and Shimakawa, K. (2002). *Phil. Mag. B* **82**, 855.

Spear, W.E. (1988). Transport and tail state interactions in amorphous silicon. In: Fritzsche, H. (ed.), *Amorphous Silicon and Related Materials*. World Scientific, Singapore, p. 721.

Street, R.A. (1991). *Hydrogenated Amorphous Silicon*. Cambridge University Press, Cambridge

Toyozawa, Y. (1959). *Prog. Theor. Phys.* **26**, 29.

Toyozawa, Y. (1962). In: Kuper, C.G. and Whitfield, G.D. (eds), *Polarons and Excitons*. Oliver and Boyd, Edinburgh.

4 Optical properties

One of the most important properties of any amorphous solid (a-solid) is its optical properties. There are two reasons for this: (1) optical properties of crystalline semiconductors (c-semiconductors) are very well studied, so the optical properties of amorphous semiconductors (a-semiconductors) can easily be related and compared and (2) optical properties are directly related to the structural and electronic properties of solids, and hence very important in device applications. In this regard, two very important and related optical properties are the absorption coefficient and density of charge carrier states. A detailed knowledge of these can provide a huge amount of information on materials about their structure, opto-electronic behavior, transport of charge carriers, etc. Figure 4.1 shows the plots of log of absorption coefficient, $\log \alpha$, as a function of the photon energy, $E = \hbar\omega$, for four different samples of amorphous silicon (a-Si) (Demichellis *et al.*, 1986) and in the inset are also shown the same plots at four different temperatures in crystalline silicon (c-Si). What is illustrated in Fig. 4.1 is that the absorption coefficient decreases systematically in both a-Si : H and c-Si. In the former, the systematic decrease is due to the increase in the hydrogen concentration from 0 to 29 at.%, whereas in the latter it is due to decrease in sample temperature. This clearly illustrates that the energy gap is influenced by increasing the H-concentration in a-Si : H.

Figures 4.2(a) and (b) show schematically typical features of density of states (DOS) of a c-semiconductor and a-semiconductor, respectively. In c-semiconductor, there is no DOS within the free electron energy gap, E_g, whereas in the a-semiconductor, there exists a non-zero DOS within the mobility gap. It is now well established that these distinct features of the absorption coefficient and DOS of a-semiconductors arise due to the presence of tail states, which are localized states, and which do not exist in c-semiconductors (Mott and Davis, 1979; Cody, 1984; Overhof and Thomas, 1989; Adachi, 1999; Morigaki, 1999). It is also well established that the origin of tail states is the absence of long range order in a-semiconductors, or in other words, the presence of disorders like weak bonds. To our knowledge, there is no systematic theory developed for a-semiconductors to demonstrate that the DOS within the gap is like what is shown in Fig. 4.2(b).

In Section 4.1, we will discuss the DOS in a-semiconductors, and then we will describe the absorption coefficient in Section 4.2.

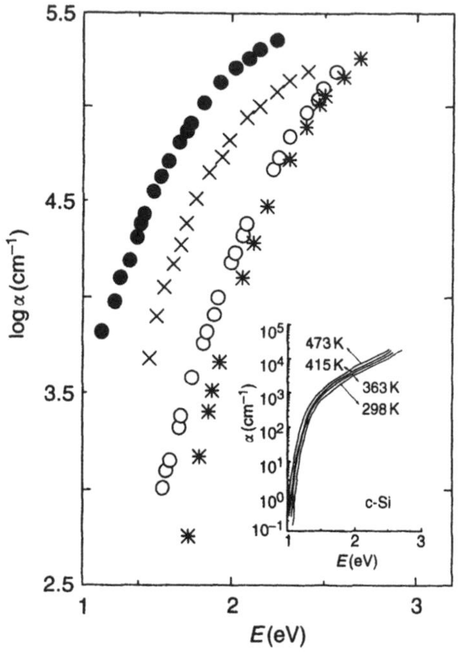

Figure 4.1 The absorption coefficient plotted as a function of photon energy near the region of the energy gap for four different samples of a-Si : H with hydrogen concentration of 0 at. % (●), 9 at. % (×), 23 at. % (✱) and 29 at. % (○). These results have been compared with those in crystalline silicon measured at four different temperatures (Demichellis *et al.*, 1986).

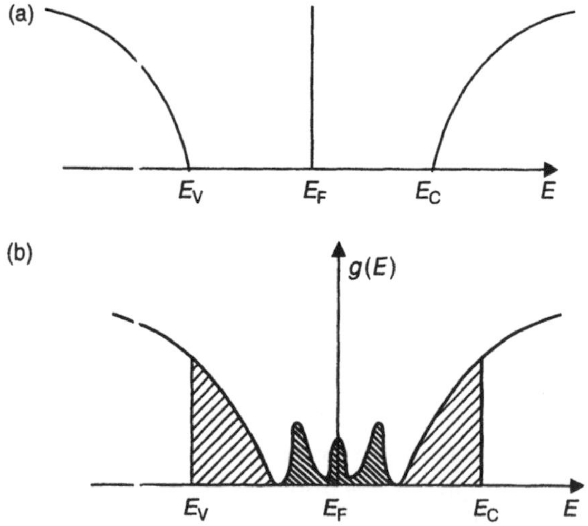

Figure 4.2 (a) Schematic representation of DOS in c-semiconductors, and (b) DOS in a-semiconductors. Localized states are shaded. E_F is the Fermi level (Shklovskii and Efros, 1984).

4.1 Density of states

The concept of free electron DOS can be applied to the extended states, using the effective mass approach developed in Chapter 3. The effective mass concept can be applied to transform each ionic potential into a square well potential, and then a fully coordinated amorphous network can be regarded as an effective medium of a Fermi free electron gas. In this effective medium, each charge carrier can be regarded as free and independent with an effective mass, calculated in Chapter 3. For an electron in this gas, the Schrödinger equation is well known as

$$-\frac{\hbar^2}{2m_e^*}\nabla^2\psi(r) = E\psi(r), \tag{4.1}$$

where m_e^* is the effective mass of electron. The energy eigenvalue is obtained as

$$E = \frac{\hbar^2\rho^2}{2m_e^*}, \tag{4.2}$$

where ρ is a function of energy as defined in Chapter 3, and the eigenfunctions are obtained as

$$\psi(\mathbf{r}) = A\exp(\mathrm{i}\boldsymbol{\rho}\cdot\mathbf{r}) + B\exp(-\mathrm{i}\boldsymbol{\rho}\cdot\mathbf{r}). \tag{4.3}$$

It is to be noted that $\boldsymbol{\rho}$ can be direction dependent, although in a fully coordinated network as the energy levels are expected to be quasi continuous, it may be assumed to be isotropic. The confinement condition essentially means that at the edges of the sample there exists an infinitely high potential barrier. Thus all eigenfunctions must vanish at the sample boundaries. Applying the condition that the probability of incidence and reflection at the boundaries of the square well is the same, we get $|A|^2 = |B|^2$, which leads to the standing wave solutions. Then, the normalizing constant can be determined by the boundary condition that the electron is confined in a finite size sample, and thus setting

$$\int \psi^*(\mathbf{r})\psi(\mathbf{r})\,\mathrm{d}^3\mathbf{r} = 1. \tag{4.4}$$

For a cubic sample of length L (from 0 to L), the standing wave solutions are then obtained as

$$\psi(x, y, z) = \left(\frac{8}{L^3}\right)^{1/2} \sin\left(n_x\frac{\pi}{L}x\right)\sin\left(n_y\frac{\pi}{L}y\right)\sin\left(n_z\frac{\pi}{L}z\right), \tag{4.5}$$

which is well known. Here $n_i = 1, 2, \ldots, i = x, y, z$, are integral quantum numbers, and the allowed energy values are then obtained from Eq. (4.1) as

$$E = \frac{\hbar^2\rho_0^2}{2m_e^*}(n_x^2 + n_y^2 + n_z^2), \quad \rho_0 \equiv \frac{\pi}{L}. \tag{4.6}$$

For macroscopic samples with large L, the separation between the allowed energy eigenvalues in ρ-space, or in other words in energy-space, becomes infinitesimally small. Thus the number of allowed energy states in a sample for ρ values within ρ and $\rho + d\rho$ will be

$$g(\rho)\,d\rho = \frac{V\rho^2}{\pi^2}\,d\rho. \tag{4.7}$$

Here V is the volume of the sample. The result of Eq. (4.7) is well known (see, e.g. Elliott, 1998). (This is equal to the volume of the spherical shell of radius ρ and thickness $d\rho$ in the positive octant $= \frac{1}{8}4\pi\rho^2\,d\rho$, multiplied by the density of standing waves $= V/\pi^3$ and multiplied by the electron spin degeneracy $= 2$). The DOS in the energy space can then be written as

$$g(E) = g(\rho)\frac{d\rho}{dE}. \tag{4.8}$$

Using Eqs (4.2) and (4.7) in Eq. (4.8), we get

$$g(E) = \frac{V}{2\pi^2}\left(\frac{2m_e^*}{\hbar^2}\right)^{3/2} E^{1/2}. \tag{4.9}$$

This well-known result of the DOS of Fermi free electron gas can be applied to both valence and conduction extended states, provided the respective effective masses of charge carriers are used. The Fermi energy and Fermi sphere can also be defined in a similar manner. Following the Pauli exclusion principle, one state can accommodate only one electron, and hence at $T = 0$ K, N electrons will fill the lowest N states, up to the Fermi energy $E = E_F$, as

$$N = \int_0^{E_F} g(E)\,dE = \frac{V}{3\pi^2}\left(\frac{2m_e^*}{\hbar^2}E_F\right)^{3/2} = \frac{2}{3}E_F\,g(E_F), \tag{4.10}$$

where $g(E_F)$ is the DOS at the Fermi energy obtained from Eq. (4.9). The Fermi energy is then obtained from Eq. (4.10) as

$$E_F = \frac{\hbar^2}{2m_e^*}\left(\frac{3\pi^2 N}{V}\right)^{2/3}, \tag{4.11}$$

which depends on the density of electrons, $n = N/V$. One can also define radius of the Fermi sphere in ρ-space as

$$\rho_F = \left[\frac{2m_e^*}{\hbar^2}E_F\right]^{1/2}, \tag{4.12}$$

and the Fermi temperature as

$$T_F = \frac{E_F}{\kappa_B}, \tag{4.13}$$

where κ_B is the Boltzmann constant. These results illustrate the point that the form of the DOS, obtained in Eq. (4.9), does not depend on the translational symmetry.

It can be applied to any solid crystalline or non-crystalline, as long as it is a free electron system, because the above derivations are true only for free electron gas. If the electrons are not free, they may be made so by using the effective mass approach and then they will be free with that effective mass. Systems with different effective masses of their charge carriers will have different density of electron states. Thus as shown in Chapter 3, charge carriers in a-semiconductors have usually a different effective mass than is used for crystalline solids (c-solids). Hence, the DOS is expected to be different. For energies, $E > E_C$, in the conduction states, and $E < E_V$ in the valence states, the dependence of DOS in Eq. (4.9) on the energy agrees with what has been observed in most a-semiconductors. However, Eq. (4.9) does suggest a sharp drop in the DOS at the mobility edges, which is not observed in any a-solid. The sharp drop can easily be seen by replacing E by $(E - E_C)$ for conduction extended states and with $(E_V - E)$ for valence extended states. This is because so far we have only considered a fully coordinated network of atoms, which is usually not the case in any a-solid. There are also disorders present, which influence the DOS near the mobility edges.

At energies near the mobility edges, E_C or E_V and within the mobility gap in a-solids, the non-zero DOS arises from the localized tail states contributed by the disorders. Of course, the influence of Franz–Keldysh effect (Dow and Redfield (1970), also see a discussion in Mott and Davis (1979)) is also a contributing factor to the DOS near the mobility edges. It is, therefore, impossible to develop a single approach or theory that can describe the DOS near the mobility edges, below E_C in the conduction states, and above E_V in the valence states, which are the regions of localized tail states in a-semiconductors. For example, as one moves down in energy from the electron mobility edge, E_C, the individual contribution from the above four factors to the DOS is expected to change. It is very difficult, if not impossible, to determine which of the above four components become more important than others in which energy region. One needs to study other optical quantities such as the absorption coefficient for understanding the distribution of electronic energy states within the mobility gap. This will be described in the next section.

4.2 Absorption coefficient

There are three distinct regions, A, B and C, observed in the absorption coefficient of a-semiconductors near the electron mobility edge (schematically shown in Fig. 4.3). Above the mobility edge, in the region of strong absorption (region A in Fig. 4.3), the dependence of the absorption coefficient on photon energy can be described as (Wood and Tauc, 1972; Davis, 1993):

$$\alpha\hbar\omega \propto (\hbar\omega - E_0)^2, \tag{4.14}$$

where α is the absorption coefficient, $\hbar\omega$ the energy of the absorbed photon, and E_0 the optical gap. Usually Eq. (4.14) is written as

$$(\alpha\hbar\omega)^{1/2} = C(\hbar\omega - E_0), \tag{4.15}$$

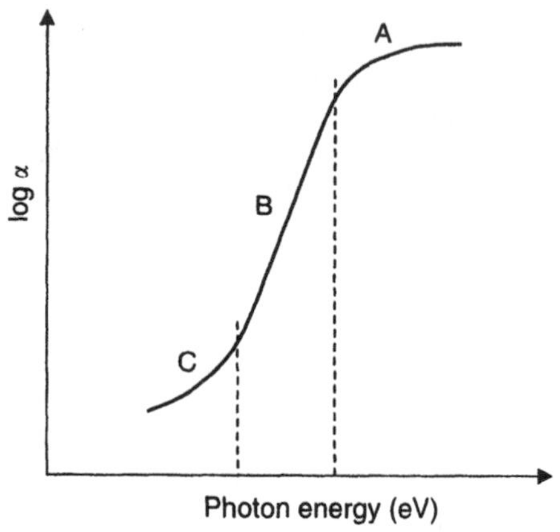

Figure 4.3 Three principal regions, A, B and C of optical absorption coefficient in a-semiconductors.

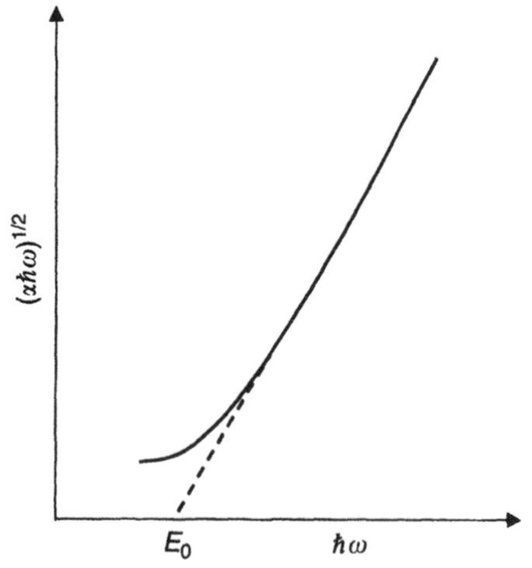

Figure 4.4 Schematic representation of the Tauc plot in a-semiconductors.

where C is independent of the photon energy. A plot of $(\alpha\hbar\omega)^{1/2}$ as a function of the photon energy is called the Tauc plot, and is schematically shown in Fig. 4.4. Examples of the Tauc plot, as observed in a few a-semiconductors are shown in Fig. 4.5 (Morigaki, 1999).

In the low absorption region (B region in Fig. 4.3), α increases exponentially with energy. In this region it can be written as

$$\alpha \propto \exp(\hbar\omega/E_{\mathrm{U}}), \tag{4.16}$$

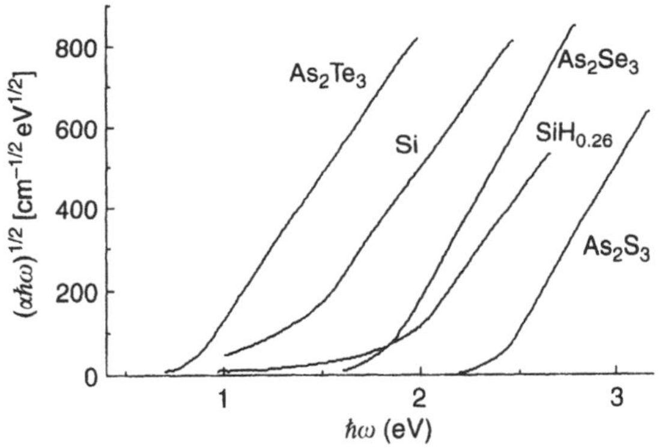

Figure 4.5 Tauc plots for different a-semiconductors (Morigaki, 1999).

where E_U is the width of the localized tail states. The exponential tail of region B in Fig. 4.3 is usually referred to as the Urbach tail.

Likewise, in the region C of Fig. 4.3, the absorption coefficient is written as another exponential function of the photon frequency

$$\alpha \propto \exp(\hbar\omega/E_d), \tag{4.17}$$

where E_d is the width of the defect states, and usually it is found that E_d is larger than E_U. It is established that the region C of Fig. 4.3 is rather sensitive to the structural properties of materials.

While there seems to be a general consensus (Mott and Davis, 1979; Fritzsche, 1980; Cody *et al.*, 1982, 1984) about the Tauc plot (Eq. (4.15)) of the absorption coefficient from valence extended states to conduction extended states, there are many views about Eqs (4.16) and (4.17) (Demichellis *et al.*, 1985). Even for Eq. (4.15) some experimental data in a-Si : H (Klazes *et al.*, 1982; Vorlicek *et al.*, 1981) fit much better to the following relation (Fagen and Fritzsche, 1970):

$$(\alpha\hbar\omega)^{1/3} = C_F(\hbar\omega - E_0), \tag{4.18}$$

and have therefore used it to determine the optical gap E_0. Here C_F is another constant. This makes it very difficult to determine the unique value for the optical gap for any a-semiconductors. This also raises a question of whether all amorphous samples of the same size and same material, but prepared at different times or by different methods, should have the same optical gap. Probably not. The ratio of the number of defect and weak bond states to that of the extended states should influence the ratio of their widths. According to the results derived in Chapter 3, the ratio of the effective mass of charge carriers in the tail states to that in extended states depends on the ratio of width of the tail states to that of extended states. Hence the position of E_0 should depend on such ratio.

Before proceeding further, it is desirable to present the quantum mechanical derivation of the absorption coefficient and how it is related to the imaginary part of the dielectric constant of solids.

4.2.1 Derivation of absorption coefficient

For transitions due to absorption of photons, the electron–photon interaction Hamiltonian in a solid is given by (e.g. Yu and Cardona, 1996):

$$\hat{H}_\text{I} = -\frac{e}{m_\text{e}^* c}\mathbf{A}\cdot\mathbf{p}, \tag{4.19}$$

where $\mathbf{A}$ is the vector potential of the electromagnetic radiation, $\mathbf{p}$ momentum operator of electron with effective mass m_e^* and c speed of light.

4.2.1.1 In crystalline solids

The case of c-solids will be presented first for comparison, because this theory is very well established. Applying the Fermi golden rule, the rate of absorption of photons per unit volume, due to transitions from valence to conduction bands, can be written as

$$R_\text{t} = \frac{2\pi}{\hbar V}\sum_{\mathbf{k}_\text{c},\mathbf{k}_\text{v}} |\langle 1,\mathbf{k}_\text{c}|\hat{H}_\text{I}|0,\mathbf{k}_\text{v}\rangle|^2\,\delta(E_\text{c}(\mathbf{k}_\text{c}) - E_\text{v}(\mathbf{k}_\text{v}) - \hbar\omega), \tag{4.20}$$

where the right-hand side of Eq. (4.20) has been divided by V, which is the illuminated volume of crystal, so that the rate, R_t, can be expressed in the unit of per second per unit volume.[1] $|t,\mathbf{k}\rangle$ represents the Bloch eigenvector for a charge carrier in band $t = 1$ (conduction) or $t = 0$ (valence) with wavevector $\mathbf{k}$, given by

$$|t,\mathbf{k}\rangle = |\psi_{t\mathbf{k}}(\mathbf{r})\rangle,\quad \psi_{t\mathbf{k}}(\mathbf{r}) = V_0^{-1/2}\text{e}^{(\text{i}\mathbf{k}\cdot\mathbf{r})}u_{t\mathbf{k}}(r). \tag{4.21}$$

Here V_0 is the total volume of the crystal, and $u_{t\mathbf{k}}(r)$ is the Bloch function. It should be noted here that the volume V_0 used in Eq. (4.21) is the normalization constant, and therefore it has to be the total volume of the crystal. But the rate in Eq. (4.20) is divided by a volume V, in which the charge carriers are excited, and therefore it is the volume that has been illuminated by the electromagnetic radiation, as used by Cody (1984). The delta function $\delta(E_\text{c}(\mathbf{k}_\text{c}) - E_\text{v}(\mathbf{k}_\text{v}) - \hbar\omega)$ ensures the energy conservation, $E_\text{c}(\mathbf{k}_\text{c}) = E_\text{v}(\mathbf{k}_\text{v}) + \hbar\omega$. $E_\text{v}(\mathbf{k}_\text{v})$ is the initial energy of the electron in the valence band with wavevector $\mathbf{k}_\text{v}$, $E_\text{c}(\mathbf{k}_\text{c})$ is its final energy in the conduction band at $\mathbf{k}_\text{c}$ after the excitation and $\hbar\omega$ is the photon energy. The

1 Equation (4.20) can be multiplied by a factor of 2 for spin degeneracy, but this will be taken into account through the DOS in Eq. (4.9) later on.

momentum conservation is applied in evaluating the transition matrix element, that is, $\mathbf{k}_c = \mathbf{k}_v$ for direct transitions.

The vector potential can be expressed as

$$\mathbf{A} = A\hat{\mathbf{e}}, \quad A = -\frac{F}{2q}\{\exp[\mathrm{i}(\mathbf{q}\cdot\mathbf{r} - \omega t)] + c.c.\}, \tag{4.22}$$

where $\hat{\mathbf{e}}$ is the unit vector in the direction of the vector potential, F and $\mathbf{q}$ are electric field strength and wavevector of the electromagnetic radiation (photons), respectively. Expanding $\exp(\mathrm{i}\mathbf{q}\cdot\mathbf{r})$ in series, as the photon wavevector is usually very small neglecting all terms dependent on $\mathbf{q}$, one can write Eq. (4.22) as

$$A \approx -\frac{F}{2q}[\exp(-\mathrm{i}\omega t) + c.c.]. \tag{4.23}$$

Using Eqs (4.22) and (4.23), the square of the transition matrix element, used in the rate (Eq. (4.20)), becomes

$$|\langle 1, \mathbf{k}_c|\hat{H}_\mathrm{I}|0, \mathbf{k}_v\rangle|^2 = \left[\frac{eF}{2m_e^* cq}\right]^2 |\langle 1, \mathbf{k}_c|\hat{\mathbf{e}}\cdot\mathbf{p}|0, \mathbf{k}_v\rangle|^2. \tag{4.24}$$

Using $cq = \omega$ in Eq. (4.24) and then using Eq. (4.24) in Eq. (4.20), the rate becomes

$$R_\mathrm{t} = \frac{2\pi}{\hbar V}\left(\frac{e}{m_e^*\omega}\right)^2\left(\frac{F}{2}\right)^2 \sum_{\mathbf{k}_c,\mathbf{k}_v} |\langle 1, \mathbf{k}_c|\mathbf{e}\cdot\mathbf{p}|0, \mathbf{k}_v\rangle|^2 \delta(E_c(\mathbf{k}_c) - E_v(\mathbf{k}_v) - \hbar\omega). \tag{4.25}$$

The rate of loss of energy due to absorption per second per unit volume (or rate of power loss per unit volume), W, is given by

$$W = R_\mathrm{t}\hbar\omega, \tag{4.26}$$

which can also be calculated from the rate of loss of incident energy per unit volume, as

$$W = -\frac{\mathrm{d}I}{\mathrm{d}t} = \frac{\mathrm{d}I}{\mathrm{d}x}\frac{\mathrm{d}x}{\mathrm{d}t}. \tag{4.27}$$

Using $I = I_0\exp(-\alpha x)$, where α is the absorption coefficient and x thickness of sample absorbing the light, one gets from Eq. (4.27):

$$W = \alpha\frac{c}{n}I = \frac{\varepsilon_2\omega}{n^2}I, \tag{4.28}$$

where n is the real part of the refractive index of the material. The second part of the above equaion is obtained using $\alpha = \varepsilon_2\omega/nc$, where ε_2 is the imaginary part of the dielectric constant. Using the average incident energy (intensity) per unit volume as

$$I = \frac{n^2}{8\pi} F^2, \tag{4.29}$$

in Eq. (4.28), and then comparing Eq. (4.28) with Eq. (4.26), we get

$$\varepsilon_2 = \left(\frac{2\pi e}{m_e^*}\right)^2 \left(\frac{1}{V\omega^2}\right) \sum_{\mathbf{k}_c, \mathbf{k}_v} |\langle 1, \mathbf{k}_c | \mathbf{e} \cdot \mathbf{p} | 0, \mathbf{k}_v \rangle|^2 \delta(E_c(\mathbf{k}_c) - E_v(\mathbf{k}_v) - \hbar\omega), \tag{4.30}$$

or

$$\alpha = \frac{1}{nc} \left(\frac{2\pi e}{m_e^*}\right)^2 \left(\frac{1}{V\omega}\right) \sum_{\mathbf{k}_c, \mathbf{k}_v} |\langle 1, \mathbf{k}_c | \mathbf{e} \cdot \mathbf{p} | 0, \mathbf{k}_v \rangle|^2 \delta(E_c(\mathbf{k}_c) - E_v(\mathbf{k}_v) - \hbar\omega). \tag{4.31}$$

Equations (4.30) and (4.31) are the key equations used for studying the behavior of the absorption coefficient in c-solids. Different methods have been applied by different groups (Mott and Davis, 1979; Yu and Cardona, 1996; Elliott, 1998) to evaluate the matrix element $\langle 1, \mathbf{k}_c | \mathbf{e} \cdot \mathbf{p} | 0, \mathbf{k}_v \rangle$. However, in most cases it is assumed to be independent of energy and wavevectors $\mathbf{k}_c$ and $\mathbf{k}_v$, and is written as

$$\langle 1, \mathbf{k}_c | \mathbf{e} \cdot \mathbf{p} | 0, \mathbf{k}_v \rangle = p_{cv} \delta_{\mathbf{k}_c, \mathbf{k}_v}, \tag{4.32}$$

where p_{cv} has the dimensions of linear momentum. The assumption that the matrix element is independent of photon energy will be addressed later on. Substituting Eq. (4.32) into Eq. (4.31) gives

$$\alpha = \frac{1}{nc} \left(\frac{2\pi e}{m_e^*}\right)^2 \left(\frac{1}{V\omega}\right) |p_{cv}|^2 \sum_{\mathbf{k}_c, \mathbf{k}_v} \delta(E_c(\mathbf{k}_c) - E_v(\mathbf{k}_v) - \hbar\omega) \delta_{\mathbf{k}_c, \mathbf{k}_v}. \tag{4.33}$$

Note that the application of momentum conservation in Eq. (4.33) corresponds to a direct transition and it reduces the double summation (over $\mathbf{k}_c$ and $\mathbf{k}_v$) into a single summation (over $\mathbf{k}_v$ or $\mathbf{k}_c$). Replacing the summation by an integration over the energy variable using

$$\sum_{\mathbf{k}} = \sum_{E} = \int g(E)\, \mathrm{d}E, \tag{4.34}$$

where $g(E)$ is the DOS, one can write Eq. (4.33) as

$$\alpha = \frac{1}{nc} \left(\frac{2\pi e}{m_e^*}\right)^2 \left(\frac{\hbar}{V\hbar\omega}\right) |p_{cv}|^2 \int_{E_g}^{\hbar\omega} g_j(E_{cv}) \delta(E_{cv} - \hbar\omega)\, \mathrm{d}E_{cv}. \tag{4.35}$$

It should be noted that in Eq. (4.34), $g_j(E_{cv})$ is the joint DOS used for direct transitions in c-solids and it is calculated from the energy of both states,

$E_{cv} = E_c(k) - E_v(k)$, $E_v(k)$ of the valence band (from which the electron is excited by creating a hole) and $E_c(k)$ of the conduction band (to which the electron is excited). It is interesting to note that the DOS gets into the expression of the absorption coefficient only when the summation in Eq. (4.33) is converted into an integration in Eq. (4.35).

One of the ways of deriving an expression for the joint DOS is to write the excitation energy, $E_c(k) - E_v(k)$, using the effective mass approximation as (Singh, 1994):

$$E_c(k) - E_v(k) \approx E_c(0) - E_v(0) + \frac{\hbar^2 k^2}{2m_e^*} + \frac{\hbar^2 k^2}{2m_h^*}, \tag{4.36}$$

where $E_c(0) - E_v(0) \equiv E_g$ is the minimum direct band gap energy, and m_e^* and m_h^* are the effective masses of thus excited electron and hole, respectively. Thus, for a direct band solid, Eq. (4.36) becomes

$$E_{cv} \equiv E_c(k) - E_v(k) \approx E_g + \frac{\hbar^2 k^2}{2\mu}, \tag{4.37}$$

where μ is the reduced effective mass $[\mu^{-1} = (m_e^*)^{-1} + (m_h^*)^{-1}]$ of the excited pair of charge carriers. Equation (4.37) is also a parabolic and therefore its DOS can be written in the same form as Eq. (4.9), replacing m_e^* with μ, and the energy variable E_{cv} with $E_{cv} - E_g$ as

$$g_j(E) = \frac{V}{2\pi^2}\left(\frac{2\mu}{\hbar^2}\right)^{3/2}(E_{cv} - E_g)^{1/2}. \tag{4.38}$$

From Eq. (4.38), it is obvious that the DOS vanishes for $E_{cv} \leq E_g$, and therefore the lower limit of integration in Eq. (4.35) is E_g, not zero. For solids, where this is not true (i.e. DOS below E_g is non-zero), like in a-solids, the concept of joint DOS cannot be used.

Substituting Eq. (4.38) into Eq. (4.35), gives

$$\alpha = \frac{2}{nc\omega}\left(\frac{e}{m_e^*}\right)^2\left(\frac{2\mu}{\hbar^2}\right)^{3/2}|p_{cv}|^2(\hbar\omega - E_g)^{1/2}. \tag{4.39}$$

Equation (4.39) demonstrates that plotting $(\hbar\omega\alpha)^2$ as a function of the photon energy $\hbar\omega$ gives a straight line that intersects the energy axis at $\hbar\omega = E_g$ and thus produces a sharp band edge for direct transitions in c-solids. For a-solids, such a sharp drop of the absorption coefficient to zero at a band edge is not observed. The derivation of the absorption coefficient in a-solids will be presented in the next subsection.

4.2.1.2 Amorphous solids

For a-solids, a theory of transition can be developed in the same way as presented above. However, as stated in the previous subsection, the concept of the joint DOS cannot be used for a-solids because there is no well-defined energy band gap as obtained for a c-solid through Eq. (4.36). Also, the DOS is non-zero within the gap. For a-solids, one should consider the transitions in different energy ranges separately. Therefore, let us first consider those transitions in which electrons are excited from the valence extended states to conduction extended states, and refer to these as extended–extended transitions. Other transitions involving the tail states, that is, from valence extended to conduction tail states, from valence tail to conduction extended states and from valence tail to conduction tail states, will be treated later on.

4.2.1.2.1 EXTENDED–EXTENDED STATES TRANSITIONS

For transitions from valence extended states to conduction extended states, the theory presented in the previous section can be applied as follows. As **k** is not a good quantum number its conservation cannot be applied in a-solids, but the momentum and energy conservations are also applicable for transitions in a-solids.[2] Under this condition the Fermi golden rule (Eq. (4.25)), should be written in the energy space, as (Singh, 2002a,b, 2003)

$$R_{\mathrm{t}} = \frac{2\pi}{\hbar V}\left(\frac{e}{m_{\mathrm{e}}^{*}\omega}\right)^{2}\left(\frac{F}{2}\right)^{2}\sum_{E_{\mathrm{c}}',E_{\mathrm{v}}'}|\langle 1, E_{\mathrm{c}}' \mid \hat{\mathbf{e}}\cdot\mathbf{p} \mid 0, E_{\mathrm{v}}'\rangle|^{2}\delta(E_{\mathrm{c}}' - E_{\mathrm{v}}' - \hbar\omega), \tag{4.40}$$

where the summations over E_{v}' and E_{c}' are to cover all states in the valence and conduction extended states, respectively, excluding any tail states. The eigenvector $|t, E\rangle = |\psi_{tE}(r)\rangle (t = 0, 1)$, and ψ_{tE} is the normalized eigenfunction of an electronic state in the valence (0) or conduction (1) extended states with energy E. It is not really important to worry about the mathematical form of these functions at this stage. Using Eq. (4.40), the whole formulation of the c-solids presented in the preceding section can be adopted analogously, and then Eqs (4.30) and (4.31) can be written for a-solids, respectively, as

$$\varepsilon_2 = \left(\frac{2\pi e}{m_{\mathrm{e}}^{*}}\right)^{2}\left(\frac{1}{V\omega^{2}}\right)|p_{\mathrm{cv}}|^{2}\sum_{E_{\mathrm{c}}',E_{\mathrm{v}}'}\delta(E_{\mathrm{c}}' - E_{\mathrm{v}}' - \hbar\omega), \tag{4.41}$$

and

$$\alpha = \frac{1}{nc}\left(\frac{2\pi e}{m_{\mathrm{e}}^{*}}\right)^{2}\left(\frac{1}{V\omega}\right)|p_{\mathrm{cv}}|^{2}\sum_{E_{\mathrm{c}}',E_{\mathrm{v}}'}\delta(E_{\mathrm{c}}' - E_{\mathrm{v}}' - \hbar\omega). \tag{4.42}$$

2 This point has been stated frequently in the literature that the momentum conservation is relaxed in a-solids, which is not true. **k** represents momentum only in c-solids, but not in a-solids. However, fortunately as the photon momentum is very small, this cannot be expected to cause any quantitative worry.

Here also, in accordance with Eq. (4.32), it is assumed that the matrix element is independent of the energies in the conduction and valence extended states. Now, using Eq. (4.34), the double summations over energies E'_v and E'_c in Eqs (4.41) and (4.42) can be changed into double integrations as

$$\varepsilon_2 = \left(\frac{2\pi e}{m_e^*}\right)^2 \left(\frac{1}{V\omega^2}\right) |p_{cv}|^2 I, \tag{4.43}$$

and

$$\alpha = \frac{1}{nc}\left(\frac{2\pi e}{m_e^*}\right)^2 \left(\frac{1}{V\omega}\right) |p_{cv}|^2 I, \tag{4.44}$$

where

$$I = \int_{E_c}^{E_v+\hbar\omega} \int_{E-E_v}^{E_v} g_c(E'_c) g_v(E'_v) \delta(E'_c - E'_v - \hbar\omega)\, dE'_c\, d(E'_v). \tag{4.45}$$

$g_c(E)$ and $g_v(E)$ are the DOS of the conduction and valence extended states, respectively. E_v and E_c denote the energy of the hole and electron mobility edges, respectively. Using the form of the DOS of valence and conduction extended states as given in Eq. (4.9), the integral in Eq. (4.45) can be evaluated (as shown in Appendix A). We then get ε_2 in Eq. (4.43), and absorption coefficient in Eq. (4.44), respectively, as

$$\varepsilon_2 = \left(\frac{2\pi e}{m_e^*}\right)^2 \left(\frac{V(m_e^* m_h^*)^{3/2}}{4\pi^3\omega^2\hbar^6}\right) |p_{cv}|^2 [\hbar\omega - E_0]^2, \tag{4.46}$$

and

$$\alpha = \frac{4\pi}{nc}\left(\frac{e}{m_e^*}\right)^2 \left(\frac{V(m_e^* m_h^*)^{3/2}}{2^2\pi^2\omega\hbar^6}\right) |p_{cv}|^2 [\hbar\omega - E_0]^2, \tag{4.47}$$

where $E_0 = E_c - E_v$, is the optical gap. The absorption coefficient α in Eq. (4.47) can be rearranged to

$$(\hbar\omega\alpha)^{1/2} = \left[\frac{4\pi}{nc}\left(\frac{e}{m_e^*}\right)^2 \left(\frac{V(m_e^* m_h^*)^{3/2}}{2^2\pi^2\hbar^5}\right)\right]^{1/2} |p_{cv}|[\hbar\omega - E_0]. \tag{4.48}$$

Thus plotting $(\hbar\omega\alpha)^{1/2}$ as a function of the photon energy $\hbar\omega$ enables one to determine the optical gap of an a-semiconductor. Equation (4.48) is a well known result first obtained by Tauc (1968). The optical gap is estimated by extrapolating the straight line to the energy axis (x-axis) as shown in Fig. 4.4. The expression derived in Eq. (4.48) has been successful in determining the optical gap for several

Table 4.1 Optical gap E_0 and constant B determined experimentally for some a-semiconductors using Eq. (4.49)

Material	E_0 (eV)	B (cm^{-1} eV^{-1})
Si[a]	1.26	5.2×10^5
$SiH_{0.26}$[a]	1.82	4.6×10^5
a-Ge : H[b]	1.05	6.7×10^5
As_2S_3[a]	2.32	4.0×10^5
As_2Se_3[a]	1.76	8.3×10^5
As_2Te_3[a]	0.83	4.7×10^5

Notes
a Mott and Davis (1979).
b Aoki *et al.* (1999).

a-solids, as shown in Fig. 4.5. In fact, for analyzing the experimental data, one usually applies Eq. (4.47) as

$$\alpha = B\frac{[\hbar\omega - E_0]^2}{\hbar\omega}. \tag{4.49}$$

Both the optical gap E_0 and the constant B can thus be estimated experimentally and for a few a-semiconductors, E_0 and B are listed in Table 4.1. However, as the above absorption coefficient in Eq. (4.47) is derived only for transitions from valence extended to conduction extended states, (without the involvement of any tail states) the optical gap thus obtained should correspond to the situation where there are no tail states, like in c-solids.

Nevertheless, before we proceed any further with calculating the absorption coefficient involving tail states, it is desirable to inspect the assumption of p_{cv} being independent of energy and momentum of the excited charge carriers, as used above from Eq. (4.32) onward.

Matrix element p_{cv}

The matrix element p_{cv} has the dimension of momentum, and therefore it does not appear to be justified to assume it as independent of momentum and energy. However, as described below, there are two methods to evaluate the matrix element. In one way, as shown by Mott and Davis (1979) it appears to be justified to assume it is independent, whereas in the other way, as demonstrated by Cody (1984), it is not independent.

Method 1 The momentum operator used by Mott and Davies (1979) is

$$\mathbf{p} = -i\hbar\nabla, \tag{4.50}$$

which gives the matrix element as

$$\langle 1, E_{c'}|\mathbf{e}\cdot\mathbf{p}|0, E_{v'}\rangle = -i\hbar D, \tag{4.51}$$

D is the resulting matrix element given by

$$D \equiv \langle 1, E_{c'} | \hat{\mathbf{e}} \cdot \nabla | 0, E' \rangle = \pi (a/V)^{1/2}, \tag{4.52}$$

and the parameter a is used as the average inter-atomic spacing in the sample for matching the dimension of D (Mott and Davis, 1979). One uses the same result for both c- and a-solids. In this form, the matrix element in Eq. (4.49) becomes independent of the energy of excited charge carriers, $E_{c'}$ and $E_{v'}$. Using Eq. (4.52) in Eq. (4.47), one gets the well known form for Tauc's plot in a-solids as

$$(\hbar\omega\alpha) = \left[\frac{4\pi}{nc} \left(\frac{e}{m_e^*} \right)^2 \left(\frac{a(m_e^* m_h^*)^{3/2}}{2^2 \hbar^3} \right) \right] [\hbar\omega - E_0]^2. \tag{4.53}$$

Method 2 The other way of writing the momentum operator is

$$\mathbf{p} = \mathrm{i} \frac{m_e^*}{\hbar} [\hat{H}, \mathbf{r}], \tag{4.54}$$

where $\mathbf{r}$ is the conjugate of the momentum operator $\mathbf{p}$. Using Equation (4.54), as shown in Appendix B, the matrix element for c-solids is obtained as

$$\langle 1, \mathbf{k}_c | \mathbf{e} \cdot \mathbf{p} | 0, \mathbf{k}_v \rangle = \mathrm{i} \frac{m_e^*}{\hbar} [E(\mathbf{k}_c) - E(\mathbf{k}_v)] Q_c \delta_{\mathbf{k}_c, \mathbf{k}_v}, \tag{4.55}$$

which clearly depends on the energy difference $E(\mathbf{k}_c) - E(\mathbf{k}_v)$. This depends on the photon's energy and wavevector $\mathbf{k}$. Q_c is the average separation between the excited electron and hole pair in a c-semiconductor and it is probably more justified to assume it is independent of energy and momentum of the excited charge carriers than the whole matrix element. Using Eq. (4.55) in Eq. (4.31) one gets

$$\alpha = \frac{(2\pi e)^2}{nc} \left(\frac{\omega}{V} \right) Q_c^2 \sum_{\mathbf{k}_c, \mathbf{k}_v} \delta(E_c(\mathbf{k}_c) - E_v(\mathbf{k}_v) - \hbar\omega) \delta_{\mathbf{k}_c, \mathbf{k}_v}, \tag{4.56}$$

where $E(\mathbf{k}_c) - E(\mathbf{k}_v) = \hbar\omega$ is used to get the above form. Equation (4.56) can be evaluated in the same way as Eq. (4.33) and one gets the absorption coefficient for direct c-semiconductors as

$$\alpha = \frac{2e^2}{nc} \left(\frac{2\mu}{\hbar^2} \right)^{3/2} Q_c^2 \omega (\hbar\omega - E_g)^{1/2}, \tag{4.57}$$

and ε_2 is obtained as

$$\varepsilon_2 = 2e^2 \left(\frac{2\mu}{\hbar^2} \right)^{3/2} Q_c^2 (\hbar\omega - E_g)^{1/2}. \tag{4.58}$$

Equation (4.57) is different from Eq. (4.39) derived using method 1.

A similar deviation is found in the results for a-semiconductors as well, and Eqs (4.43) and (4.44), respectively, become as

$$\varepsilon_2 = (2\pi e)^2 \left(\frac{V(m_e^* m_h^*)^{3/2}}{2^2 \pi^3 \hbar^6} \right) Q_a^2 [\hbar\omega - E_0]^2, \tag{4.59}$$

and

$$\alpha = \frac{(2\pi e)^2}{nc} \left(\frac{V(m_e^* m_h^*)^{3/2}}{2^2 \pi^3 \hbar^6} \right) Q_a^2 \omega [\hbar\omega - E_0]^2 \tag{4.60}$$

where Q_a is the average separation between the excited electron and hole pair in a-semiconductors. Q_a may be different from the corresponding Q_c in c-solids used in Eqs (4.57) and (4.58), but it may be assumed to be independent of energy and momentum of the excited charge carriers. It is obvious from Eq. (4.60) that $[\alpha\hbar\omega]^{1/2}$ does not have a linear relation with the photon energy $\hbar\omega$, as obtained in the Tauc's plot (Eq. (4.53)) by using the first form of the momentum operator. However, $[\varepsilon_2]^{1/2}$ derived from Eq. (4.59), or $[\alpha/\hbar\omega]^{1/2}$ derived from Eq. (4.60), depends linearly on the photon energy as

$$[\varepsilon_2]^{1/2} = (2\pi e) \left[\frac{V(m_e^* m_h^*)^{3/2}}{2^2 \pi^3 \hbar^6} \right]^{1/2} Q_a [\hbar\omega - E_0], \tag{4.61}$$

and

$$\left[\frac{\alpha}{\hbar\omega} \right]^{1/2} = \frac{(2\pi e)}{\sqrt{nc}} \left(\frac{V(m_e^* m_h^*)^{3/2}}{2^2 \pi^3 \hbar^7} \right)^{1/2} Q_a [\hbar\omega - E_0]. \tag{4.62}$$

In this case Eq. (4.61) or (4.62) should be used for a Tauc's plot. Cody (1984) has plotted $[\varepsilon_2]^{1/2}$ as a function of photon energy for 30 films of a-Si : H_x ($x = 0.09$) and found excellent agreement with the experimental data.

Thus, it may be concluded that there are two different ways of evaluating the transition matrix element. Applying one approach it is found to be independent of the energy and momentum of the excited charge carriers, but the second approach shows that it depends on the photon energy and hence on energy of charge carriers. If one uses the first approach, then $[\alpha\hbar\omega]^{1/2}$ is found to give the correct Tauc's plot (i.e. it is linear with the photon energy). However, if the second method was used, then $[\varepsilon_2]^{1/2}$ or $[\alpha/\hbar\omega]^{1/2}$ would give the correct Tauc's plot. As the objective of Tauc's plot is to determine the optical gap, one may expect both approaches will produce the same value for the optical gap.

It may be noted that in using Eq. (4.61) or (4.62) for the Tauc's plot, there are two unknowns, the volume V and Q_a. These unknowns occur through the first approach as well, but then the inter-atomic spacing a is used in Eq. (4.52) to get around the problem. For applying the second approach Cody (1984) has defined as

$$\frac{2N_0}{V} = \nu\rho_A, \tag{4.63}$$

where N_0 is the number of single spin states in the valence/conduction extended states, $2N_0$ number of total valence electrons occupying N_0 states, ν number of valence electrons per atom and ρ_A atomic density per unit volume. Assuming that the wavefunctions of valence and conduction extended states of a-semiconductors can be expanded in terms of Bloch waves of a virtual crystal, Cody has been able to replace the volume V by V/N_0 in Eq. (4.61). Then substituting Eq. (4.63) in Eqs (4.59), and (4.60) gives

$$\varepsilon_2 = (\pi e)^2 \left[\frac{2(m_e^* m_h^*)^{3/2}}{\pi^3 \hbar^6 \nu\rho_A}\right] Q_a^2 [\hbar\omega - E_0]^2, \tag{4.64}$$

and

$$\left[\frac{\alpha}{\hbar\omega}\right] = \frac{(\pi e)^2}{nc} \left(\frac{2(m_e^* m_h^*)^{3/2}}{\pi^3 \hbar^7 \nu\rho_A}\right) Q_a^2 [\hbar\omega - E_0]^2. \tag{4.65}$$

Plotting $[\alpha\hbar\omega]^{1/2}$ in Eq. (4.53) or $[\alpha/\hbar\omega]^{1/2}$ in Eq. (4.65) as a function of the photon energy $\hbar\omega$, one can determine the energy of the optical gap E_0, as well as Tauc's constant, slope of the straight line thus obtained. However, as stated above, it is usually Eq. (4.53) that is used to fit the experimental data and then E_0 is determined from that fit.

Theoretically, these equations have seldom been used to estimate either the slope or optical gap of a-semiconductors. As a matter of exercise, we may as well try to estimate these quantities using both Eqs (4.53) and (4.65). Writing Eq. (4.53) in the form of Eq. (4.49), we get the constant B as (Singh, 2002a,b, 2003)

$$B = \frac{1}{nc\varepsilon_0} \left(\frac{e}{m_e^*}\right)^2 \left(\frac{a(m_e^* m_h^*)^{3/2}}{2^2 \hbar^3}\right). \tag{4.66}$$

For calculating B in SI units, 4π is replaced by $1/\varepsilon_0$, $\varepsilon_0 = 8.8542 \times 10^{-14}\,\mathrm{F\,cm^{-1}}$ is the vacuum permittivity. As the absorption coefficient is measured in $\mathrm{cm^{-1}}$, the value of speed of light should also be used in $\mathrm{cm\,s^{-1}}$ in Eq. (4.66). Using then the effective masses for carriers in the extended states given in Table 3.1, the constant B can easily be calculated and the values thus obtained for a-Si : H and a-Ge : H are listed in Table 4.2. For a-Si : H we get $B = 6.0 \times 10^6\,\mathrm{cm^{-1}\,eV^{-1}}$ (Singh, 2002), and for a-Ge : H, $B = 4.1 \times 10^6\,\mathrm{cm^{-1}\,eV^{-1}}$, which are an order of magnitude higher than those given in Table 4.1 estimated from experiments (Mott and Davis, 1979; Aoki *et al.*, 1999). It may also be remarked here that the values given in Table 4.1, estimated for different materials from experiments, do vary by an order in their magnitude.

Table 4.2 Theoretical values of B and B' calculated using Eqs (4.66) and (4.68), respectively, for a-Si : H (Singh, 2002) and a-Ge : H. The same refractive index is used for both a-Si : H and a-Ge : H, but different values may be used if desired

	$m_e^* = m_h^*$	a (nm)	n	B ($cm^{-1}\,eV^{-1}$)	B' ($cm^{-1}\,eV^{-3}$)
a-Si : H	$0.34\,m_e$	0.235	4.0	6.0×10^6	4.6×10^3
a-Ge : H	$0.22\,m_e$	0.245	4.0	4.1×10^6	1.3×10^3

We can also write Eq. (4.65) in the form of Eq. (4.49) as (Singh, 2002):

$$\frac{\alpha}{\hbar\omega} = B'[\hbar\omega - E_0]^2, \tag{4.67}$$

where B' is obtained by comparing Eq. (4.67) with Eq. (4.65) as

$$B' = \frac{e^2}{nc\varepsilon_0}\left[\frac{(m_e^* m_h^*)^{3/2}}{2\pi^2\hbar^7 \nu \rho_A}\right] Q_a^2, \tag{4.68}$$

where again the vacuum permittivity is used as in Eq. (4.66). The problem in calculating B' is that Q_a is not known. Using the atomic density of c-Si and four valence electrons per atom, Cody (1984) has estimated $Q_a^2 = 0.9\,\text{Å}^2$, which gives $Q_a \approx 0.095$ nm. This value is less than half of the inter-atomic separation of 0.235 nm in a-Si : H, but in the same range of order of magnitude. Using $\nu = 4$, $\rho_A = 5 \times 10^{28}\,m^{-3}$, $Q_a^2 = 0.9\,\text{Å}^2$ and extended state effective masses from Table 3.1, we get $B' = 4.6 \times 10^3\,cm^{-1}\,eV^{-3}$ for a-Si : H (Singh, 2002). Cody has estimated an optical gap, $E_0 = 1.64$ eV for a-Si : H. Substituting this into Eq. (4.67), we get $\alpha = 1.2 \times 10^3\,cm^{-1}$ at a photon energy of, $\hbar\omega = 2$ eV. This agrees reasonably well with $\alpha = 6.0 \times 10^2\,cm^{-1}$ used by Cody. If we use inter-atomic spacing in place of Q_a in Eq. (4.68), we get $B' = 2.8 \times 10^4\,cm^{-1}\,eV^{-3}$, and then the corresponding absorption coefficient becomes $7.3 \times 10^3\,cm^{-1}$. This suggests that for an estimate one may be able to use the inter-atomic spacing in place of Q_a, if the latter is unknown. Calculated values of B' for a-Si : H and a-Ge : H (using the same Q_a as above for a-Si : H) are also listed in Table 4.2.

Variation from Tauc's plot

In some a-solids, experimental data exhibit deviation from the square relations derived in Eqs (4.53) or (4.67). A cubic dependence on the photon energy given by

$$[\alpha\hbar\omega] \propto (\hbar\omega - E_0)^3, \tag{4.69}$$

has been found to fit better with some experimental data. For example, the data obtained by Aspnes *et al.* (1984) for a-Si, appear to fit Eq. (4.69) better than Eq. (4.59) and produce $E_0 = 1.35$ eV with $[\alpha\hbar\omega]^{1/3}$ plot and $E_0 = 1.80$ eV with $[\alpha\hbar\omega]^{1/2}$ plot. Klazes *et al.* (1982) have also found that their data for a-Si fit better with the cubic root than square root. Likewise Khawaja and Hogarth (1988) have found relatively better agreement with the cubic root fit for their experimental data from binary vanadate glasses, like a-Si, a-Ge a-Ta_2O_5 and a-As_2S_3, than with the square root fit.

As shown in Appendix A, the cubic dependence on photon energy can be obtained only when the DOS depends linearly on energy, provided of course the assumption of constant transition matrix is valid for every a-solid. On the basis that the DOS may depend linearly on energy near the optical gap, the cubic root dependence was suggested by Mott and Davis (1979). Using Eq. (4.49), Sokolov *et al.* (1991) have modeled the cubic dependence on photon energy by considering the fluctuations in the optical band gap energy due to structural disorders and assuming the constant transition matrix element. For getting the cubic root dependence they finally assume that the fluctuations are constant over the range of integration and then the integration of Eq. (4.49) over the optical gap energy produces a cubic root dependence on the photon energy. Although their theory shows a way of getting the cubic root dependence, as the integration over the optical gap is carried out by assuming constant fluctuations, Sokolov *et al.*'s model is little different from the linear density of states model suggested by Mott and Davis. There is no other more appropriate theoretical development on the cubic root dependence of $[\alpha\hbar\omega]$ on photon energy.

According to Eq. (4.34), it should also be noted that the DOS, as stated earlier, plays a role in the absorption coefficient only when the summation over energy states is converted into an integration. It is difficult to decide whether the integration is applicable near the optical gap or not. Moreover, according to Section 4.1, the DOS can be made to have parabolic dependence on energy (Eq. (4.9)) for any particle by associating it with an appropriate effective mass. Then, the resulting absorption coefficient, obtained within the assumption of constant transition matrix element, will have the square root dependence on energy involving a different effective mass. In this view, a deviation from Tauc's plot cannot be explained by assuming the constant matrix element.

On the other hand, using the energy dependent matrix element obtained in Eq. (4.55), we get $[\alpha\hbar\omega]$ from Eq. (4.67) as

$$[\alpha\hbar\omega] = B'(\hbar\omega)^2[\hbar\omega - E_0]^2. \tag{4.70}$$

The right-hand side of Eq. (4.70) is a fourth-order polynomial. Depending on which term of the polynomial may dominate, one can obtain a deviation from Tauc's plot. Therefore in a material in which a deviation from Tauc's plot is observed the transition matrix element may not be constant, as has recently been pointed out by Singh *et al.* (2002).

In chalcogenides the situation gets even more complicated, because the DOS in the valence and conduction states are expected to have different forms. This can also lead to a deviation from the Tauc relation, which will be described below.

It is observed in amorphous chalcogenides that $(\alpha\hbar\omega)^n \propto (\hbar\omega - E_0)$ and $n \neq \frac{1}{2}$. A linear dependence ($n = 1$) on photon energy has been observed in amorphous

Se (a-Se) and a cubic dependence ($n = 3$) in the multicomponent of chalcogenide glasses (Mott and Davis, 1979). The deviation from the simple Tauc relation may be a reflection of the deviation of the dependence of DOS on energy from the free electron DOS given in Eq. (4.9). In these materials, DOS need to be described by taking into account the fractals, which are known to dominate in many physical properties of a-semiconductors (Zallen, 1983).

Let us express DOS of the conduction and valence extended states in chalcogenides as: $g_c(E_c) = G_c(E - E_c)^s$ and $g_v(E_v) = G_v(E_v - E)^p$, where G_c and G_v are constants. Substituting these into Eq. (4.45), we get

$$\alpha\hbar\omega = C'(\hbar\omega - E)^{p+s+1}, \tag{4.71}$$

where C' is the constant term obtained including the matrix element p_{cv}. This can be written as

$$[\alpha\hbar\omega]^n = C'^n(\hbar\omega - E), \tag{4.72}$$

where $n = 1/(p+s+1)$. If the form of both g_c and g_v is parabolic, then we obtain $p = s = \frac{1}{2}$, and $n = \frac{1}{2}$ in three-dimensional (3D) systems as described above. Let us now analyze the experimental results to find possible values of p and s in a-chalcogenides.

Meherun-Nessa *et al.* (2000) have measured the optical transmittance in the interband absorption region in films (0.5–1 μm) of several amorphous chalcogenides of Ge-based (a-GeS_2 (Fig. 4.6) and a-$GeSe_2$), As-based (a-As_2S_3 (Fig. 4.7) and a-As_2Se_3) and a-Se and deduced their optical absorption coefficient. These measurements are performed before and after thermal annealing for 2 h in vacuum below the glass transition temperature T_g.

In Figs 4.7(a) and (b) some examples of the experimental results for As-based chalcogenides before and after annealing, respectively, are shown for obliquely deposited a-As_2S_3, As_2Se_3 and Se. The data are fitted to the plot of $(\alpha\hbar\omega)^n$ versus $(\hbar\omega - E_0)$ giving $n = 0.7$ before annealing and 0.59 after annealing for a-As_2S_3. The values of n and optical gap E_0 thus deduced from measurements are listed in Table 4.3.

First, we will discuss the simple case of a-Se, where $n = 1$ is always observed. Applying Eq. (4.72), the sum of $(p + s)$ should be zero for $n = 1$ in a-Se. This is only possible if the product of DOS is independent of the energy. Although the origin of such DOS was argued a long time ago (Mott and Davis, 1979), it has not been fully understood. A chain-like structure is basically expected in a-Se. The top of the valence states is known to be formed by p-lone pair (LP) orbitals (lone pair interaction) of Se atoms. The interaction between lone pair electrons should be of 3D nature and therefore the parabolic DOS near the valence extended states edge can be expected giving $p = \frac{1}{2}$. The bottom of the conduction states, on the other hand, is formed by anti-bonding states of Se. If the interaction between chains is ignored, the DOS near the conduction states may have 1D nature, that is, $s = -\frac{1}{2}$. We thus obtain $n = 1/(p + s + 1) = 1$, producing a linear dependence in the absorption coefficient on energy, that is, $(\alpha\hbar\omega) \propto (\hbar\omega - E_0)$ as observed experimentally.

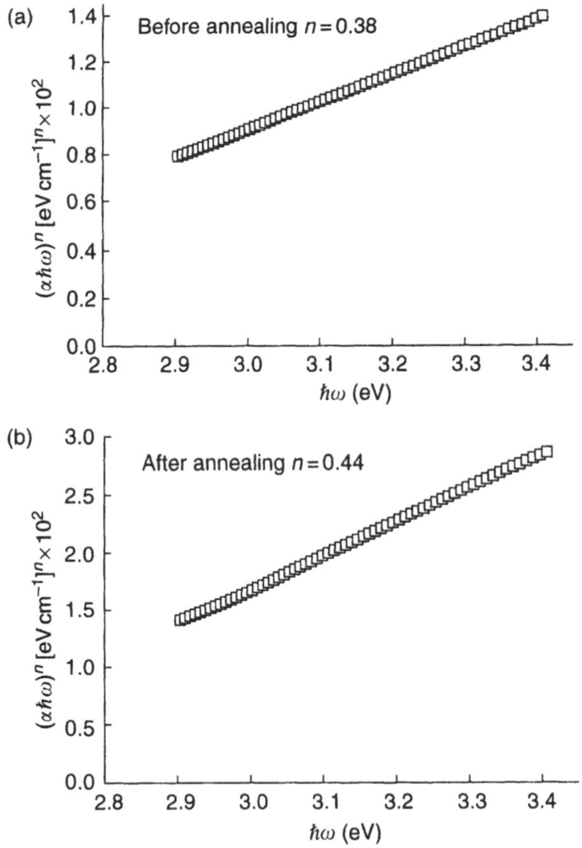

Figure 4.6 Optical absorption spectrum measured in obliquely deposited a-GeS_2 fitted to $(\alpha\hbar\omega)^n \propto (\hbar\omega - E_0)$. (a) $n = 0.38$ before annealing and (b) $n = 0.44$ after annealing at 200°C for 2 h.

Next, we discuss As- and Ge-based binary systems. Although there is no systematic deviation from $n = \frac{1}{2}$ in these systems, it is larger for the oblique systems in comparison with that observed for normal systems. Therefore DOS for the normal (flat) films, before and after thermal annealing, appears to be closer to 3D in nature. For example, in a-As_2Se_3 (Table 4.3) one finds $n = 0.73$ and 0.58 before and after annealing, respectively, whereas the corresponding values for a-GeS_2 are 0.38 and 0.44; n approaching to 0.5 after annealing.

Structurally, As-based systems are suggested to have layers (Elliott, 1990; Tanaka, 1998). The top of valence states is formed by the LP so the parabolic DOS near valence states can also be expected in these systems, since LP–LP interactions occur in 3D space as already mentioned. Unlike a-Se, the bottom of conduction extended states in these solids should arise from a 2D structure in nature, if the layer–layer interactions can be ignored in the anti-bonding states.

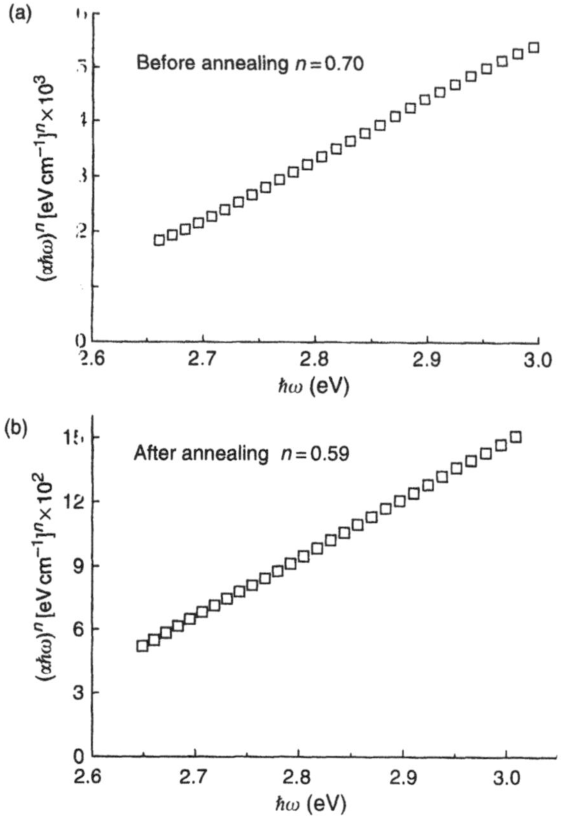

Figure 4.7 Optical absorption spectrum measured in obliquely deposited a-As_2S_3 and fitted to $(\alpha\hbar\omega)^n \propto (\hbar\omega - E_0)$. (a) $n = 0.70$ before annealing and (b) $n = 0.59$ after annealing at 170°C for 2 h.

That means the corresponding DOS is independent of energy ($s = 0$). The value of n, in this case, should be $\frac{2}{3}$, because $p = \frac{1}{2}$ and $s = 0$ are on the basis of space dimensions of DOS. This is close to some of the observed values for as-deposited oblique films of As-based chalcogenides (see Table 4.3). Note, however, that the layer–layer interactions cannot be ignored for the DOS of valence states (Watanabe *et al.*, 1988).

On the basis of space dimension argument we have achieved $n = \frac{1}{2}, \frac{2}{3}$ and 1, which cannot be applied to all binary systems (see Table 4.3), because according to Table 4.3 the value of n in these solids lies between 0.73 and 0.38. The deviation from $n = 1, \frac{2}{3}$ or $n = \frac{1}{2}$ may be attributed to the fractal nature in the DOS; that is, p and s cannot only be assigned with values of $\frac{1}{2}$ (3D), 0 (2D), and $-\frac{1}{2}$ (1D). In obliquely deposited a-As_2Se_3, for example, $n = 0.73$ (before annealing) produce $p+s = 0.37$. The DOS for the extended states on d-dimensional space is given by

$$N(E)\,\mathrm{d}E \propto \rho^{d-1}\,\mathrm{d}\rho, \qquad (4.73)$$

Table 4.3 Values of n and optical bandgap E_0 before and after annealing, and annealing temperature T_a (°C) for obliquely deposited systems

Sample	*Annealing temperature* (°C)	*Before annealing* n	*After annealing* n	*Before annealing* E_0 (eV)	*After annealing* E_0 (eV)
a-As_2S_3	170	0.70	0.59	2.50	2.46
a-As_2Se_3	160	0.73	0.58	1.90	1.80
a-GeS_2	200	0.38	0.44	2.24	2.43
a-Se	30	1.00	1.00	2.05	2.06

where ρ is defined as $(2m^*E)^{1/2}/\hbar$, instead of the wave vector **k**, since it is not a good quantum number in disordered matters, and m^* is the effective mass of a charge carrier (Singh *et al.*, 2002).

On fractal space, on the other hand, a fractal dimension D is introduced, instead of d (Mandelbrot, 1982). The DOS for the extended states on fractal space of dimension D can be written as (He, 1990):

$$N(E)\,\mathrm{d}E \propto \rho^{D-1}\,\mathrm{d}\rho \propto E^{(D-2)/2}\,\mathrm{d}E. \tag{4.74}$$

Note that D is introduced as $m(r) \propto r^D$, where $m(r)$ is the "mass" in a space, and hence D can take any fractional value (even larger than 3). As discussed above, the energy dependence of DOS for the conduction states is expected to be different from that for the valence states in amorphous chalcogenides, but usually the space dimensionality for valence states is larger than that for conduction states. Thus, denoting the dimensionality of valence and conduction states by D_v and D_c, respectively, $p + s + 1$ in Eq. (4.71) is replaced by

$$p + s + 1 = \frac{D_v + D_c + 1}{2}, \tag{4.75}$$

for fractional dimension systems. From the value of $n = 2/(D_v + D_c) - 2$, $D_v + D_c$ can be deduced. In As_2Se_3, for example, $D_v + D_c$ is 4.74. After thermal annealing, the value of n tends to 0.5, which gives $D_v + D_c = 6$, indicating that each of D_v and D_c approaches its 3D value. This is due to the fact that thermal annealing produces more ordered and dense structural network. For a cubic energy dependence, $n = \frac{1}{3}$, observed for multicomponents (e.g. Ge-As-Te-Si), $D_v + D_c = 8$ is obtained. This higher fractal dimension may be related to the "branching" or "cross-linking" between Te chains by introducing As, Ge and Si atoms (Zallen, 1983). The "branching" may be equivalent to "Bethe lattice" (or Cayley tree), resulting in an increase in space dimension (Zallen, 1983).

It may therefore be concluded that the fundamental optical absorption, empirically presented by the relation $(\alpha\hbar\omega)^n \propto (\hbar\omega - E_0)$, where $0.3 < n < 1$, in amorphous chalcogenides, can be interpreted by introducing the DOS on fractals. The energy-dependent DOS form is not the same for both conduction and valence states. The presence of disorder can greatly influence the nature of the electronic

DOS even for the extended states. However, in these cases the transition matrix element is assumed to be constant.

4.2.1.2.2 TRANSITIONS IN TAIL STATES

Let us now consider optical transitions involving the tail states. As described in the previous section, an exponential tail in the absorption coefficient (see parts B and C of Fig. 4.3) of a-solids is usually observed for photon energies below the optical gap. It is usually referred to as exponential tail or Urbach tail and it was first observed in alkali halides and silver halides. As suggested by Urbach (1953), at higher temperatures the optical absorption coefficient follows an empirical rule given by

$$\alpha = \alpha_0 \exp[-\sigma\beta(E_1 - E)]. \tag{4.76}$$

Here α_0, σ and E_1 are nearly independent of the photon energy E. $\beta^{-1} = k_B T$ depends inversely on temperature. The expression shows that a plot of $\ln \alpha$ as a function of photon energy gives straight lines for various temperatures, which converge at an energy $E = E_1$, usually referred to as the focus point. The slope of the straight line is given by $\sigma\beta$. In alkali halides as the tail was found to be temperature dependent, its origin was expected to be the exciton–phonon interaction. Therefore, in alkali halides, where the slope tends to a constant value as $T \to 0$, the temperature parameter $\beta^{-1} = k_B T$ is replaced by an effective temperature in the low temperature region given by

$$k_B T_{\text{eff}} = \frac{\hbar\omega}{2} \coth\left(\frac{\beta\hbar\omega}{2}\right), \tag{4.77}$$

where $\hbar\omega$ is of the order of phonon energy. Some theoretical developments to explain the Urbach tail in molecular crystals have also been made (Singh, 1981). The empirical rule given in Eq. (4.76) is also called the Urbach rule, and it holds over several decades of the absorption coefficient in a-semiconductors.

Toyozawa's group has studied the tail absorption for more than two decades. First, they developed a theory for the temperature-dependent Urbach tail (Toyozawa, 1959; Sumi and Toyozawa, 1971), and then another theory for the Urbach tail due to static disorder (Abe and Toyozawa, 1981) applicable to a-solids. For the temperature dependent Urbach tail in alkali halide crystals, they have found that the exciton–phonon interaction causing momentarily exciton self-trapping, produces a shift and broadening in the exciton absorption leading to an exponential tail (Sumi and Toyozawa, 1971). They found that the thermal distribution of the energy of localized exciton coupled linearly to lattice vibrations is Gaussian with a variance W^2 given by

$$W^2 = S k_B T, \tag{4.78}$$

where S is the lattice relaxation energy. The significance of the Gaussian shape will be discussed in detail with the static disorder case given in the next section.

Another suggested cause for the Urbach tail was the electric field broadening of the absorption edge, known as the Franz–Keldysh effect. Accordingly, in the presence of an electric field a red shift in the edge of free electron absorption occurs due to tunneling of Bloch waves into the gap region and an exponential tail is introduced in the absorption. The origin of electric field in a-solids has been proposed to arise from atomic vibrations in the amorphous network, which also implies that the magnitude of the effect depends on the strength of electron–phonon interaction.

Dow and Redfield (1971, 1972) have followed another approach for exciton absorption and have found that the effect of a uniform electric field causes broadening in the direct exciton transitions and gives rise to a more accurate exponential tail. In this case, the Coulomb interaction between the excited electron and hole is considered, which is not taken into account in Franz–Keldysh effect.

A further development was then made by Abe and Toyozawa (1981). By calculating the inter-band absorption spectra of a crystal with Gaussian site diagonal disorder and using the coherent potential approximation (CPA), they found out that the Urbach tail occurs due to static disorders. Accordingly, the tail states in a-solids contribute to an exponential absorption tail as a result of interplay between transfer energy and Gaussian distributed site energies.

Based on the experimental results, the current stage of thinking is that the Urbach tails in a-semiconductors are caused by both thermal effects and static disorders (Cody, 1984). The work of Abe and Toyozawa will be reviewed in the next subsection. However, as a considerable part of the work is computational and cannot be explained analytically, readers may refer to the original paper for details.

Urbach tail due to static disorder

We can write the electronic Hamiltonian in Eq. (3.26), dropping the spin index and considering only two bands, conduction and valence, for a crystal in which the diagonal site energies vary randomly as

$$\hat{H}^j = \sum_n E_n^j a_{jn}^+ a_{jn} + \sum_{n \neq m}\sum T_{nm}^j a_{jn}^+ a_{jm}. \tag{4.79}$$

The creation and annihilation operators are used as $a_{jn}^+ \equiv |\, j, n\rangle$ and $a_{jn} \equiv \langle j, n|$, respectively. E_n^j and T_{nm}^j are random site energy and regular site transfer energy, respectively, of the band $j =$ c, v; c denotes the conduction and v valence bands, given by

$$H^j|\, jn\rangle = E_n^j|\, jn\rangle. \tag{4.80}$$

The most important assumption is that of the probability distribution function for the random site energies to be Gaussian given by

$$P(\{E_n^j\}) = \prod_n p(x_n, y_n), \tag{4.81}$$

where $x_n = E_n^{\mathrm{c}} - \bar{E}_n^{\mathrm{c}}$, $y_n = E_n^{\mathrm{v}} - \bar{E}_n^{\mathrm{v}}$ and $\bar{E}_n^j$, $j = \mathrm{c}, \mathrm{v}$ are the mean values of the random site energies associated with conduction and valence bands. Also,

$$p(x_n, y_n) = \frac{1}{2\pi\sqrt{\det D}} \exp\left[-\frac{1}{2}(xy)D^{-1}\begin{pmatrix} x \\ y \end{pmatrix}\right], \tag{4.82}$$

where D is a positive definite symmetric 2×2 matrix given by

$$D = \begin{pmatrix} W_{\mathrm{c}}^2 & \gamma W_{\mathrm{c}} W_{\mathrm{v}} \\ \gamma W_{\mathrm{c}} W_{\mathrm{v}} & W_{\mathrm{v}}^2 \end{pmatrix}. \tag{4.83}$$

Here W_{c}, W_{v} and γ are three parameters used to characterize the randomness in the system. The first two of these represent the root mean square standard deviations in the random site energies in the two bands as

$$W_j = \sqrt{\langle\langle (E_n^j - \bar{E}_n^j)^2 \rangle\rangle}, \tag{4.84}$$

and γ represents the correlation coefficient between the two bands given by

$$\gamma = \frac{\langle\langle (E_n^{\mathrm{c}} - \bar{E}_n^{\mathrm{c}})(E_n^{\mathrm{v}} - \bar{E}_n^{\mathrm{v}}) \rangle\rangle}{W_{\mathrm{c}} W_{\mathrm{v}}}. \tag{4.85}$$

The notation $\langle\langle \cdots \rangle\rangle$ denotes the ensemble average.

The electronic excitation operator is used as the dipole moment operator given by

$$\prod^{+} = \sum_n |c, n\rangle p \langle v, n|, \tag{4.86}$$

where p is the dipole transition matrix element as given in Eq. (4.B.7) in Appendix B. The important point to note here is that the momentum operator matrix element is not used. Instead the momentum operator is used in the form of Eq. (4.54), and then the integral I in Eq. (4.45) can be expressed as

$$I = \int_{E_{\mathrm{c}}}^{E_{\mathrm{v}}+\hbar\omega} \int_{E-E_{\mathrm{v}}}^{E_{\mathrm{v}}} M(E'_{\mathrm{c}}, E'_{\mathrm{v}}) D_{\mathrm{c}}(E'_{\mathrm{c}}) D_{\mathrm{v}}(E'_{\mathrm{v}}) \delta(E'_{\mathrm{c}} - E'_{\mathrm{v}} - \hbar\omega)\, \mathrm{d}E'_{\mathrm{c}}\, \mathrm{d}(E'_{\mathrm{v}}), \tag{4.87}$$

where $M(E'_{\mathrm{c}}, E'_{\mathrm{v}}) = |p_{\mathrm{cv}}|^2$ is the squared dipole matrix element. This is not assumed to be independent of energy, as in Eq. (4.45), and $D_j(E'_j)$ is the DOS of the band j given by:

$$D_j(E'_j) = \langle\langle \mathrm{Tr}_j \delta(E'_j - H^j) \rangle\rangle. \tag{4.88}$$

As will become clear later on, the form of this DOS is not the same as given in Eq. (4.9). This is why it is denoted by a different symbol.

As the absorption coefficient is proportional to $I(E)$ (see Eq. (4.44)), the evaluation of $I(E)$ can provide information about the absorption spectrum. For evaluating $I(E)$, Eq. (4.86) is expressed as

$$I(E) = \iint dE'_c\, dE'_v \delta(E'_c - E'_v - E) S(E'_c, E'_v), \tag{4.89}$$

where $S(E'_c, E'_v)$ is used in place of the square of the transition matrix element and the product of the two DOS and it is defined by

$$S(E'_c, E'_v) = \left\langle\left\langle \mathrm{Tr}_v \left[\prod \delta(E'_c - H^c) \prod^{+} (E'_v - H^v) \right] \right\rangle\right\rangle. \tag{4.90}$$

Comparing Eq. (4.89) with Eq. (4.88) we find that $M(E'_c, E'_v)$ can be expressed as

$$M(E'_c, E'_v) = \frac{S(E'_c, E'_v)}{D_c(E'_c) D_v(E'_v)}. \tag{4.91}$$

Here H^j, $j = \mathrm{c, v}$ is the Hamiltonian given in Eq. (4.79), and Tr_j is defined within the subspace of each band as

$$\mathrm{Tr}_j[\cdots] = \sum_n \langle j, n| \cdots |j, n\rangle. \tag{4.92}$$

Substituting Eqs (4.86) and (4.92) into Eq. (4.89), we get

$$S(E'_c, E'_v) = \langle\langle |p|^2 \langle c, n|\delta(E'_c - H^c)|c, n\rangle \langle v, n(E'_v - H^v)|\, v, n\rangle |\rangle\rangle. \tag{4.93}$$

The normalized densities of states $D_j(E'_j)$ and $S(E'_c, E'_v)$ can be written as

$$\begin{aligned} D_j(E'_j) &= \left\langle\left\langle \frac{1}{N} \sum_n \langle jn|\delta(E'_j - H^j)|j, n| \right\rangle\right\rangle \\ &= \langle j, n|\langle\langle \delta(E'_j - H^j)\rangle\rangle |j, n|. \end{aligned} \tag{4.94}$$

$$\begin{aligned} S(E'_c, E'_v) &= |p|^2 \left\langle\left\langle \frac{1}{N} \sum_n \langle c, n|\langle\langle \delta(E'_c - H^c)|c, n\rangle \langle v, n(E'_v - H^v)|v, n\rangle \right\rangle\right\rangle \\ &= |p|^2 \langle c, n|\langle\langle \delta(E'_c - H^c)|c, n\rangle \langle v, n(E'_v - H^v)|\rangle\rangle |v, n\rangle]. \end{aligned} \tag{4.95}$$

Equations (4.94) and (4.95) are obtained by applying the following normalization conditions

$$\int dE\, D_j(E) = 1, \quad \text{and} \quad \int dE\, I(E) = 1, \tag{4.96}$$

in Eqs (4.88) and (4.89), respectively.

Introducing the single particle Green function as

$$G^j = (E'_j - H^j)^{-1}, \tag{4.97}$$

and two particle Green function as

$$K = \left\langle G^{\mathrm{c}} \sum_m |cm\rangle\langle vm| G^{\mathrm{v}} \right\rangle, \tag{4.98}$$

in Eq. (4.95), $S(E'_{\mathrm{c}}, E'_{\mathrm{v}})$ can be written as

$$S(E'_{\mathrm{c}}, E'_{\mathrm{v}}) = \frac{|p|^2}{(2\pi \mathrm{i})^2}[K(E'^{+}_{\mathrm{c}}, E'^{+}_{\mathrm{v}}) - K(E'^{+}_{\mathrm{c}}, E'^{-}_{\mathrm{v}})$$
$$- K(E'^{-}_{\mathrm{c}}, E'^{+}_{\mathrm{v}}) + K(E'^{-}_{\mathrm{c}}, E'^{-}_{\mathrm{v}})], \tag{4.99}$$

where $K(Z_{\mathrm{c}}, Z_{\mathrm{v}})$ is the site diagonal matrix element of the two particle Green function in Eq. (4.98) given by

$$K(Z_{\mathrm{c}}, Z_{\mathrm{v}}) = \langle c, n|K|v, n\rangle, \tag{4.100}$$

and $E^{\pm} = E \pm \mathrm{i}0$. The CPA is finally applied to evaluate $K(Z_{\mathrm{c}}, Z_{\mathrm{v}})$.

The regular transfer energy part of the Hamiltonian in Eq. (4.79) is written in **k**-space as

$$\sum_{n\neq} \sum_{m} E^j_{nm} a^{+}_{jn} a_{jm} = \sum_k E^j_k a^{+}_{jk} a_{jk}, \tag{4.101}$$

where E^j_k is the energy dispersion relation for the band $j = \mathrm{c, v}$. The electronic DOS is assumed to be elliptic, as used for molecular crystals, given by

$$D_j(E) = \frac{1}{N}\sum_k \delta(E - E_j(k)) = \frac{1}{\pi B_j^2}(B_j^2 - E^2)^{1/2}, \quad \text{for } |B_j| \leq |E|, \tag{4.102}$$

and

$$g_j(E) = 0, \quad \text{for } |B_j| > |E|.$$

For numerical calculations, it is also assumed that valence and conduction bands have similarity in dispersion, which is expressed by

$$\frac{E_{\mathrm{v}}(k)}{B_{\mathrm{v}}} = \mp \frac{E_{\mathrm{c}}(k)}{B_{\mathrm{c}}}, \tag{4.103}$$

where the minus sign corresponds to direct gap transitions and the plus sign to indirect gap transitions.

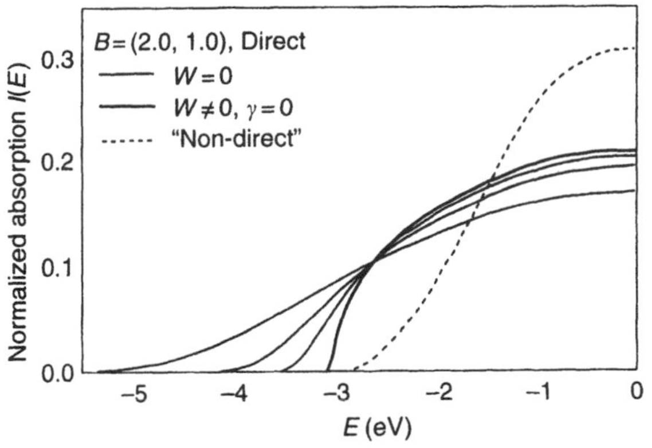

Figure 4.8 Interband absorption spectra calculated for $B_c = 2.0$, $B_v = 1.0$ and $\gamma = 0$ with $W_c = W_v = W$ as a parameter for direct gap. The broken curve shows the convolution of the DOS for $W = 0$ (Abe and Toyozawa, 1981).

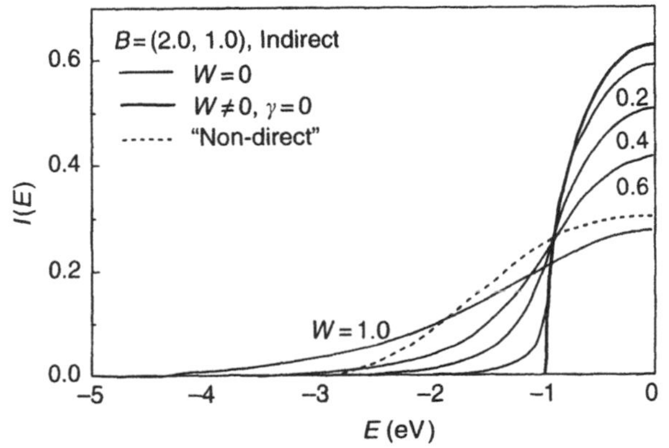

Figure 4.9 Interband absorption spectra calculated for $B_c = 2.0$, $B_v = 1.0$ and $\gamma = 0$ with $W_c = W_v = W$ as a parameter for indirect gap. The broken curve shows the convolution of the DOS for $W = 0$ (Abe and Toyozawa, 1981).

The absorption spectra are calculated numerically for $B_v = 1$ and $B_c = 2$. Assuming that the randomness parameters are equal for both conduction and valence bands, that is, $W_c = W_v = W$, W is varied from 0 to 1. First, the case of uncorrelated conduction and valence band site energies, that is, for $\gamma = 0$, is considered. $I(E)$, thus computed, is plotted as a function of the photon energy E below the conduction band edge in Fig. 4.8 for direct gap, and in Fig. 4.9 for indirect gap transitions. In both figures, the dotted curve represents the convolution of the DOS by evaluating Eq. (4.87) for $W = 0$ which means assuming

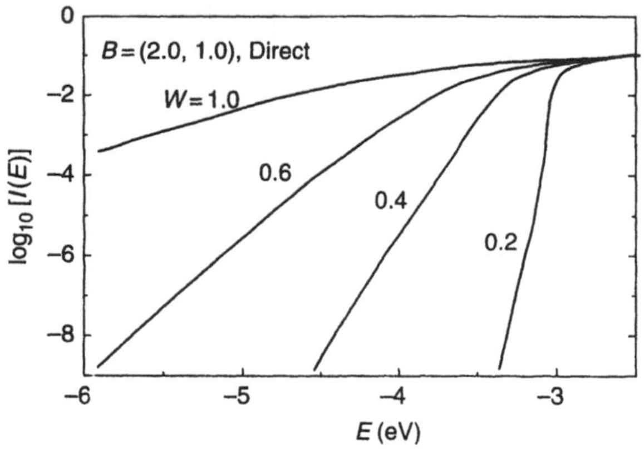

Figure 4.10 Calculated absorption edges in Fig. 4.8 plotted on a logarithmic scale (Abe and Toyozawa, 1981).

$M(E_c', E_v')$ as constant. The solid line curve with $W = 0$, represents a crystal with no disorder; and therefore it gives a sharp conduction band edge at $E = -3.0$ for the direct gap (Fig. 4.8) and at $E = -1.0$ for the indirect band gap (Fig. 4.9). It is very clear from Figs 4.8 and 4.9 that with even a slight introduction of disorder, say $W = 0.2$, the absorption spectrum broadens and the absorption edge shifts to lower energy.

The data of Fig. 4.8 are plotted in the form of $\log_{10}[I(E)]$ as a function of energy in Fig. 4.10 which exhibits a very clear linear relation representing an exponential Urbach tail. The exponential energy dependence is found to hold over several decades of absorption intensity, provided W is small in comparison with both B_c and B_v. From the numerical results it is obvious that substituting Eq. (4.89) into Eq. (4.44), we can write the absorption coefficient for transitions involving the tail states as

$$\alpha\hbar\omega = C \exp(E/E_t), \tag{4.104}$$

where E_t denotes the inverse slope of the linear relation plotted in Fig. 4.10, and C is the pre-exponential constant (see Eq. (4.44)). Following the derivation of Eq. (4.57), the exponential relation can also be expressed as

$$\frac{\alpha}{E} = C' \exp(E/E_t'), \tag{4.105}$$

where C' and E_t' are the corresponding constants as in Eq. (4.104). Neither of these constants depend on the photon energy. It is the slope parameter (E_t or E_t') that can be determined experimentally, and hence the relation in Eq. (4.104) or Eq. (4.105) is very useful. However, the experimental results of these parameters cannot be compared with any theoretical parameters for any further quantitative understanding. Therefore, the above theory is used primarily to fit the experimental data and the slope is determined from such fittings. In this way, the theory is used to interpret the experimental results qualitatively.

Another interesting result of Abe and Toyozawa's calculations is the DOS of conduction and valence bands plotted in Fig. 4.11 as a function of energy. The lower curves represent the DOS on a linear scale and the upper on a logarithmic scale. The densities of states consist of two parts: the main band part for $|E| \leq B_j$, and the tail part for $|E| \geq B_j$. In the main part, $E^{1/2}$-dependence is quite clearly retained, but then this changes in the tails and does not exhibit $E^{1/2}$ dependence. The plot on the logarithmic states shows linear dependence of densities of states in the tail region, which means exponential character like the absorption spectra in Fig. 4.9. Thus, the DOS in the tail region can be expressed as

$$D^{\text{tail}}(E) \approx \exp\left(\frac{E}{\Gamma}\right), \tag{4.106}$$

where Γ is obtained approximately as

$$\Gamma \approx C_0 \frac{W^2}{B}, \tag{4.107}$$

and C_0 is a constant close to 0.5. It is also found that the profound exponential tailing is produced when the range of parameters is such that $W/B \leq 0.7$. The similarity between the exponential densities of states and exponential absorption spectrum is interpreted as the cause of the Urbach tail to be the change in DOS from elliptic or parabolic in the band absorption region to exponential in the tail region. Abe and Toyozawa have thus found the cause of Urbach tail being the introduction of static disorder in the crystal and they have concluded their result as follows. In the main part of the DOS, the regular transfer energy plays the dominant role, while at energies far off the band region the effect of transfer of energy becomes very small so that the DOS tends to the density of site energy distribution, which is assumed to be Gaussian.

This conclusion is quite obvious from Eqs (4.30) and (4.31) for c-solids, and from Eqs (4.41) and (4.42) for a-solids. As stated above, in these equations if the absorption coefficient is evaluated by summing over all states in the conduction and valence bands, the densities of states do not appear in the expression. It is only when the summation is converted into integrations, according to Eq. (4.34), the DOS appears in the expression, as is obvious from Eqs (4.35) and (4.45). What it means in a physical sense is that, if the energy states are not quasi continuous in any energy region and states are reasonably far apart from one another, one should not convert the summation into integration. Thus, in the band region in crystals or extended states region in a-solids, where energy states are quasi continuous, the absorption spectrum should be evaluated by integration and the DOS will play a role. In the tail region, for example, in a-solids, the energy states may not be so close as to be quasi continuous and then the calculation of absorption by integration may not be valid. In this situation, according to Abe and Toyozawa, the site distribution energies behave like DOS, which can be expected to play a role only if the integration method is adopted. This means the exponential densities of states in the tail region of Fig. 4.11 obtained by Abe and Toyozawa are due to static disorders, but this form of densities of states is not the cause of the exponential tail. Both exponential densities of states and exponential Urbach tail are due to static disorders, but the latter does not occur due to former.

Abe and Toyozawa's numerical results suggest that the Urbach tail is caused by static disorder in crystalline materials, which therefore can be applied to a-solids. The results also suggest that the effect of disorder plays role in the low energy region of the absorption, where the DOS also has exponential character, but the DOS itself is not the cause of Urbach tail. Instead the cause of tail is the distribution of site energies, which is assumed to be Gaussian. What is not clear, however, is why and how the Gaussian distribution of site energies produces an exponential Urbach tail. In other words, if the Gaussian distribution itself is the cause of the Urbach tail, how does a Gaussian distribution of site energies produce an exponential shape in the tail spectra? As stated above, because the approach of calculating the absorption spectrum due to static disorder is numerical, this question cannot be answered easily in any sensible way.

Comparison with experimental results

From the absorption spectra of high energy transitions (i.e. from valence to conduction extended states) we have obtained two relations given in Eqs (4.53) and (4.65) to determine the optical gap E_0. Likewise, we have obtained two relations, Eqs (4.104) and (4.105) for the lower energy transitions involving the tail states as well. It is very important to determine both the optical gap energy, E_0, and inverse Urbach slope, E_t, of a-semiconductors. The drift mobility has been measured in a-Si : H_x and the data demonstrate the existence of an exponential density of trapped states proceeding from the valence band that controls the hole drift mobility. It has also been interpreted to contribute the exponential absorption edge (Tiedje *et al.*, 1981, 1983; Tiedje and Rose, 1981). This is because the inverse logarithmic slope of the trap distribution is the same as E_t' in (Eq. (4.105)) (Cody, 1984).

Analyzing the experimental data on measuring the change in the optical gap as a function of temperature in a-Si : H (Cody *et al.*, 1981) and in a-Ge : H (Bludau *et al.*, 1974), Cody (1981) has fitted the change to follow the following equation

$$\Delta E_0(T) = \frac{K}{[\exp(\theta_E/T) - 1]}, \tag{4.108}$$

where $K = 220 \times 10^{-3}$ eV and $\theta_E = 400$ K for a-Si : H, and $K = 96 \times 10^{-3}$ eV and $\theta_E = 200$ K for a-Ge : H. Cody also found a linear relation between the change

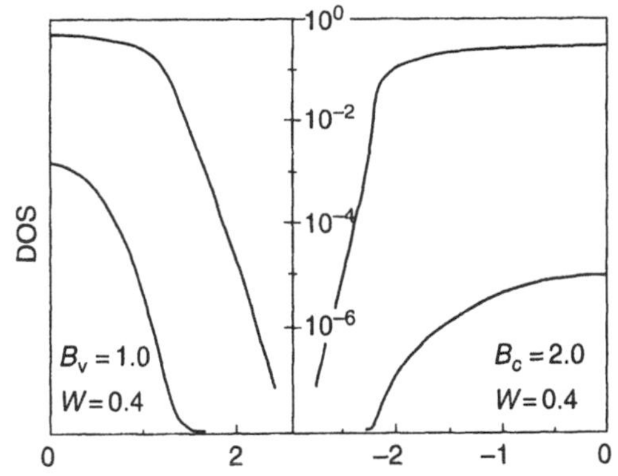

Figure 4.11 Calculated density of states for $B_c = 2.0$, $B_v = 1.0$ and $W = 0.4$ on linear (lower curves) and logarithmic (upper curves) scales (Abe and Toyozawa, 1981).

in the optical gap, ΔE_0 and associated change in the inverse Urbach slope, $\Delta E_t'$, and then fitted a straight line for a-Si : H given by

$$E_0 = -6.2E_t' + 2.0\,\text{eV}, \tag{4.109}$$

where $E_0 = 1.64\,\text{eV}$ and $E_t' = 50 \times 10^{-3}\,\text{eV}$ are used.

Cody *et al.* (1981) have measured the optical absorption coefficient α in a-Si : H_x as a function of photon energy E as shown in Fig. 4.12. The filled symbols indicate data obtained at the measurement temperature T_M below room temperature and unfilled ones denote measurements made on a similar film at $T = 293$ K, where the film had been isochronally annealed for 30 min periods at temperatures up to 625°C. It is clear from Fig. 4.12 that the data demonstrate an Urbach tail, which is also temperature dependent. On the basis of the relation of Eq. (4.109) and the results of Fig. 4.12, Cody has suggested that it would be impossible to distinguish between the effects of thermal and structural disorders on the absorption edge in a-Si : H and expressed the additive effect of thermal and static disorder in the inverse Urbach slope as (Cody, 1984):

$$E_t'(T, X) = \left(\frac{\theta_E}{\sigma}\right)\left\{\frac{1}{\exp(\theta_E/T) - 1} + \frac{1 + X}{2}\right\}, \tag{4.110}$$

where σ is the Urbach edge parameter and X is a measure of disorder. The significance of these two parameters can be understood as follows. As the temperature becomes infinitely large, $T \to \infty$, the second term of Eq. (4.110), representing the disorder, can be neglected, and then expanding the exponential gives, $E_t(T, X) \to T/\sigma$. As the thermal energy disorder is given by $k_B T$, writing T/σ as $k_B T/(\sigma k_B)$, one can clearly see that if $\sigma > k_B^{-1}$ then the thermal disorder energy $k_B T$ reduces in value. Therefore, σ represents a reduction factor for the thermal disorder energy.

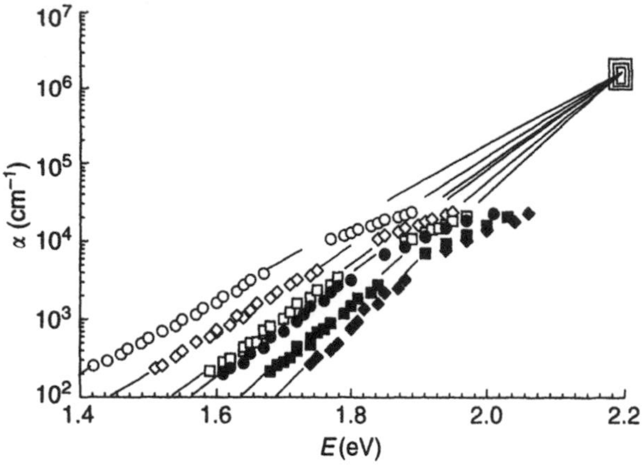

Figure 4.12 The absorption edge of a-Si : H_x as a function of isochronal annealing temperature T_A and measurement temperature T_M. Open symbols (T_M = 300 K): T_A = 575°C (○), 525°C (◇), 475°C (□). Solid symbols: T_M = 293 K (•), T_M = 151 K (■), and T_M = 12.7 K (◇) (Cody, 1984).

For a-Si : H_x, Cody has shown that $\sigma \approx 1$, and using this value in Eq. (4.110) for $E(300, X) = 50 \times 10^{-3}$ eV, he has estimated $X = 1.3$ (Fig. 4.12).

The essence of the above analysis may be summarized as follows. In the lower temperature region, the occurrence of Urbach tail may be considered to be primarily from the structural disorders. As temperature increases, the thermal vibrations become active and the associated thermal disorders due to atomic vibrations also contribute to the exponential absorption. Once a sample is prepared and annealed, the structural disorders may not vary, however, as the temperature increases high enough the effect of structural disorders may be overtaken by that of thermal disorders. The temperature at which such an overtaking occurs is called the fictive temperature T_F, above which the influence of structural disorders on the absorption spectrum becomes relatively negligibly small. The fictive temperature can be derived from Eq. (4.110) by writing it as

$$E_t(T_F, 0) \approx \left(\frac{\theta_E}{\sigma}\right)\left\{\frac{1}{\exp(\theta_E/T_F) - 1}\right\}, \tag{4.111}$$

which is obtained by neglecting the second term, associated with the structural disorders, within the curly brackets of Eq. (4.110). Rearranging Eq. (4.111), one obtains

$$T_F = 6_E\left[\ln\left(\frac{1+\sigma S}{\sigma S}\right)\right]^{-1}, \tag{4.112}$$

where $S = E(T_F, 0)/\theta_E$. The fictive temperature in Eq. (4.112) is obtained by considering that at this temperature all of the second term of Eq. (4.110) becomes negligible, not only X, therefore the expression of Eq. (4.112) is slightly different from that derived by Cody (1984). For a-Si : H_x, $T_F \approx 580\,K$ for annealing temperature $T_A \leq 700\,K$ and measurement temperature $T_M = 300\,K$ (Cody, 1984).

Appendix A: evaluation of the integral in Eq. (4.45)

The integral in Eq. (4.45) is

$$I = \int_{E_c}^{E_v+\hbar\omega} \int_{E-E_v}^{E_v} g_c(E'_c) g_v(E'_v) \delta(E'_c - E'_v - \hbar\omega)\, dE'_c\, d(E'_v). \tag{4.A.1}$$

Let us first integrate over the valence states to get

$$I = \int_{E_c}^{E_v+\hbar\omega} g_c(E'_c) g_v(E'_c - \hbar\omega)\, dE'_c. \tag{4.A.2}$$

Expressing the density of states derived in Eq. (4.9) for the valence and conduction extended states, respectively, as

$$g_v(E) = \frac{V}{2\pi^2}\left(\frac{2m_h^*}{\hbar^2}\right)^{3/2}(E_v - E)^{1/2}, \tag{4.A.3}$$

and

$$g_c(E) = \frac{V}{2\pi^2}\left(\frac{2m_e^*}{\hbar^2}\right)^{3/2}(E - E_c)^{1/2}. \tag{4.A.4}$$

where m_h^* is the effective mass of hole in the valence extended states. It is to be noted that the density of valence extended states is expressed in terms of $E_v - E$, because these states are below the hole mobility edge at E_v. Substituting (4.A.3) and (4.A.4) into (4.A.2), the integral becomes

$$I = \left(\frac{V}{2\pi^2}\right)^2 \left(\frac{8}{\hbar^6}\right)(m_e^* m_h^*)^{3/2} Y\left(\frac{1}{2}\right), \tag{4.A.5}$$

where

$$Y\left(\frac{1}{2}\right) = \int_{E_c}^{E_v+\hbar\omega} (E - E_c)^{1/2}(E_v + \hbar\omega - E)^{1/2}\, dE. \tag{4.A.6}$$

Let us evaluate the integral in (4.A.6) in a general form as

$$Y(r) = \int_{E_c}^{E_v+\hbar\omega} (E - E_c)^r (E_v + \hbar\omega - E)^r\, dE, \tag{4.A.7}$$

where r can be a fraction or integer. Divide and multiply (4.A.7) by $[\hbar\omega + E_v - E_c]^{2r}$ to write it as

$$Y(r) = [\hbar\omega + E_v - E_c]^{2r} \int_{E_c}^{E_v+\hbar\omega} \left[\frac{E - E_c}{\hbar\omega + E_v - E_c}\right]^r \left[\frac{\hbar\omega + E_v - E}{\hbar\omega + E_v - E_c}\right]^r dE.$$

(4.A.8)

Defining a new variable of integration x as

$$x = \frac{\hbar\omega + E_v - E}{\hbar\omega + E_v - E_c}, \qquad x = 0 \text{ for } E = \hbar\omega + E_v, \quad \text{and } x = 1 \text{ for } E = E_c,$$

we can write (4.A.8) as

$$Y(r) = [\hbar\omega - (E_c - E_v)]^{2r+1} \int_0^1 x^r (1 - x)^r \, dx,$$

and then the integral can be evaluated in terms of Γ functions as

$$Y(r) = [\hbar\omega - (E_c - E_v)]^{2r+1} \frac{\Gamma(r+1)\Gamma(r+1)}{\Gamma(2r+2)}. \tag{4.A.9}$$

For $r = \frac{1}{2}$, Eq. (4.A.9) gives

$$Y\left(\frac{1}{2}\right) = \frac{\pi}{8}[\hbar\omega - (E_c - E_v)]^2, \tag{4.A.10}$$

and substituting (4.A.10) into (4.A.2) we get

$$\alpha = \frac{4\pi}{nc}\left(\frac{e}{m_e}\right)^2 \left(\frac{V(m_e^* m_h^*)^{3/2}}{2\pi^2 \omega \hbar^6}\right) |p_{cv}|^2 [\hbar\omega - (E_c - E_v)]^2, \tag{4.A.11}$$

which is the same as Eq. (4.47).

It may also be pointed out here that using $r = 1$ in (4.A.9) the value of the integral becomes

$$I(1) = \tfrac{1}{6}[\hbar\omega - (E_c - E_v)]^3, \tag{4.A.12}$$

which shows that the absorption coefficient can depend on the photon energy as $[\hbar\omega - (E_c - E_v)]^3$, if the DOS depends linearly on energy.

Appendix B: derivation of the matrix element $p_{cv} = \langle 1, \mathbf{k}_c | \hat{\mathbf{e}} \cdot \mathbf{p} | 0, \mathbf{k}_v \rangle$ in Eqs (4.30) and (4.31)

Using the effective mass approximation the single particle electronic Hamiltonian in a crystal is just the kinetic energy operator given by

$$\hat{H} = -\frac{\hbar^2 \nabla^2}{2m_e^*}, \tag{4.B.1}$$

where m_e^* is the effective mass of an electron. In this form, the eigenvalues of the Hamiltonian for conduction and valence Bloch states are

$$\hat{H}|1, \mathbf{k}_c\rangle = E_c(\mathbf{k}_c)|1, \mathbf{k}_c\rangle, \tag{4.B.2}$$

and

$$\hat{H}|1, \mathbf{k}_v\rangle = E_v(\mathbf{k}_v)|1, \mathbf{k}_v\rangle. \tag{4.B.3}$$

The momentum operator can be written as

$$\mathbf{p} = \mathrm{i}\frac{m_e^*}{\hbar}[\hat{H}, \mathbf{r}], \tag{4.B.4}$$

where $\mathbf{r}$ is the complex conjugate coordinate operator associated with $\mathbf{p}$. Using (4.B.4), we evaluate the matrix element

$$\begin{aligned} p_{cv} &= \langle 1, \mathbf{k}_c|\hat{\mathbf{e}}\cdot\mathbf{p}|0, \mathbf{k}_v\rangle = \mathrm{i}\frac{m_e^*}{\hbar}\langle 1, \mathbf{k}_c|\hat{H}, \hat{\mathbf{e}}\cdot\mathbf{r}|0, \mathbf{k}_v\rangle \\ &= \mathrm{i}\frac{m_e^*}{\hbar}[\langle 1, \mathbf{k}_c|\hat{H}\hat{\mathbf{e}}\cdot\mathbf{r}|0, \mathbf{k}_v\rangle - \langle 1, \mathbf{k}_c|\hat{\mathbf{e}}\cdot\mathbf{r}\hat{H}|0, \mathbf{k}_v\rangle]. \end{aligned} \tag{4.B.5}$$

Using Eqs (4.B.2) and (4.B.3), we can write Eq. (4.B.5) as

$$\begin{aligned} p_{cv} &= \mathrm{i}\frac{m_e^*}{\hbar}[E_c(\mathbf{k}_c)\langle 1, \mathbf{k}_c|\hat{\mathbf{e}}\cdot\mathbf{r}|0, \mathbf{k}_v\rangle - \langle 1, \mathbf{k}_c|\hat{\mathbf{e}}\cdot\mathbf{r}|0, \mathbf{k}_v\rangle E_v(\mathbf{k}_v)] \\ &= \mathrm{i}\frac{m_e^*}{\hbar}[E_c(\mathbf{k}_c) - E_v(\mathbf{k}_v)]\langle 1, \mathbf{k}_c|\hat{\mathbf{e}}\cdot\mathbf{r}|0, \mathbf{k}_v\rangle. \end{aligned} \tag{4.B.6}$$

Thus, the objective of writing the momentum operator in the form of (4.B.4) is that the momentum transition matrix element gets converted into the dipole transition matrix element, given by

$$\langle 1, \mathbf{k}_c|\hat{\mathbf{e}}\cdot\mathbf{r}|0, \mathbf{k}_v\rangle. \tag{4.B.7}$$

Assuming that the average separation between the excited electron and hole is independent of the photon energy and momentum, and denoting it by Q_c, we can write (4.B.7) as

$$\langle 1, \mathbf{k}_c|\hat{\mathbf{e}}\cdot\mathbf{r}|0, \mathbf{k}_v\rangle = Q_c\delta_{\mathbf{k}_c,\mathbf{k}_v}, \tag{4.B.8}$$

where the subscript c on Q_c denotes the average separation in crystals.

For a-semiconductors similar results will be obtained, but we do not use $\mathbf{k}$-dependent wavefunctions. Following the same procedure as above, but expressing the wavefunctions in terms of energy, the matrix element for a-semiconductors is obtained as

$$p_{cv} = \langle 1, E_c|\hat{\mathbf{e}}\cdot\mathbf{p}|0, E_v\rangle = \mathrm{i}\frac{m_e^*}{\hbar}[E_c - E_v]Q_a\delta_{E_c,E_v}, \tag{4.B.9}$$

where the subscript a on Q_a represents the average separation between electron and hole in a-semiconductors.

References

Abe, S. and Toyozawa, Y. (1981). *J. Phys. Soc. Jpn*, **50**, 2185.

Adachi, S. (1999). *Optical Properties of Crystalline and Amorphous Semiconductors: Materials and Fundamental Principles*. Kluwer Academic, Boston.

Aoki, T., Shimada, H., Hirao, N., Yoshida, N., Shimakawa, K. and Elliott, S.R. (1999). *Phys. Rev. B* **59**, 1579.

Aspnes, D.E., Studna, A.A. and Kinsborn, E. (1984). *Phys. Rev. B* **29**, 768.

Blundau, W., Onton, A., and Heinke, H. (1974). *J. Appl. Phys.* **45**, 1846.

Cody, G.D. (1984). In: Pankov, J.I (ed.), *Hydrogenated Amorphous Silicon*, Vol. 21, part B, Academic Press, New York, p. 11.

Cody, G.D., Tiedje, T., Abeles, B., Brooks, B. and Goldstein, Y. (1981). *Phys. Rev. Lett.* **47**, 1480.

Cody, G.D., Brooks, B.G. and Abeles, B. (1982). *Sol. Energy Mat.* **8**, 231.

Davis, D.A. (1993). *Jpn. J. Appl. Phys.* **32**, 178.

Demichellis, F., Minetti-Mezzetti, E., Tagliaferro, A., Tresso, E., Rava, P. and Ravindra, N.M. (1986). *J. Appl. Phys.* **59**, 611.

Dow, J.D. and Redfield, D. (1971). *Phys. Rev. Lett.* **26**, 762.

Dow, J.D. and Redfield, D. (1972). *Phys. Rev. B* **5**, 594.

Elliott, S.R. (1990). *Physics of Amorphous Materials*, 2nd edn. Longman Scientific & Technical, London.

Elliott, S.R. (1998). *The Physics and Chemistry of Solids*. John Wiley & Sons, Sussex.

Fagen, E.A. and Fritzsche, H.J. (1970). *J. Non-Cryst. Solids* **4**, 480.

Fritzsche, H. (1980). *Sol. Energy Mat.* **3**, 447.

He, X.F. (1990). *Phys. Rev. B* **42**, 11751.

Khawaja, E.E and Hogarth, C.A. (1988). *J. Phys. C. Solid State Phys.* **21**, 607.

Klazes, R.H., Van den Broek, M.H.L.M., Bezemer, J. and Radelaar, S. (1982). *Phil. Mag. B* **45**, 377.

Mandelbrot, B.B. (1982). *The Fractal Geometry of Nature*. Freeman, New York.

Mehern-Nessa, Shimakawa, K., Ganjoo, A. and Singh, J. (2000). *J. Optoelect. Adv. Mater.* **2**, 133.

Morigaki, K (1999). *Physics of Amorphous Semiconductors*. World Scientific, Singapore.

Mott, N.F. and Davis, E.A. (1979). *Electronic Processes in Non-Crystalline Materials*. Clarendon Press, Oxford.

Overhof, H. and Thomas, P. (1989). *Electronic Transport in Hydrogenated Amorphous Semiconductors*. Springer-Verlag, Berlin.

Shklovskii, B.I. and Efros, A.L. (1984). *Electronic Properties of Doped Semiconductors*. Springer-Verlag, Berlin Heidelberg.

Singh, J. (1981). *Phys. Rev. B* **23**, 2011.

Singh, J. (1994). *Excitation Energy Transfer Processes in Condensed Matter*. Plenum, New York.

Singh, J (2002a). *J. Non-Cryst. Solids* **299–302**, 444.

Singh, J. (2002b). *Nonlinear Optics*. **29**, 115.

Singh, J. (2003). *J. Mat. Science* (in press).

Singh, J., Aoki, T. and Shimakawa, K. (2002). *Phil. Mag. B* **82**, 855.

Sokolov, A.P., Shebanin, A.P., Golikova, O.A. and Mezdrogina, M.M. (1991). *J. Phys. Cond. Matter.* **3**, 9887.

Sumi, H. and Toyozawa, Y. (1971). *J. Phys. Soc. Jpn.* **31**, 342.

Tanaka, Ke. (1998). *Jpn J. Appl. Phys.* **37**, 1747.

Tauc, J. (1968). *Mat. Res. Bull.* **3**, 37.
Tauc, J. (1976). *Amorphous and Liquid Semiconductors*. Plenum, New York.
Tauc, J. (1979). In: Abeles, F. (ed.), *The Optical Properties of Solids*. North Holland, Amsterdam, p. 277.
Tiedje, T. and Rose, A. (1981). *Solid State Commun.* **37**, 39.
Tiedje, T., Cebulka, J.M., Morel, D.L. and Abeles, B. (1981). *Phys. Rev. Lett.* **46**, 1425.
Tiedje, T., Abeles, B., Cebulka, J.M. and Pelz, J. (1983a). *Appl. Phys. Lett.* **42**, 712.
Tiedje, T., Abeles, B. and Cebulka, J.M. (1983b). *Solid State Commun.* **47**, 493.
Toyozawa, Y. (1959). *Prog. Theor. Phys.* **26**, 29.
Urbach, F. (1953). *Phys. Rev.* **92**, 1324.
Vorlicek, V., Zavetova, M., Pavlov, S.K. and Pajasova, L. (1981). *J. Non-Cryst. Solids* **45**, 289.
Watanabe, Y., Kawazoe, H. and Yamane, M. (1988). *Phys. Rev. B* **38**, 5677.
Wood, D.L. and Tauc, J. (1972). *Phys. Rev. B* **5**, 3144.
Yu, P.Y. and Cardona, M. (1996). *Fundamentals of Semiconductors*. Springer-Verlag, Berlin-Heidelberg.
Zallen, R. (1983). *The Physics of Amorphous Solids*. John Wiley & Sons, New York.

5 Photoluminescence

The description of charge carrier states in amorphous solids (a-solids) was introduced in Chapter 1 using the chemical bonding approach of molecular orbitals formed by the linear combination of atomic orbitals. In Chapter 3, we introduced the concept of effective mass of charge carriers applicable to a-solids. In this chapter, the effective mass approach will be extended to formulate the excited charge carrier energy states or excitonic states with a view to study the optical properties like photoluminescence (PL) in a-solids. Although atomic vibrations giving rise to phonons are an intrinsic part of any solid, the effect of phonons will not be considered as yet. We will consider first a rigid amorphous network, and the effect of carrier–phonon interaction will be presented in Chapter 6.

Photoluminescence occurs when an excited pair of charge carriers (e and h) in a solid recombines radiatively. As introduced in Chapter 1, a-solids contain extended and tail states. Therefore, if the excitation of charge carriers occurs at low temperatures, the excited pairs of e and h will go through rapid thermalization in the extended states, which means they will relax down to their respective tail states very rapidly, typically within about 10^{-13} s. From the tail states they will either recombine radiatively or non-radiatively. The radiative recombination mechanism in the tail states tends to dominate at low temperature, and the non-radiative processes dominate at higher temperature above about 100 K (Street, 1991; Morigaki, 1999a). Thus at low temperatures PL occurs dominantly from the radiative recombinations in the tail states. In the tail states excited charge carriers, usually localized on different sites, must move to the same site for recombination. This is the essence of the radiative tunneling (RT) model suggested by Tsang and Street (1979), and it is briefly described below.

5.1 RT model

As the excited carriers in the extended states rapidly relax down to their tail states, it is expected that all recombinations, both radiative and non-radiative, occur after the excited carriers are trapped in the localized tail states. Carriers in localized states means that in an excited pair of e and h, e can be localized on one site and h on another, but they must be on the same site at the instant of recombination.

Thus one or both must travel to the same site before a recombination can occur. At low temperatures, as mentioned briefly in Chapter 1, the transport of charge carriers can only occur in the localized tail states through quantum tunneling. For radiative tunneling, the recombination time between a pair of e and h localized in the tail states and separated by a distance R is given by

$$\tau = \tau_0 \exp(2R/R_0), \tag{5.1}$$

where R_0 is the localization length and τ_0 is a time constant expected to be about 10^{-8} s. As the separation between excited e and h in the tail states is expected to vary randomly, the RT model suggests a distribution of lifetime for the occurrence of PL. As the earlier PL experiments (Tsang and Street, 1979) did exhibit luminescence decay time extending from 10^{-8} to 10^{-2} s, the RT model has been widely accepted for interpreting the PL spectra in a-Si : H.

However, RT model has failed to explain more recent PL lifetime distribution spectra observed in a-Si : H (Boulitrop and Dunstan, 1985; Searle *et al.*, 1987; Ambros *et al.*, 1991; Stachowitz *et al.*, 1991, 1998); using the frequency resolved PL spectroscopy technique (FRS) (see Figs 5.1 and 5.5). These measurements reveal that the PL lifetime distribution in a-Si : H is not a single broad distribution from 10^{-8} to 10^{-2} s as observed earlier, but instead it consists of a double peak structure, a slow peak with peak intensity at 10^{-3} s and a fast peak with peak intensity at 10^{-6} s. This indicates that there are two different radiative recombination channels, which are well separated in time. As the only variable in Eq. (5.1) is the separation R between e and h, it is impossible to find two different channels in the RT model on the basis of separation.

The other problem with the RT model is that it cannot explain the observed dependence of the time corresponding to the peak positions and corresponding quantum efficiencies of both slow and fast peaks on the temperature and generation rate. At low generation rates, the lifetime and quantum efficiency of both slow and fast peaks remain constant and at higher generation rates both start decreasing proportional to $G^{-0.5}$, G being the generation rate. At higher generation rates, it may be expected that the average separation between e and h would decrease, which will decrease the recombination time according to Eq. (5.1). However, any decrease in separation should not affect the PL quantum efficiency or PL intensity, which is also observed to decrease in the same way as the recombination time.

In order to explain the observed results, two kinds of recombination processes have been suggested in the last two decades: (1) Geminate pair recombination, and (2) distant pair recombination. A geminate pair is the electron–hole (e–h) pair excited by the same photon, and a distant pair is that e and h in an initially excited pair has moved far away from each other and are no longer aware of their initial correlation by the Coulomb interaction. A distant pair is also associated with excited e and h not necessarily created by the same photon. As will become clear later on, a geminate pair is the same as an e and h bound in an excitonic state and a distant pair is the same as a pair of e and h created by dissociation of excitonic states.

As the concept of excitons is not regarded to be valid in a-solids, particularly if the charge carriers are excited in the tail states, the term geminate pair, first suggested by Onsager, was used. The recombination kinetics of distant pairs or dissociated excitonic charge carriers will be described in the chapter on transport of charge carriers, because their recombination involves bringing together a pair of e and h localized on different sites.

However, before we can address the problem of PL in a-solids, it is important to develop a theory of excitation in a-solids. This will be described in the next section.

5.2 Electronic excitation in a-solids

Here, we consider the case of an amorphous insulator/semiconductor, such that all valence extended states are fully occupied and conduction states are completely empty. If such a solid is exposed to light, with photons of energy equal or greater than the optical energy gap of the solid, light will be absorbed by exciting electrons from the valence to conduction extended states. This creates a hole in the valence extended state as well, corresponding to every electronic excitation. Thus, the absorption of every photon creates a pair of electron (e) and hole (h) in any insulator/semiconductor. The Coulomb interaction between e and h in a pair thus created gets activated as soon as they are excited, and therefore the separation between e and h plays a very important role in the optical and opto-electronic properties of a solid.

Following Chapter 3, we set the vacuum state of an a-solid, denoted by $|0\rangle$, to the fully occupied valence states and completely empty conduction states. Let us now consider that the solid is hit by a photon, which is absorbed by exciting a hole in the valence and an electron in the conduction extended states. At the instant of excitation both e and h are excited from the same site, but once they are excited it is difficult to identify which sites they came from because both the valence and conduction extended states are quasi continuum. Therefore, for the purpose of generalization we will consider that the excited e and h are created at sites $\mathbf{l}_e$ and $\mathbf{l}_h$ with spins σ_e and σ_h, respectively. Then considering the two bands approximation and denoting the valence and conduction extended states by 0 and 1, respectively, we can write the state vector for such an excited state as

$$|1,0\rangle = \sum_{\mathbf{l}_e,\mathbf{l}_h} C_{10}(\mathbf{l}_e,\mathbf{l}_h) \sum_{\sigma_e,\sigma_h} a_{1\mathbf{l}_e}^{+}(\sigma_e) d_{0\mathbf{l}_h}^{+}(\sigma_h)|0\rangle, \tag{5.2}$$

where $C_{10}(\mathbf{l}_e,\mathbf{l}_h)$ is the probability amplitude coefficient, and $a_{1\mathbf{l}_h}^{+}(\sigma_h)$ and $d_{0\mathbf{l}_h}^{+}(\sigma_h)$ are the creation operators of electron and hole in the conduction and valence states, 1 and 0, respectively, as defined in Sections 3.2 and 3.3. The summation over spins in Eq. (5.2) corresponds to two different possibilities of excited spin states, a singlet or a triplet, given by

$$\sum_{\sigma_e,\sigma_h} a^+_{1\mathbf{l}_e}(\sigma_e) d^+_{0\mathbf{l}_h}(\sigma_h)|0\rangle = \frac{1}{\sqrt{2}}\left[a^+_{1\mathbf{l}_e}\left(\frac{1}{2}\right) d^+_{0\mathbf{l}_h}\left(-\frac{1}{2}\right) + a^+_{1\mathbf{l}_e}\left(-\frac{1}{2}\right) d^+_{0\mathbf{l}_h}\left(\frac{1}{2}\right)\right]\Big|0\Big\rangle, \tag{5.3}$$

for a singlet, and

$$\sum_{\sigma_e,\sigma_h} a^+_{1\mathbf{l}_e}(\sigma_e) d^+_{0\mathbf{l}_h}(\sigma_h)|0\rangle = \frac{1}{\sqrt{3}}\left[a^+_{1\mathbf{l}_e}\left(\frac{1}{2}\right) d^+_{0\mathbf{l}_h}\left(\frac{1}{2}\right) + a^+_{1\mathbf{l}_e}\left(-\frac{1}{2}\right) d^+_{0\mathbf{l}_h}\left(-\frac{1}{2}\right)\right.$$
$$\left. + \frac{1}{\sqrt{2}}\left\{a^+_{1\mathbf{l}_e}\left(\frac{1}{2}\right) d^+_{0\mathbf{l}_h}\left(-\frac{1}{2}\right) - a^+_{1\mathbf{l}_e}\left(-\frac{1}{2}\right) d^+_{0\mathbf{l}_h}\left(\frac{1}{2}\right)\right\}\right]\Big|0\Big\rangle \tag{5.4}$$

for a triplet. Using Eqs (5.2)–(5.4), we want to determine the energy eigenvalue W of the following Schrödinger equation:

$$\hat{H}|1,0\rangle = W|1,0\rangle, \tag{5.5}$$

where $\hat{H}$ is the electronic Hamiltonian consisting of two terms given in Eqs (3.22) and (3.23). To solve the Schrödinger equation in Eq. (5.5) we will follow the approach used in Section 3.3 for calculating the energy eigenvalue of a single charge carrier, and evaluate the following:

$$\langle 1,0|\hat{H}|1,0\rangle = W\langle 1,0|1,0\rangle. \tag{5.6}$$

The easiest way to evaluate Eq. (5.6) is: first convert all hole operators into electron operators using the relations given in Eq. (3.17), and then use the anti-commutation relations of fermion operators given in Eq. (3.12). We thus obtain from Eq. (5.6) as

$$W = W_0 + \sum_{\mathbf{l}_e,\mathbf{l},\mathbf{l}_h} [C^*(\mathbf{l}_e,\mathbf{l}_h)C(\mathbf{l},\mathbf{l}_h)E_{1\mathbf{l}_e\mathbf{l}} - C^*(\mathbf{l}_e,\mathbf{l}_h)C(\mathbf{l}_e,\mathbf{l})E_{0\mathbf{l}_h\mathbf{l}}]$$
$$- \sum_{\mathbf{l}_e,\mathbf{l}_1,\mathbf{l}_h,\mathbf{l}_2} [C^*(\mathbf{l}_e,\mathbf{l}_2)C(\mathbf{l}_1,\mathbf{l}_h)E_b(\mathbf{l}_e,\mathbf{l}_2,\mathbf{l}_1,\mathbf{l}_h;S), \tag{5.7}$$

where

$$E_{1\mathbf{l}_e\mathbf{l}} = \left\langle 1,\mathbf{l}_e\left| -\frac{\hbar^2\nabla^2}{2m_e} + V(r)\right| 1,\mathbf{l}_1\right\rangle + \sum_{\mathbf{l}_1}\{2\langle 1,\mathbf{l}_e;0,\mathbf{l}|U|0,\mathbf{l};1,\mathbf{l}_1\rangle$$
$$- \langle 1,\mathbf{l}_e;0,\mathbf{l}|U|1,\mathbf{l}_1;0,\mathbf{l}\rangle\}, \tag{5.8}$$

$$E_{0\mathbf{l}_h\mathbf{l}} = \left\langle 0,\mathbf{l}_h\left| -\frac{\hbar^2\nabla^2}{2m_h} + V(r)\right| 0,\mathbf{l}\right\rangle + \sum_{\mathbf{l}_1}\{2\langle 0,\mathbf{l}_h;g,\mathbf{l}_1|U|0,\mathbf{l}_1;0,\mathbf{l}\rangle$$
$$- \langle 0,\mathbf{l}_h;0,\mathbf{l}_1|U|0,\mathbf{l};0,\mathbf{l}_1\rangle\}, \tag{5.9}$$

$$E_b(\mathbf{l}_e,\mathbf{l}_2,\mathbf{l}_1,\mathbf{l}_h,S) = \langle 1,\mathbf{l}_e;0,\mathbf{l}_2|U|0,\mathbf{l}_h;1,\mathbf{l}_1\rangle$$
$$- (1-S)\langle 1,\mathbf{l}_e;0,\mathbf{l}_2|U|1,\mathbf{l}_1;0,\mathbf{l}_h\rangle, \tag{5.10}$$

where $S = 0$ for singlets and $S = 1$ for triplets and W_0 is the total energy of all the electrons in the valence states before the excitation. It is obtained as [see also Eq. (3.19)]

$$W_0 = \sum_{\mathbf{l}_1} \left[2\left\langle 0, \mathbf{l}_1 \left| -\frac{\hbar^2 \nabla^2}{2m_e} + V(r) \right| 0, \mathbf{l}_1 \right\rangle + \sum_{\mathbf{l}_2} \{2\langle 0, \mathbf{l}_1; 0, \mathbf{l}_2 \,|\, U \,|\, 0, \mathbf{l}_2; 0, \mathbf{l}_1 \rangle - \langle 0, \mathbf{l}_1; 0, \mathbf{l}_2 |\, U \,| 0, \mathbf{l}_1; 0, \mathbf{l}_2 \rangle\} \right]. \tag{5.11}$$

The factor of 2 in Eqs (5.8), (5.9) and (5.11) comes from the two possible spins. It is to be noted that Eq. (5.7) is obtained by applying the following condition:

$$\sum_{\mathbf{l}_e, \mathbf{l}_h} [C^*(\mathbf{l}_e, \mathbf{l}_h) C(\mathbf{l}_e, \mathbf{l}_h) = 1, \tag{5.12}$$

which ensures that the excited pair of charge carriers is within the solid. The energy matrix elements $E_{1\mathbf{l}_e\mathbf{l}}$ (Eqs (5.7) and (5.8)) and $E_{0\mathbf{l}_h\mathbf{l}}$ (Eqs (5.7) and (5.9)), represent the energy transfer matrix element between sites $\mathbf{l}_e$ and $\mathbf{l}$ for an electron excited in the conduction states and between sites $\mathbf{l}_h$ and $\mathbf{l}$ for a hole in the valence states, respectively. The last term of Eq. (5.7) with the energy matrix element $E_b(\mathbf{l}_e, \mathbf{l}_2, \mathbf{l}_1, \mathbf{l}_h; S)$, represents the interaction between the excited e and h as e moves from site $\mathbf{l}_e$ to $\mathbf{l}_1$, and h from site $\mathbf{l}_h$ to $\mathbf{l}_2$, as is obvious from Eq. (5.10). Depending on the relative magnitude of this energy matrix element, an excited pair of e and h may or may not form a bound state like an exciton. Therefore, the magnitude of this term of Eq. (5.7) contributes to the binding energy of an excited e and h pair. It is obvious from Eq. (5.10) that the binding energy depends on the spin of the excited pair. The first term of the binding energy is due to the Coulomb interaction, and the second is due to the exchange interaction. The exchange interaction between e and h vanishes for a triplet spin state ($S = 1$), and that means the binding energy in a singlet state is usually less than in a triplet state. In other words a singlet excitonic state is usually formed at a higher energy than the triplet excitonic state.

Before moving any further with the derivation of energy of the excited pair of e and h, it is desirable to have more information about the formation of an excitonic state in amorphous semiconductors (a-semiconductors). Is it an exciton state? If the answer is "yes," then is it a Wannier exciton or Frenkel exciton? We will try to address these questions in the following section.

5.2.1 *Frenkel or Wannier excitons in a-solids?*

In drawing a line between Wannier and Frenkel excitons, the most important part is played by the overlap of inter-atomic electronic wavefunctions (Singh, 1994). The theory of Wannier excitons is applicable in crystalline solids (c-solids) with a large overlap of inter-atomic electronic wavefunctions and most inorganic semiconductors like crystalline Si and Ge fulfill this condition. On the other hand, c-solids with a small overlap of inter-atomic/molecular wavefunctions fulfill the requirement of Frenkel excitons. These are usually organic crystals like those of anthracene, naphthalene, etc. In addition to the inter-atomic/molecular overlap of electronic wavefunctions, the other important point to realize is that the energy of a Wannier exciton, $E_n(K)$, consists of a kinetic energy term associated with its center of mass motion and an internal state energy term due to binding of the e–h pair in a hydrogenic state with principal quantum number n, given by

$$E_n(K) = \frac{\hbar^2 K^2}{2M_x} + E_g - E_b(n), \tag{5.13}$$

where K is the wavevector associated with the momentum of the center of mass motion, and for Wannier excitons it is the sum of the wavevectors of both electron and hole. E_g is the energy of the band gap of the c-solid, and $E_b(n)$ is the exciton binding energy in its internal hydrogenic energy state with $n = 1, 2, 3, \ldots$. For Frenkel excitons also the exciton energy can be written in the same form (Singh, 1994), but no hydrogenic bound state is formed, because a Frenkel exciton usually represents the excited molecular state. Therefore, the concept of exciting e and h pairs is not valid for Frenkel excitons. Also the wavevector **K** or the momentum of the center of mass and the effective mass, M_x, of of Frenkel excitons are defined differently. Nevertheless, in both cases, the wavevector **K** arises from the long range symmetry of crystalline materials (c-materials), which does not exist in the amorphous form. This illustrates quite clearly that in principle the concept of excitons cannot be directly applied to study the properties of excitations created in a-solids.

Let us look at some details of the theory of Wannier and Frenkel excitons. In inorganic crystals, where Wannier excitons are usually created, because of the translational symmetry and large overlap of inter-atomic electronic wavefunctions, the theory is developed in the reciprocal lattice vector space (**k**-space) using Bloch functions of single charge carriers. The excited e in the conduction band and h in the valence band, may first relax non-radiatively down to their respective band edges, and then form excitons or form an exciton first, which then relaxes to its lowest state non-radiatively. Although, the details of the formation of excitons may be different in different materials, generally the above processes follow. In the real crystal space, the separation between e and h of a Wannier exciton can be quite large and for that reason it is also called the large radii orbital exciton.

In the case of Frenkel excitons, inter-atomic/molecular overlap of electronic functions is relatively small, so the excitation in these solids occurs mainly as excitation of individual atoms/molecules. In fact in the original theory of Frenkel excitons there is no binding between e and h in hydrogenic states, and experimentally also such bound states are not observed. It is only when the Frenkel exciton theory is developed in terms of second quantization (Davydov, 1971; Singh, 1994), a Frenkel exciton can be represented as an excited pair of e and h created on the same atom/molecule. Then such an excited pair of e and h moves throughout the crystal together by hopping, which means that the excited pair of e and h on a molecule recombine and transfer that energy to excite another pair on a neighboring molecule. Thus, unlike the case of Wannier excitons, individual charge carriers do not move from one site to another. This is the reason that the theory of Frenkel excitons is formulated in the real crystal space, which has led people to believe that the theory of Frenkel excitons can be applied to study excitations in a-solids (Mott and Davis, 1979; Stutzmann and Brandt, 1992). Of course, as there is no translational symmetry in a-solids, the theory of excitations in such solids must be formulated in the real coordinate space, but that does not make these excitations as Frenkel excitons. We need to develop the theory of Wannier excitons in the real coordinate space, as done above, for studying the excitation properties of a-solids. The theory presented above can be applied to a-solids and it can also be applied to Frenkel excitons, but as shown below, the Frenkel exciton's picture is inadequate to study excitations in a-solids.

In Eq. (5.7), consider that both e and h are excited on the same site, as considered in the theory of Frenkel excitons. Then separate the terms that do not contribute to the energy of Frenkel excitons to write it as

$$W = W_0 + \sum_{\mathbf{l}}[C^*(\mathbf{l},\mathbf{l})C(\mathbf{l},\mathbf{l})E_{1\mathbf{l}\mathbf{l}} - C^*(\mathbf{l},\mathbf{l})C(\mathbf{l},\mathbf{l})E_{0\mathbf{l}\mathbf{l}}] - \sum_{\mathbf{l},\mathbf{m}}[C^*(\mathbf{l},\mathbf{l})C(\mathbf{m},\mathbf{m})E_b(\mathbf{l},\mathbf{m},\mathbf{m},\mathbf{l};S) - D_{eh}, \tag{5.14}$$

where $E_{1\mathbf{l}\mathbf{l}}$ and $E_{0\mathbf{l}\mathbf{l}}$ are obtained by setting $\mathbf{l}_e = \mathbf{l}_h = \mathbf{l}$ in Eqs (5.8) and (5.9) respectively, and $E_b(\mathbf{l},\mathbf{m},\mathbf{m},\mathbf{l};S)$ is obtained by setting $\mathbf{l}_e = \mathbf{l}_h = \mathbf{l}$ and $\mathbf{l}_1 = \mathbf{l}_2 = \mathbf{m}$ in Eq. (5.10). The rest of the terms are collected in D_{eh} given by

$$D_{eh} = \sum_{\mathbf{l}_e,\mathbf{l},\mathbf{l}_h(\mathbf{l}_e\neq\mathbf{l})} C^*(\mathbf{l}_e,\mathbf{l}_h)C(\mathbf{l},\mathbf{l}_h)E_{1\mathbf{l}_e\mathbf{l}} - \sum_{\mathbf{l}_e,\mathbf{l},\mathbf{l}_h(\mathbf{l}_h\neq\mathbf{l})} C^*(\mathbf{l}_e,\mathbf{l}_h)C(\mathbf{l}_e,\mathbf{l})E_{0\mathbf{l}_h\mathbf{l}} - \sum_{\mathbf{l}_e,\mathbf{l}_1,\mathbf{l}_h,\mathbf{l}_2(\mathbf{l}_e\neq\mathbf{l}_h,\mathbf{l}_1\neq\mathbf{l}_2)} C^*(\mathbf{l}_e,\mathbf{l}_2)C(\mathbf{l}_1,\mathbf{l}_h)E_b(\mathbf{l}_e,\mathbf{l}_2,\mathbf{l}_1,\mathbf{l}_h;S) \tag{5.15}$$

In the theory of Frenkel excitons, only terms without D_{eh} in Eq. (5.14) are considered, because the contribution of D_{eh} is not that significant if the overlap of the inter-atomic/molecular electronic wavefunctions is negligible. Neglecting D_{eh}, and assuming the translational symmetry in c-solids, the probability amplitude coefficients can be written as

$$C(\mathbf{l}_e,\mathbf{l}_h) = N^{-1}\exp(i\mathbf{k}_e\cdot\mathbf{l}_e + \mathbf{k}_h\cdot\mathbf{l}_h), \tag{5.16}$$

where N is the number of atoms in the crystal, and $\mathbf{k}_e$ and $\mathbf{k}_h$ are electron and hole wavevectors, respectively. Using Eq. (5.16) in Eq. (5.14), without D_{eh}, the energy eigenvalue W_F of a Frenkel exciton is obtained as

$$W_F = W_0 + E_1 - E_0 - E_b(S) - N^{-1}\sum_{\mathbf{l},\mathbf{m}\neq\mathbf{l}}\exp[i\mathbf{K}\cdot(\mathbf{m}-\mathbf{l})]M_{\mathbf{lm}}(S), \tag{5.17}$$

where $\mathbf{k}_e + \mathbf{k}_h = \mathbf{K}$, $E_1 = N^{-1}\sum_{\mathbf{l}} E_{1\mathbf{l}\mathbf{l}}$, $E_0 = N^{-1}\sum_{\mathbf{l}} E_{0\mathbf{l}\mathbf{l}}$, $E_b(S) = N^{-1}\sum_{\mathbf{l}} E_b(\mathbf{l},\mathbf{l},\mathbf{l},\mathbf{l};S)$, and $M_{\mathbf{lm}}(S) = E_b(\mathbf{l},\mathbf{m},\mathbf{m},\mathbf{l};S)$ is obtained from

Eq. (5.14) as

$$M_{\mathbf{lm}}(S) = \langle 1, \mathbf{l}; 0, \mathbf{m}|U|0, \mathbf{l}; 1, \mathbf{m}\rangle - (1 - S)\langle 1, \mathbf{l}; 0, \mathbf{m}|U|1, \mathbf{m}; 0, \mathbf{l}\rangle. \tag{5.18}$$

It is to be noted that only the last term of Eq. (5.17) depends on the exciton wavevector **K** and hence contributes to the kinetic energy of Frenkel excitons. That means a Frenkel exciton moves only due to the Coulomb interaction between charge carriers. Denoting the last term of Eq. (5.17) by $L_1(\mathbf{K}, S)$, and then applying the effective mass approximation we can write it as

$$L_1(K, S) \approx L_1(0, S) + \frac{\hbar^2 K^2}{2M^*(S)}, \tag{5.19}$$

where $M^*(S)$ is the effective mass of a Frenkel exciton given by (Singh, 1994)

$$[M^*(S)]^{-1} = \frac{1}{\hbar^2}\sum_{\mathbf{l}} l^2 M_{\mathbf{l},0}(S). \tag{5.20}$$

It should also be noted that $L_1(0, S)$ in Eq. (5.19) is independent of the location of the excited site in a crystal, so it can be set to zero for convenience. Thus, the definition of effective mass for a Frenkel exciton is very different from that of a Wannier exciton. The former depends primarily on the Coulomb interaction between the excited e–h pair in a crystal. In Eq. (5.17), writing the band gap energy as $E_g = E_1 - E_0$ and using Eq. (5.19) we get the exciton energy eigenvalue as

$$W_F = W_0 + E_g + \frac{\hbar^2 K^2}{2M^*(S)} - E_b(S), \tag{5.21}$$

which depends on K and the spin, $S = 0$ (singlet) or 1 (triplet) of the exciton. The binding energy of a Frenkel exciton then becomes $E_b(S)$, which is not the same as the binding energy of a hydrogenic state used for Wannier excitons. The energy eigenvalue of a Wannier exciton is not derived here, as the theory is developed in **k**-space, and hence not relevant in the context of a-solids. However, interested readers may like to refer to Knox (1965) and Singh (1994).

As stated above, the reason for giving details about the theory of Frenkel excitons in this chapter is to illustrate the point that it cannot be applied to study the excitation related properties of amorphous inorganic solids. The following points must be noted in particular:

1 In inorganic a-semiconductors, any excitation creating an e and h pair occurs in a similar way as it occurs in their crystalline counter parts, not like molecular solids.
2 Any excited e and h move separately as individual particles in inorganic a-solids, and they have their own individual effective masses as in a Wannier

exciton, but not like in a Frenkel exciton where e and h do not have much meaning as separate particles and the exciton's effective mass is defined quite differently, as given in Eq. (5.20).

3 In a Frenkel exciton, the kinetic energy for its center of mass motion is due primarily to the Coulomb interaction within the exciton, which is not the case in inorganic a-solids where both the kinetic energy of individual particles and ionic potential energy of a solid contribute to the motion of the center of mass of an excited e and h pair.

Therefore, for studying excitations in inorganic a-solids one has to develop the theory of Wannier excitons in the real coordinate representation as described above. Then one has to determine the energy eigenvalue of the excited state of an a-solid using Eq. (5.6).

5.3 Excitonic states in a-semiconductors

To solve Eq. (5.7), we will follow the effective mass approach described in Chapter 3, and extend it to write the probability amplitude coefficient $C_{10}(\mathbf{l}, \mathbf{m})$ as (Singh *et al.*, 2002)

$$C_{10}(\mathbf{l}, \mathbf{m}) = N^{-1} \exp[i(\boldsymbol{\rho}_e \cdot \mathbf{l} + \boldsymbol{\rho}_h \cdot \mathbf{m})], \tag{5.22}$$

where

$$\rho_e = \frac{\sqrt{2m_e^*(E_e - E_c)}}{\hbar} \quad \text{and} \quad \rho_h = \frac{\sqrt{2m_h^*(E_v - E_h)}}{\hbar} \tag{5.23}$$

with m_e^* and m_h^* being the effective masses of electron in the conduction states and hole in the valence states, respectively. E_c and E_v are energies of electron and hole mobility edges, and E_e and E_h are electron and hole energies, respectively. As described in Chapter 3, the choice of the above form of probability coefficient is based on the following basic results of quantum mechanics. A particle moving from one side of a constant barrier of height V and finite width to the other side has a wavefunction like a plane wave, if its energy $E > V$, but it becomes exponentially decreasing with the distance within the barrier, if $E < V$ (see, e.g. Schiff, 1968). Here it is assumed that the mobility edge defines the height of the barrier, which exists between two atoms for a carrier to move from one site to another, and each ionic potential is a square well. The depths and inter well separations are usually not the same for every well in a-solids. However, in quasi-continuos extended, as well as tail states, such a distinction may be assumed to make little difference.

It may be remembered here again that the above choice of the probability amplitude coefficient given in Eq. (5.22) enables one to separate out extended states from tail states in both valence and conduction states. For instance, an electron with energy $E > E_c$ on any site can move to another site like a free wave, but if $E < E_c$, the electron can only tunnel through the barrier between atomic sites and the probability of transfer will decrease exponentially with the distance. This is quite consistent with Mott's condition that extended and localized tail states cannot coexist.

Substituting Eq. (5.22) in Eq. (5.6) we get

$$W = W_0 + N^{-1} \sum_{\mathbf{l}_e, \mathbf{l}} \exp[i\boldsymbol{\rho}_e \cdot (\mathbf{l} - \mathbf{l}_e)] E_{1\mathbf{l}_e\mathbf{l}} - N^{-1} \sum_{\mathbf{l}_h, \mathbf{l}} \exp[i\boldsymbol{\rho}_h \cdot (\mathbf{l} - \mathbf{l}_h)] E_{0\mathbf{l}_h\mathbf{l}}$$
$$- N^{-2} \sum_{\mathbf{l}_e, \mathbf{l}_1, \mathbf{l}_h, \mathbf{l}_2} \exp[i\boldsymbol{\rho}_e \cdot (\mathbf{l}_1 - \mathbf{l}_e) + i\boldsymbol{\rho}_h \cdot (\mathbf{l}_2 - \mathbf{l}_h)] E_b(\mathbf{l}_e, \mathbf{l}_2, \mathbf{l}_1, \mathbf{l}_h; S). \quad (5.24)$$

In the real coordinate space, Eq. (5.24) is similar to that obtained from the theory of Wannier excitons (Elliott, 1962; Singh, 1994) in the reciprocal lattice vector space for c-solids. However, there are two marked differences: (1) From the definition of ρ_e and ρ_h given in Eq. (5.23), it is obvious that these are functions of energy, and thus the exponentials which are functions of wavevectors **k** in c-solids become functions of energy in a-solids. (2) In the theory of Wannier excitons for c-solids, the exchange interaction between the excited e–h pair, which is the second term in Eq. (5.10), is neglected for solving Eq. (5.24) analytically. This is done on the assumption that the exchange interaction is of short range (Elliott, 1962) and hence it can be neglected for large radii orbital excitons as the Wannier excitons are considered in c-solids. This approximation makes Eq. (5.4) independent of spin, which is regarded to be quite valid for c-solids, but it is not so for a-solids. In a-solids, because of the lack of translational symmetry, the situation may be quite different. The charge carriers in an excited pair, in relative terms, may not be able to move far apart from each other, and hence the large radii model cannot be applied for a-solids. Thus a-solids offer a kind of confinement to the transport of charge carriers.

However, Eq. (5.24) can still be solved in a way similar to that applied for c-solids (Singh, 1994). The last three terms, which depend on ρ's, can be expanded in Taylor's series about $\boldsymbol{\rho} = \mathbf{0}$. Applying the effective mass approximation for both the charge carriers, Eq. (3.44) for electron and Eq. (3.54) for hole, in Eq. (5.24) we get

$$W = W_0 + E_1 - E_0 + \frac{p_e^2}{2m_e^*} + \frac{p_h^2}{2m_h^*}$$
$$- N^{-2} \sum_{\mathbf{l}_e, \mathbf{l}_1, \mathbf{l}_h, \mathbf{l}_2} \exp[i\boldsymbol{\rho}_e \cdot (\mathbf{l}_1 - \mathbf{l}_e) + i\boldsymbol{\rho}_h \cdot (\mathbf{l}_2 - \mathbf{l}_h)] E_b(\mathbf{l}_e, \mathbf{l}_2, \mathbf{l}_1, \mathbf{l}_h; S), \quad (5.25)$$

where m_e^* and m_h^* are the effective masses of excited electron and hole as obtained in Eqs (3.41) and (3.56), respectively. If the excited charge carriers are in their tail states one should use the corresponding effective masses given in Eqs (3.43) and (3.57), respectively. E_1 and E_0 are the average energy of excited electron and hole in the conduction and valence states, given respectively by

$$E_1 = N^{-1} \sum_{\mathbf{l}_e, \mathbf{l}} E_{1\mathbf{l}_e\mathbf{l}}, \tag{5.26a}$$

and

$$E_0 = N^{-1} \sum_{\mathbf{l}_h, \mathbf{l}} E_{0\mathbf{l}_h\mathbf{l}}. \tag{5.26b}$$

The average optical energy gap, E_{opt}, of a-solid can be defined as: $E_{\text{opt}} = E_1 - E_0$. In Eq. (5.25), $\mathbf{p}_e = \hbar\boldsymbol{\rho}_e$ and $\mathbf{p}_h = \hbar\boldsymbol{\rho}_h$ are linear momenta of the electron and hole, respectively.

The last term of Eq. (5.25), which is the interaction energy term between the excited electron and hole pair, depends on both ρ_e and ρ_h, and therefore it needs to be expanded in Taylor's series with respect to both. Let us denote it by $M(\rho_e, \rho_h, S)$, and expand it about $\boldsymbol{\rho}_e = \boldsymbol{\rho}_h = \mathbf{0}$ up to the second-order term to get

$$\begin{aligned}
M(\rho_e, \rho_h, S) &= -N^{-2} \sum_{\mathbf{l}_e, \mathbf{l}_1, \mathbf{l}_h, \mathbf{l}_2} \exp[i\boldsymbol{\rho}_e \cdot (\mathbf{l}_1 - \mathbf{l}_e) + i\boldsymbol{\rho}_h \cdot (\mathbf{l}_2 - \mathbf{l}_h)] E_b(\mathbf{l}_e, \mathbf{l}_2, \mathbf{l}_1, \mathbf{l}_h; S) \\
&\approx -N^{-2} \sum_{\mathbf{l}_e, \mathbf{l}_1, \mathbf{l}_h, \mathbf{l}_2} E_b(\mathbf{l}_e, \mathbf{l}_2, \mathbf{l}_1, \mathbf{l}_h; S) - N^{-2} i \sum_{\mathbf{l}_e, \mathbf{l}_1, \mathbf{l}_h, \mathbf{l}_2} [\boldsymbol{\rho}_e \cdot (\mathbf{l}_1 - \mathbf{l}_e) \\
&\quad + \boldsymbol{\rho}_h \cdot (\mathbf{l}_2 - \mathbf{l}_h)] E_b(\mathbf{l}_e, \mathbf{l}_2, \mathbf{l}_1, \mathbf{l}_h; S) + N^{-2} \sum_{\mathbf{l}_e, \mathbf{l}_1, \mathbf{l}_h, \mathbf{l}_2} [\{\boldsymbol{\rho}_e \cdot (\mathbf{l}_1 - \mathbf{l}_e)\}^2 \\
&\quad + \{\boldsymbol{\rho}_h \cdot (\mathbf{l}_2 - \mathbf{l}_h)\}^2] E_b(\mathbf{l}_e, \mathbf{l}_2, \mathbf{l}_1, \mathbf{l}_h; S).
\end{aligned} \tag{5.27}$$

In Eq. (5.27), again the first-order term is zero, and the second-order term provides additional kinetic energy to the motion of the excited electron and hole pair due to their mutual Coulomb interaction. It is interesting to note that this additional kinetic energy term does not seem to appear when the theory of Wannier excitons is developed in the **k**-space for crystals (Singh, 1994). As there is no translational symmetry in a-solids, the additional kinetic energy may be attributed to a kind of confinement to the motion of charge carriers offered by the amorphous nature of a-solids, as mentioned above. This supports further that the Wannier type of excitons in a-solids may not necessarily have large radii orbitals. However, the contribution of this term to the kinetic energy of the excited e–h pair is expected to be small because it is due to the contribution of the Coulomb and exchange interaction energy between the excited pair, and does not involve any interaction with ions. Keeping this in mind, this term can be neglected for simplification and thus only the zeroth order term of Eq. (5.27) will be used for solving Eq. (5.25), which then becomes

$$W = W_0 + E_{\text{opt}} + \frac{p_e^2}{2m_e^*} + \frac{p_h^2}{2m_h^*} - N^{-2} \sum_{\mathbf{l}_e, \mathbf{l}_1, \mathbf{l}_h, \mathbf{l}_2} E_b(\mathbf{l}_e, \mathbf{l}_2, \mathbf{l}_1, \mathbf{l}_h; S). \tag{5.28}$$

The last term of Eq. (4.28) is difficult to evaluate analytically. However, some simplifications can help understanding the main features of the problem. According to Eq. (5.10), $E_b(\mathbf{l}_e, \mathbf{l}_1, \mathbf{l}_2, \mathbf{l}_h; S)$ involves up to four center integrals, which are not only difficult to evaluate but their magnitude also may not be very significant. Therefore, let us neglect three and four center integrals and consider only terms with $\mathbf{l}_e = \mathbf{l}_1$ and $\mathbf{l}_h = \mathbf{l}_2$ (two center integrals only). With this simplification, Eq. (5.28) reduces to

$$W = W_0 + E_{\text{op}} + \frac{p_e^2}{2m_e^*} + \frac{p_h^2}{2m_h^*} - N^{-2} \sum_{\mathbf{l}_e, \mathbf{l}_h} E_b(\mathbf{l}_e, \mathbf{l}_h; S), \tag{5.29}$$

where $E_b(\mathbf{l}_e, \mathbf{l}_h; S)$ is obtained from Eq. (5.10) by using $\mathbf{l}_e = \mathbf{l}_1$ and $\mathbf{l}_h = \mathbf{l}_2$ as

$$E_b(\mathbf{l}_e, \mathbf{l}_h; S) = \langle 1, \mathbf{l}_e; 0, \mathbf{l}_h | U | 0, \mathbf{l}_h; 1, \mathbf{l}_e \rangle - (1 - S)\langle 1, \mathbf{l}_e; 0, \mathbf{l}_h | U | 1, \mathbf{l}_e; 0, \mathbf{l}_h \rangle. \tag{5.30}$$

Following the theory of exciton, the first term Coulomb interaction between electron and hole, can be evaluated (Singh, 1994) as

$$\langle 1, \mathbf{l}_e; 0, \mathbf{l}_h | U | 0, \mathbf{l}_h; 1, \mathbf{l}_e \rangle = \iint |\phi 1\mathbf{l}_e(\mathbf{r}_1 - \mathbf{l}_e)|^2 |\phi_{0\mathbf{l}_h}(\mathbf{r}_2 - \mathbf{l}_h)|^2 \frac{\kappa e^2}{|\mathbf{r}_1 - \mathbf{r}_2|} \, d^3\mathbf{r}_1 \, d^3\mathbf{r}_{21}. \tag{5.31}$$

Transforming the coordinates as

$$\mathbf{r}_1' = \mathbf{r}_1 - \mathbf{l}_e, \quad \text{and} \quad \mathbf{r}_2' = \mathbf{r}_2 - \mathbf{l}_h, \tag{5.32}$$

the integral in Eq. (5.31) can be written as

$$\langle 1, \mathbf{l}_e; 0, \mathbf{l}_h | U | 0, \mathbf{l}_h; 1, \mathbf{l}_e \rangle = \iint |\phi 1\mathbf{l}_e(\mathbf{r}_1')|^2 |\phi_{0\mathbf{l}_h}(\mathbf{r}_2')|^2 \times \frac{\kappa e^2}{|(\mathbf{l}_e - \mathbf{l}_h)| \, |1 - (\mathbf{r}_1' - \mathbf{r}_2'/\mathbf{l}_e - \mathbf{l}_h)|} d^3\mathbf{r}_1' \, d^3\mathbf{r}_2'. \tag{5.33}$$

Considering that $|\mathbf{r}_1' - \mathbf{r}_2'|/|\mathbf{l}_e - \mathbf{l}_h| \ll 1$, the integral of Eq. (5.33) can be written as

$$\langle 1, \mathbf{l}_e; 0, \mathbf{l}_h | U | 0, \mathbf{l}_h; 1, \mathbf{l}_e \rangle \approx \iint |\phi 1\mathbf{l}_e(\mathbf{r}_1')|^2 |\phi_{0\mathbf{l}_h}(\mathbf{r}_2')|^2 \frac{\kappa e^2}{|\mathbf{l}_e - \mathbf{l}_h|} \, d^3\mathbf{r}_1' \, d^3\mathbf{r}_2'. \tag{5.34}$$

As $\mathbf{r}_1'$ and $\mathbf{r}_2'$ are electronic coordinates measured from their nuclei, and considering that all atoms in the solid are the same, $|\mathbf{r}_1' - \mathbf{r}_2'|$ will be nearly zero and therefore the above approximation is quite justified. Then using Haken's approximation applied for Wannier excitons (Singh, 1994, p. 24), we can write

$$\int |\phi_{0\mathbf{l}_e}(\mathbf{r}_2')|^2 \, d^3\mathbf{r}_1' = \int |\phi_{0\mathbf{l}_h}(\mathbf{r}_2')|^2 \, d^3\mathbf{r}_2' \approx \frac{1}{\sqrt{\varepsilon}}, \tag{5.35}$$

where ε is the static dielectric constant of the solid. In the theory of Wannier excitons, the static dielectric constant is introduced through the Bloch wave-functions in the same way as introduced in Eq. (5.35) through the localized wavefunctions for a-solids. Substituting Eq. (5.35) into Eq. (5.34) we get

$$\langle 1, \mathbf{l}_e; 0, \mathbf{l}_h | U | 0, \mathbf{l}_h; 1, \mathbf{l}_e \rangle \approx \frac{\kappa e^2}{\varepsilon |\mathbf{l}_e - \mathbf{l}_h|}, \tag{5.36}$$

Likewise, the second term of Eq. (5.30), the exchange interaction also depends on the inverse of the distance between e and h, but its magnitude is usually less than that of the Coulomb interaction. Let us assume that the second term can be approximated by

$$\langle 1, \mathbf{l}_e; 0, \mathbf{l}_h | U | 0, \mathbf{l}_e; 1, \mathbf{l}_h \rangle = \frac{\kappa e^2}{\alpha \varepsilon |\mathbf{l}_e - \mathbf{l}_h|}, \tag{5.37}$$

where α is a parameter representing the ratio of the magnitude of the Coulomb interaction to that of the exchange interaction, and it will be determined later on.

It should be quite clear that when an excited e–h pair forms a bound state due to their Coulomb interaction, their separation in a particular energy state (orbital radius) will be fixed and will not change as they move in a bound hydrogenic orbital. Therefore, we can assume that $|\mathbf{l}_e - \mathbf{l}_h|$ does not change much during the motion of e–h pair inside the solid. Then using Eqs (5.30), (5.36) and (5.37), the binding energy term in Eq. (5.29) can be written as

$$N^{-2} \sum_{\mathbf{l}_e \, \mathbf{l}_h} E_b(\mathbf{l}_e, \mathbf{l}_h; S) \approx \frac{\kappa e^2}{\varepsilon |\mathbf{l}_e - \mathbf{l}_h|} \left[1 - \frac{(1 - S)}{\alpha} \right]. \tag{5.38}$$

Substituting Eq. (5.38) into (5.29) one gets (Singh, 2002, 2003; Singh *et al.*, 2002)

$$W = W_0 + E_{op} + \frac{p_e^2}{2m_e^*} + \frac{p_h^2}{2m_h^*} - \frac{\kappa e^2}{\varepsilon'(S) |\mathbf{l}_e - \mathbf{l}_h|}, \tag{5.39}$$

where

$$\varepsilon'(S) = \varepsilon \left[1 - \frac{(1 - S)}{\alpha} \right]^{-1}. \tag{5.40}$$

The last three terms of Eq. (5.39) represent the energy operator of an e–h pair interacting through their Coulomb interaction, exactly in the same way as that of a Wannier exciton in crystalline semiconductors (c-semiconductors) (Singh, 1994).

Replacing the linear momentum by the corresponding operators, $p \rightarrow -i\hbar\nabla$, and then transforming the coordinates into the center of mass coordinate **R** and relative coordinate **r** according to

$$\mathbf{R} = \frac{m_{\mathrm{e}}^{*}\mathbf{l}_{\mathrm{e}} + m_{\mathrm{h}}^{*}\mathbf{l}_{\mathrm{h}}}{M}, \quad \text{and} \quad \mathbf{r} = \mathbf{l}_{\mathrm{e}} - \mathbf{l}_{\mathrm{h}}, \tag{5.41}$$

the last three terms of Eq. (5.39) transform into:

$$-\frac{\hbar\nabla_{\mathrm{R}}^{2}}{2M} - \frac{\hbar^{2}\nabla_{\mathrm{r}}^{2}}{2\mu_{x}} - \frac{\kappa e^{2}}{\varepsilon'(S)r}, \tag{5.42}$$

where ∇_R and ∇_r are the differential operators associated with R and r, respectively. $M = m_{\mathrm{e}}^{*} + m_{\mathrm{h}}^{*}$, and μ_x represents the reduced mass of the excited e–h pair given by: $\mu_x^{-1} = (m_{\mathrm{e}}^{*})^{-1} + (m_{\mathrm{h}}^{*})^{-1}$. The first term of Eq. (5.42) represents the kinetic energy operator associated with the center of mass of the excited e–h pair, and last two terms give the energy operator of an e–h pair bound in hydrogenic state. The eigenfunction of the operator in Eq. (5.42) can be written as (Singh, 1994):

$$\psi(r, R) = V^{-1/2} \exp\left(\mathrm{i}\frac{\mathbf{P}\cdot\mathbf{R}}{\hbar}\right)\phi_n(r), \tag{5.43}$$

where V is the volume of crystal, **P** is the linear momentum associated with the center of mass motion, and $\phi_n(r)$ is the eigenfunction of a hydrogenic energy state with principal quantum number n. Substituting Eqs (5.42) and (5.43) into Eq. (5.39), the energy eigenvalue of an excited pair of electron and hole bound in a hydrogenic state is obtained as

$$W = W_0 + E_{\mathrm{op}} + \frac{P^{2}}{2M} + E_n(S), \tag{5.44}$$

where $E_n(S)$ is the energy eigenvalue of a hydrogenic energy state with principal quantum number n, obtained from the following Schrödinger equation:

$$\left[-\frac{\hbar^{2}\nabla^{2}}{2\mu_{x}} - \frac{\kappa e^{2}}{\varepsilon'(S)r}\right]\phi_n(r) = E_n(S)\phi_n(r), \tag{5.45}$$

which gives

$$E_n(S) = -\frac{\mu_x e^{4}\kappa^{2}}{2\hbar^{2}\varepsilon'(S)^{2}}\frac{1}{n^{2}}. \tag{5.46}$$

The form of the energy eigenvalue of an excited e–h pair in an a-semiconductor, as obtained in Eq. (5.44), is analogous to that of an exciton in c-solids (Singh, 1994). This demonstrates quite clearly that excitonic states in a-semiconductors are indeed similar to those in c-semiconductors, but with a difference that in a-semiconductors these are not necessarily large radii orbital excitons. This also means that in a-semiconductors, one should usually expect the excitonic states to have a larger binding energy than in their crystalline counterparts. However, before we talk more about the binding energy of excitonic states in a-semiconductors, let us first determine the value of the parameter α introduced in Eq. (5.37).

5.3.1 Determination of α

We may consider determining α applying the variational approach. From Eq. (5.46) we get the energy eigenvalue for the singlet state as

$$E_n(S=0) = -\frac{(\alpha-1)^2\mu_x e^4\kappa^2}{2\alpha^2\hbar^2\varepsilon^2}\frac{1}{n^2}. \tag{5.47}$$

In order to determine α, we may minimize $E_n(S=0)$ with respect to α by setting $\partial E_n(S=0)/\partial\alpha = 0$, which produces $\alpha = 1$ corresponding to the maximum not minimum of energy. Therefore, no binding between the excited pair of charge carriers is possible as the binding potential vanishes at $\alpha = 1$. It is necessary for the binding that α must be greater than unity ($\alpha > 1$). It may also be noted that for very large values of α, the contribution of the exchange interaction becomes negligible and the binding energy reaches the large radii orbital limit at which the energy difference between singlet and triplet states vanishes. One way of determining the Coulomb and exchange interactions would be to evaluate the integrals involved numerically and thus determine α numerically. However, the accuracy of such numerical evaluation will depend on the sample size and kind of electronic wavefunctions chosen. An alternative way to estimate α can be from the experimental data, as described below.

For this purpose, we first derive the singlet–triplet splitting from Eq. (5.46) for the excitonic ground state ($n = 1$) as

$$\Delta E_{\text{ex}} = E_1(S=0) - E_1(S=1) = \left[1 - \frac{(\alpha-1)^2}{\alpha^2}\right] C_{\text{M}}, \tag{5.48}$$

where

$$C_{\text{M}} = \frac{\mu_x e^4\kappa^2}{2\hbar^2\varepsilon^2}, \tag{5.49}$$

and compare it with the observed one experimentally. Let us denote the observed splitting by ΔE_{xp}, and then equate it with the splitting derived in Eq. (5.48), which gives α as

$$\alpha = \left[1 - \sqrt{1 - \frac{\Delta E_{\text{xp}}}{C_{\text{M}}}}\right]^{-1}, \tag{5.50}$$

where only one of the signs of the square root is applicable for obtaining $\alpha > 1$.

The splitting between singlet and triplet states has been observed in a-Si : H and a-Ge : H, and therefore we can determine α for these two solids as follows.

For calculating C_{M} we need the reduced effective mass of exciton μ_x, which, according to Chapter 3, depends on whether the excited charge carriers are in their extended or tail states. There are four possibilities: (a) both excited electron and hole are in their extended states, that is, electron in the conduction and hole in the valence extended states, (b) electron is in conduction extended and hole in valence tail states, (c) electron is in conduction tail and hole in valence extended states, and (d) both are in tail states, that is, electron in conduction and hole in valence tail states. Let us consider the first possibility that an exciton is created by exciting an electron in the conduction and hole in valence extended states. Using Table 3.1, we get $\mu_x = 0.17m_{\mathrm{e}}$ for a-Si : H and $\mu_x = 0.11m_{\mathrm{e}}$ for a-Ge : H. The static dielectric constant for a-Si : H is 12 (Cody, 1984), which gives $C_{\mathrm{M}} = 16.1$ meV from Eq. (5.49). For a-Ge : H, with $\varepsilon = 16$ (Conwell and Pawlik, 1976) we get $C_{\mathrm{M}} = 5.9$ meV. The observed singlet–triplet splitting, ΔE_{xp} in a-Si : H can be rounded to be 3 meV, and 1 meV in a-Ge : H (Aoki, 2001). Using these values in Eq. (5.50), we get $\alpha = 10$ for a-Si : H, and 11 for a-Ge : H. As there are uncertainties in the experimental data, it is considered reasonable here to use $\alpha = 10$ for both a-Si : H and a-Ge : H.

5.3.2 *Energy difference between singlet and triplet excitonic states*

Substituting $\alpha = 10$ in Eq. (5.47) for a-Si : H and a-Ge : H, we get the energy $E_n(S = 0)$ of the singlet state as

$$E_n(S = 0) = -\frac{0.81C_{\mathrm{M}}}{n^2}. \tag{5.51}$$

According to Eq. (5.44), the singlet excitonic state will be formed below the optical energy gap by an energy given by $E_n(S = 0)$. That means the binding energy, $E_{\mathrm{b}}(n; S = 0)$, of singlet excitonic state with principal quantum number n is

$$E_{\mathrm{b}}(n; S = 0) = -E_n(S = 0) = \frac{0.81C_{\mathrm{M}}}{n^2}. \tag{5.52}$$

Likewise the energy of the triplet excitonic state, $E_n(S = 1)$, can be obtained from Eq. (5.46) as

$$E_n(S = 1) = -\frac{C_{\mathrm{M}}}{n^2}. \tag{5.53}$$

This gives the triplet state binding energy, $E_{\mathrm{b}}(n; S = 1)$ as

$$E_{\mathrm{b}}(n; S = 1) = -E_n(S = 1) = \frac{C_{\mathrm{M}}}{n^2}. \tag{5.54}$$

Comparing Eqs (5.52) and (5.54), we find that $E_{\mathrm{b}}(n; S = 0) = 0.81E_{\mathrm{b}}(n; S = 1)$ in a-Si : H and a-Ge : H, and the energy difference, ΔE_{ex}, between singlet and

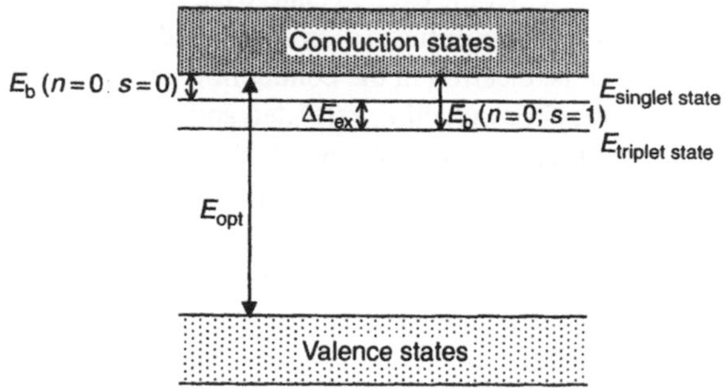

Figure 5.1 Schematic illustration of singlet and triplet excitonic states in a-solids. ΔE_{ex} represents the energy difference between singlet and triplet excitonic states.

triplet in the ground state ($n = 1$) is obtained as

$$\Delta E_{ex} = E_1(S = 0) - E_1(S = 1) \approx 0.2C_M. \tag{5.55}$$

The energy of singlet and triplet excitonic states are schematically illustrated in Fig. 5.1. It may be emphasized here that as Eq. (5.55) is obtained after applying several approximations, it should be regarded as an estimate rather than a very accurate result. The crucial point is to determine α; without that ΔE_{ex} cannot be estimated for other solids.

The Bohr radius of singlet [$a_{ex}(S = 0)$] and triplet [$a_{ex}(S = 1)$] excitonic states can also be obtained from Eqs (5.51) and (5.53), respectively, for a-Si : H and a-Ge : H. We thus find that

$$a_{ex}(S = 0) = \frac{5}{4} a_{ex}(S = 1), \tag{5.56}$$

and

$$a_{ex}(S = 1) = \frac{\mu\varepsilon}{\mu_x} a_0; \quad a_0 = \frac{\hbar^2}{\kappa\mu e^2}, \tag{5.57}$$

where a_0 is the Bohr radius, and μ is the reduced mass of elec ron in hydrogen atom. Thus the Bohr radius of singlet excitonic state is four times larger than that of triplet excitonic state, and the binding energy of a triplet state is four times larger than that of a singlet state in a-Si : H and a-Ge : H.

Once α is known, the theory developed above can be applied to study the details of PL in any a-solid. Here we will focus mainly on a-Si : H and a-Ge : H, which are reasonably well studied and in which valence and conduction states are obtained from sp^3-hybridization. Using $\alpha = 10$, and $\mu_x = 0.17m_e$ for a-Si : H and $0.11\,m_e$ for a-Ge : H for an exciton created through the first possibility that both electron and hole are excited in their extended states, we get the energy difference between

Table 5.1 Values of ΔE_{ex} calculated from Eq. (5.55) for a-Si : H and a-Ge : H. Experimental values given in the last column are by Aoki (2001)

	μ_x	ε	ΔE_{ex} (meV)	ΔE_{ex} (meV) *experimental*
a-Si : H	$0.17\, m_e$	12[a]	3.2	3.0[c]
a-Ge : H	$0.11\, m_e$	16[b]	1.2	0.8[c]

Notes
a Cody, 1984.
b Conwell and Pawlik, 1976.
c Aoki, 2001.

singlet and triplet excitonic states as $\Delta E_{ex} = 3.2$ meV for a-Si : H, and 1.2 meV for a-Ge : H. The calculated values of ΔE_{ex} from possibility (a) are listed in Table 5.1 along with the corresponding experimental values in a-Si : H and a-Ge : H. It is clear from Table 5.1 that the calculated values agree reasonably well with the experimentally estimated values for a-Si : H and a-Ge : H as described in the next section. This is expected because α is determined by comparing the theoretical and experimental results. However, the important point is that the value of α is not expected to depend on the energy of exciting photons or which states electrons and holes are excited. Therefore, once α is known the PL observed through the other three possibilities can also be studied and analyzed by applying the above theory. This will be described in the next section.

5.4 Experimental study of PL in a-solids

Distinct PL peaks of singlet and triplet excitons have not been observed in crystalline silicon (c-silicon). In a-Si : H, however, Boulitrop and Dunstan (1985), Ambros *et al.* (1991) and Aoki *et al.* (2002, 2003) have observed a double peak PL lifetime distribution by quadrature frequency resolved spectra (QFRS) technique. Ambros *et al.* (1991) have deconvoluted the double peak spectra, shown in Fig. 5.2, and estimated two PL lifetimes; one is at about 10 μs, much shorter than the other at about 1 ms. They have also estimated the corresponding PL quantum efficiencies. As described in Section 5.1, the double peak lifetime distribution appearing at a temperature of 2 K, cannot be explained on the basis of the radiative tunneling (RT) model (Tsang and Street, 1979). If the PL lifetime depends on the distance the excited pair has to travel before their radiative recombination in the tail states occurs, as the RT model suggests, then there would be a wide range of time distributions, not just two. As the observed double peak is in contradiction with the RT model, it also means that the PL lifetime does not depend on the random separation between the excited and localized charge carriers in the tail states.

The occurrence of the double peak structure in the PL lifetime distribution has also been observed in a-SiN$_x$: H (Boulitrop and Dunstan, 1985; Searle *et al.*, 1987), and recently in a-Ge : H by Aoki (2001) at 13 and 60 K (as shown in Fig. 4.3). This suggests that the double peak structure may be common to all a-solids, and hence there has to be a common explanation. This is going to be explained in the next

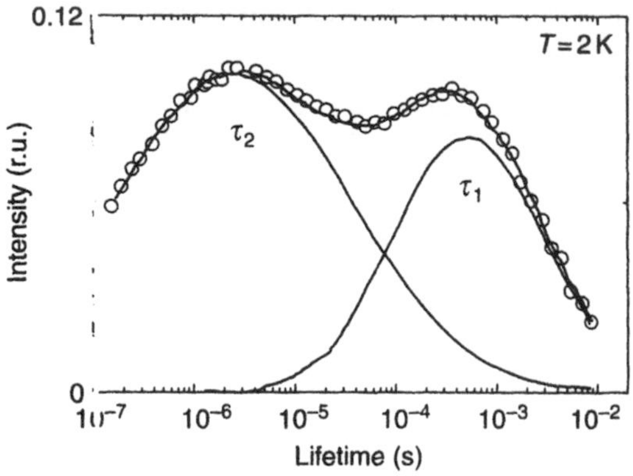

Figure 5.2 PL lifetime distribution in a-Si : H measured at a temperature of 2 K using the frequency resolved spectroscopy (Ambros *et al.*, 1991).

section on the basis of excitonic states whose theory has been developed in the preceding section.

5.4.1 PL from excitonic states in a-semiconductors

According to the theory presented in Section 5.3, an excited pair of e and h in a-semiconductors can form an excitonic state due to their attractive Coulomb interaction. Such a state is similar to exciton state in c-solids. This bonding in hydrogenic type states can take place in two different spin states, singlet and triplet, and it can take place in both extended and tail states. Depending on the material, a high energy excited pair of e and h in the extended states may or may not form an excitonic state, before relaxing down to their mobility edges. However, once the bound state is formed it will continue to be so even if the carriers relax down to the tail states, provided the temperature is low and the excitation density is not too high. At higher temperatures the excitonic states may dissociate due to thermal energy, and at higher excitation densities these can also dissociate due to the excessive repulsive Coulomb interaction.

Once the singlet and triplet excitonic states are formed, they have different binding energies as given in Eqs (5.52) and (5.54). The singlet state is formed at a higher energy than the triplet state, and a radiative recombination of the singlet state is spin allowed, whereas that of a triplet is spin forbidden. This is because without the spin–orbit interaction the oscillator strength of a triplet state is zero, whereas that of a singlet is non-zero. Hence, on the time scale the radiative recombination of a singlet should be much faster than that of a triplet. This can be used to explain the occurrence of the double peak structure in the PL time distribution. Accordingly, when the excitonic states relax from extended states to tail states, their binding energies will change because as described in Section 5.3, the effective masses will change but singlet will remain singlet and triplet will remain triplet. Therefore, only two types of radiation time distribution are possible corresponding to each of the four possibilities, stated above. In an excitonic state, the separation between e and h is decided by the excitonic Bohr radius (see Eqs (5.56) and (5.57)). As the effective mass of charge carriers increases in the tail states, the excitonic Bohr radius will be smaller than that in the extended states and the binding energy will increase. This explains why both singlet and triplet excitonic states can be observed in a-Si : H but not in c-Si.

Aoki (2001) has drawn the Arrhenius plot of η_2/η_1 as a function of the inverse of temperature ($1/T$), where η_1 and η_2 are the relative quantum efficiencies of the slow and fast PL components observed in a-Ge : H. Considering that the energy difference between slow (triplet) and fast (singlet) PL peaks is ΔE_{ex}, one can express η_2/η_1 as

$$\eta_2/\eta_1 = \eta_0 \exp(-\Delta E_{ex}/k_B T), \tag{5.58}$$

where η_0 is a constant and k_B the Boltzmann's constant. Plotting $\ln(\eta_2/\eta_1)$ as a function of $1/T$, gives the Arrhenius plot whose slope determines ΔE_{ex}. The Arrhenius plot from the data of a-Ge : H obtained by Aoki (2001) is shown in Fig. 5.4, which gives $\Delta E_{ex} \approx 0.8$ meV, close to the theoretical value of 1.17 meV obtained above from Eq. (5.55). The Arrhenius plots for a-Si : H from two different sets of experimental data are also shown in Fig. 5.4. Aoki has deconvoluted the lifetime distributions in a-Si : H observed by Searle *et al.* (1987) and

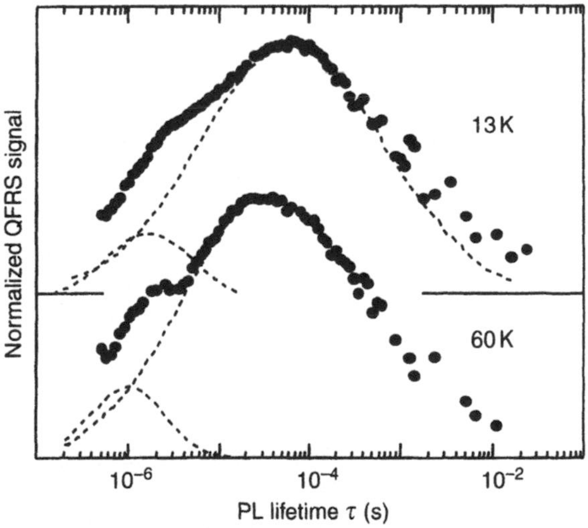

Figure 5.3 PL lifetime distribution in a-Ge : H at $T = 13$ and 60 K with a generation rate $G \approx 2 \times 10^{20}\,\mathrm{cm^{-3}\,s^{-1}}$. Dashed curves are convoluted ones (Ishii *et al.*, 2000; Aoki, 2001).

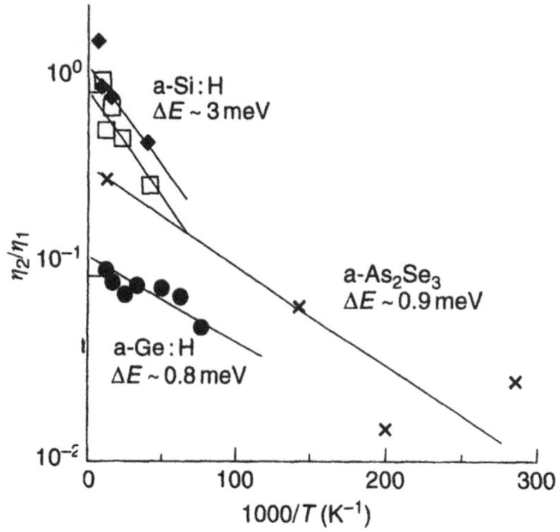

Figure 5.4 Arrhenius plots for the ratio PL quantum efficiency η_2/η_1 for a-Ge : H(•) giving an activation energy of 0.8 meV (Aoki, 2001), for a-Si : H calculated from the data of Searle *et al.* (1987) (□) and of Schubert *et al.* (1993) (■) giving the activation energy of 3 meV (Aoki, 2001) and for a-As_2Se_3 giving an activation energy of 0.9 meV (Aoki, 2001).

Schubert *et al.* (1993) and obtained the Arrhenius plot as shown in Fig. 5.4 giving $\Delta E_{ex} \approx 3$ meV in agreement with 3.22 meV calculated from Eq. (5.55) for a-Si : H. Figure 5.4 also shows the Arrhenius plot for amorphous chalcogenide (a-chalcogenide), a-As_2Se_3 with $\Delta E_{ex} \approx 0.9$ meV.

As stated above there are four distinct possibilities for the PL to originate in a-solids: (a) both e and h recombine directly from the extended states; (b) h is localized in the tail states while e remains in the extended states; (c) e may relax down and get localized in the tail states while h remains in the extended states and (d) both e and h are in their respective tail states and then recombine radiatively. In sp^3 a-semiconductors it is expected that $m_e^* = m_h^*$, therefore, both (b) and (c) possibilities will give the same value of the reduced mass and ΔE_{ex} in such semiconductors. One may thus expect to observe only a total of six PL peaks, three singlets and three triplets, in these materials. For the possibilities (b) and (c) in a-Si : H, using $m_{et}^* = m_{ht}^* = 7.1 m_e$ (see Table 3.1) and $m_e^* = m_h^* = 0.34 m_e$, we get $\mu_x = 0.32 m_e$, which gives $\Delta E_{ex} = 6.1$ meV from Eq. (5.55). This is slightly higher but of the same order of magnitude as the experimental value of 3 meV. Such variations in the effective mass and Bohr radius may be attributed to the broad PL peak observed in a-Si : H and a-Ge : H. As α is relatively large ($\approx$10) in these materials, any change in effective mass can cause only a small variation in the PL energy.

Although it may be possible to observe a maximum of six PL peaks (three singlets and three triplets) in any sp^3 a-semiconductors, all may not be observed due to the detection limit of the experimental set up. For instance, it is known that the PL from singlet excitonic states created in the extended states in a-Si : H occurs on the nanosecond time scale, and hence difficult to detect on a slower time scale detection. Nevertheless, more than one singlet and triplet peaks may be observed in any a-semiconductor, producing different values of the singlet–triplet splitting. The expression derived in Eq. (5.55) is not general. It is valid only for the situation where both singlet and triplet states are excited by one of the four possibilities of transitions described above. A general expression of singlet–triplet splitting can be derived from Eqs (5.52) and (5.54) as

$$\Delta E_{\mathrm{ex}}^{01} = \left[\mu_x^1 - \mu_x^0 \frac{(\alpha - 1)^2}{\alpha^2}\right] C_{\mathrm{M}}^0, \tag{5.59}$$

where μ_x^0 and μ_x^1 are excitonic reduced masses of singlet and triplet states, respectively, and

$$C_{\mathrm{M}}^0 = \frac{e^4 \kappa^2}{2\hbar^2 \varepsilon^2}. \tag{5.60}$$

For $\mu_x^0 = \mu_x^1 = \mu_x$, Eq. (5.59) reduces to Eq. (5.48). Thus, depending on different reduced masses one can expect to find a few different values of the singlet–triplet splitting in the same material. For instance, in a-Si : H, say, the PL is observed from a triplet state such that both excited electron and hole are in their respective tail states, and a singlet state such that electron and hole are in the extended state. In this case $\mu_x^1 = 2.85 m_{\mathrm{e}}$, and $\mu_x^0 = 0.17 m_{\mathrm{e}}$ (using Table 3.1), and then using $\alpha = 10$, we get $\Delta E_{\mathrm{ex}}^{01} \approx 257$ meV from Eq. (5.59), which is much higher than 3.22 meV calculated above when both singlet and triplet are in the extended states [possibility (a)]. It is also higher than 6.1 meV, which is obtained from possibility (b) and (c), but it is close to the experimental estimate suggested by Morigaki (1999b) for a singlet–triplet splitting of about 150 meV.

It should also be noted that ΔE_{ex} (Eq. (5.55)) is more sensitive to the value of the dielectric constant, ε, than to that of the reduced mass of the excited pair of charge carriers. However, we really need many more experimental results to address this issue any further.

It may also be desirable to clarify here that a "geminate pair" in a-solids is the same as a pair of e and h bound in an excitonic state. It has to be bound in excitonic states, otherwise as explained above, a double peak structure will not be observed experimentally. Thus the PL from a geminate pair is the same as that from the corresponding excitonic state formed in the tail states.

5.4.2 PL Stokes shift

It is useful to evaluate the binding energy of excitonic states as well, as it corresponds to the observed Stokes shift between the excitation energy and PL energy.

According to Eqs (5.52) and (5.54), the ground state binding energy of a singlet excitonic state is $0.81C_M$ and that of a triplet is C_M. The value of C_M for the possibility (a), with e and h in their extended states, is 16.12 meV for a-Si : H and 5.87 meV for a-Ge : H. For the possibilities (b) and (c), with one carrier in the extended and one in the tail states, $C_M = 30.5$ meV and for the possibility (d), with both e and h in their tail states, $C_M = 1.2$ eV for a-Si : H. The observed Stokes shift in a-Si : H is known to be 0.4–0.5 eV (Street, 1991) and in a-Ge : H about 0.2 eV (Ishii *et al.*, 2000; Aoki, 2001). The observed values in both, a-Si : H and a-Ge : H, are thus an order of magnitude higher than those for possibility (a), corresponding to which a good agreement has been found in the singlet–triplet splitting energy in the preceding section. In contrast to this, the observed value in a-Si : H agrees more closely with the theoretical estimate for the possibility (d), and hence suggests that PL occurs after the excited charge carriers have relaxed down to their corresponding tail states.

However, it is important to realize that in the above theory of calculating the singlet–triplet splitting energy, no involvement of phonons is considered. It should also be noted that usually the Stokes shift is defined to be the difference in energy between a free exciton and a self-trapped exciton states. The latter arises due to the strong exciton–lattice interaction, which is known to lower the energy of a free exciton state (e.g. Chapter 6). In the present treatment, as stated above, no lattice interaction is considered. Therefore, a direct comparison of Stokes shift obtained from the results of the present theory and those of experiments may not be considered justified.

5.4.3 *Radiative recombination of excitonic states*

The radiative recombination depends on the oscillator strength between two energy states involved in a transition. In the case of an excitonic state, the transition involves radiative recombination of an excited pair of e and h bound in a hydrogenic energy state. Let us consider that the excitonic state thus formed is in its ground state ($n = 1$), and according to Eq. (5.44) the energy of such an excited state in a solid will be

$$W = V_0 + E_{\text{opt}} + \frac{P^2}{2M} + E_1(S). \tag{5.61}$$

As $E_1(S) < 0$, the energy of this excited state is slightly less than E_{opt}, if the pair was excited by a photon of energy $\geq E_{\text{opt}}$ at low temperatures. A radiative recombination occurring from such an excitonic state, although e and h are bound in the hydrogenic state, is different from that occurring in hydrogenic atoms. In the case of a hydrogen atom such a transition does not occur from the ground state, otherwise the atom will become unstable. For an excitonic state, this transition is to be regarded as a recombination between an electron in the conduction and a hole in the valence states, but bound in a hydrogenic energy state. It is the bound hydrogenic state that fixes the spin and relative distance between e and h, because if not bound in hydrogenic states e and h will be free to be in any spin state. A recombination between a free e and h will not have distinct singlet or triplet spin character. The dipole transition matrix element between the conduction states with an electron and valence state with a hole can be written as

$$\sum_{\mathbf{l},\mathbf{m}} C_1^*(\mathbf{l}, \rho_e) C_0(\mathbf{m}, \rho_h) \langle 1, \mathbf{l}, \sigma_e | \mathbf{r} | 0, \mathbf{m}, -\sigma_h \rangle, \qquad (5.62)$$

which vanishes unless $\sigma_e = -\sigma_h$. Examining each term of the singlet and triplet eigenvectors given in Eqs (5.3) and (5.4), respectively, we find that the dipole transition matrix element in Eq. (5.62) vanishes for triplets, but is non-zero for singlets. As the oscillator strength of a transition is proportional to the square of the dipole transition matrix element, the oscillator strength of a triplet transition vanishes. Thus, a triplet transition becomes spin forbidden unless spin–orbit interaction is included. However, spin–orbit interaction is usually weak, which means that the triplet transitions will not be as efficient as singlet transitions. As a result the radiative recombination from the singlet excitonic state is faster than that from the corresponding triplet excitonic state.

5.4.4 Temperature dependence of PL

Hydrogenated amorphous Silicon and a-Ge : H are the two semiconductors in which the temperature dependence of PL has been studied in detail. Figure 5.3 shows the double peak time distribution of PL in a-Ge : H observed by Aoki (2001). It is quite obvious from the deconvoluted curves that the peak intensity of the fast peak increases when the temperature is increased from 13 to 60 K. In a-Ge : H, the experimentally estimated $\Delta E_{ex} \approx 0.8$ meV (10 K) and its theoretical value is 1.17 meV (14 K). This is lower but of the same order of magnitude as 47 K, the difference of temperature found experimentally. One can see from Fig. 5.3 that the intensity of the fast peak starts increasing, compared with that of the slow peak, when the temperature increases from 13 to 60 K.

Likewise, Fig. 5.5 shows the temperature dependent quadrature resolved frequency luminescence (QFRL) intensity in a-Si : H as a function of frequency observed by Searle *et al.* (1987). As the temperature increases from 4 to 25 K and 65 K, the long time (small frequency) peak remains dominant, but at 115 and 150 K the short time (large frequency) peak intensity increases and the long time (small frequency) peak intensity decreases. In a-Si : H, ΔE_{ex} is about 3 meV(=35 K), which is close to the observed value of 50 K above 65 K when the triplet excitonic states get enough thermal energy and get converted into singlets, and therefore the fast time component increases. However, at lower temperatures (<115 K) the thermal energy is too small to convert triplets into singlets and hence the long time component remains dominant.

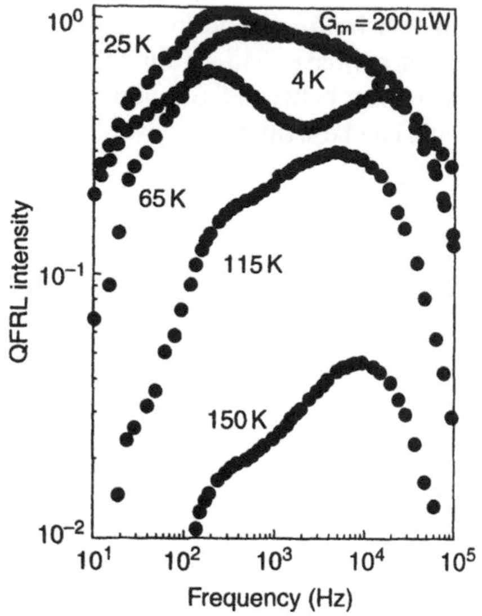

Figure 5.5 Quadrature frequency resolved luminescence (QFRL) observed at different temperatures in a-Si : H (Searle *et al.*, 1987).

In order to be conclusive about the temperature of conversion from triplets to singlets, we need more experimental results at many different temperatures of small intervals. What one can conclude at this stage, however, are: (1) The occurrence of two PL peaks in the time distribution can be associated with the radiative recombinations from singlet and triplet excitonic states in a-semiconductors. (2) The slow peak is dominant at lower temperatures, but a reversal in this behavior occurs at higher temperatures, because the thermal energy enables triplets to be converted into singlets.

5.4.5 *Dependence of PL on the generation rate*

The density of excitation is also known to influence the PL properties in a-semiconductors. The dependence of the lifetime distribution of the double peak and associated quantum efficiencies on the excitation intensity or generation rate has been studied experimentally by Ambros *et al.* (1991) in a-Si : H and Ishii *et al.* (1999) in a-Ge : H. The maximum of the lifetime distribution and associated quantum efficiency as a function of the excitation intensity (measured in the unit of $cm^{-2}\,s^{-1}$) for both the long and short living channels observed at 2 K by Ambros *et al.* in a-Si : H are shown in Fig. 5.6. The lifetime and quantum efficiency of the long time PL component are denoted by τ_1 and η_1, and those of the short time

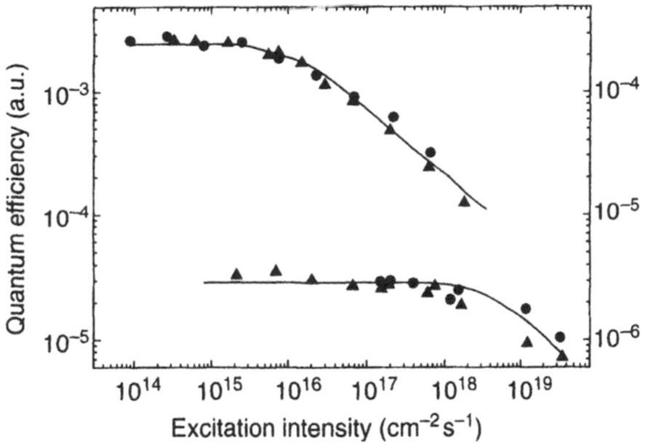

Figure 5.6 Observed photoluminescence lifetime and quantum efficiency plotted as a function of the excitation intensity in a-Si : H (Ambros *et al.*, 1991). The top curve represents the long lifetime • and the associated quantum efficiency ▲, and the bottom curve represents short lifetime • and the corresponding quantum efficiency.

components by τ_2 and η_2, respectively. The quantum efficiency is defined by

$$\eta = \frac{I_{PL}}{I_{exc}}, \tag{5.63}$$

where I_{exc} is the excitation intensity defined by the number of photons incident per unit area, per unit time and I_{PL} is the PL intensity. As can be seen from Fig. 5.6, both τ_1 and τ_2 have a region of constant lifetime, after which they start decreasing in proportion to $I_{exc}^{-0.5}$. The corresponding quantum efficiencies also show the same dependence on I_{exc}. The region of constant lifetime and quantum efficiency is shorter (up to $10^{16}\,cm^{-2}\,s^{-1}$) for the slow component and longer (up to $10^{18}\,cm^{-2}\,s^{-1}$) for the fast component.

The dependence of lifetime and corresponding quantum efficiencies as a function of the generation rate G (observed by Ishii *et al.* at 13 and 60 K in a-Ge : H) are shown in Fig. 5.7. The generation rate G (in $cm^{-3}\,s^{-1}$) is proportional to the excitation intensity I_{exc}, and it is defined as the number of photons absorbed per unit volume per unit time. The relation between G and I_{exc} is given by

$$G = AI_{exc}, \tag{5.64}$$

where A (in cm^{-1}) is the absorption coefficient. In a-Ge : H, it is found that both τ_1 and η_1 of the long time component remain constant at smaller G and then begin to decrease proportional to $G^{-0.5}$ at $G \geq 3 \times 10^{20} cm^{-3}\,s^{-1}$ (see Fig. 5.7). However, the short-time component (τ_2 and η_2) is found to remain constant within the range of measurements. As described in Section 5.1, the observed dependence of τ and η on G cannot be explained on the basis of RT model.

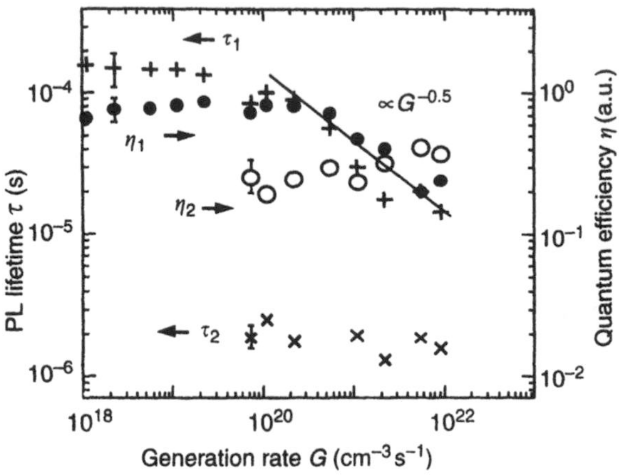

Figure 5.7 PL lifetime and quantum efficiency obtained from deconvolution of lifetime distribution as a function of the generation rate in a-Ge : H. Plots with + and × represent long (τ_1) and short (τ_2) lifetimes, respectively, and those with ● and ○ represent the corresponding quantum efficiencies, η_1 and η_2.

In order to understand the observed dependence on the generation rate (excitation intensity) of (τ_1, η_1) and (τ_2, η_2) in a-Si : H and a-Ge : H shown in Figs 5.6 and 5.7, respectively, it is important to realize the significance of radiative and non-radiative recombinations of excited pairs of charge carriers in a-semiconductors and a-insulators. Both radiative and non-radiative processes increase with the increase in generation rate, and hence they compete with each other in every situation. At very low temperatures and generation rates, it is expected that the radiative process is much more dominant than the non-radiative one, because of the following two main reasons: (1) The number of excited pairs of charge carriers is not large enough to build up the repulsive potential energy between like charge carriers to a limit that it can overcome the binding energy (attractive potential energy) of excitonic states and (2) the thermal energy is not large enough to dissociate excitonic states into free carrier states. This situation, however is reversed, at higher temperatures and higher generation rates. Both the thermal energy and repulsive energy between like excited charge carriers can dissociate excitonic states into free excited carriers, which may recombine non-radiatively. One may ask, however, why free e and h thus created cannot recombine radiatively? This question can be answered qualitatively, as described below, on the basis of the theory of excitonic states developed in Section 5.3.

Once excitonic states are ionized, e and h become free from each other. That means, although the attractive Coulomb interaction still exists between the two, they are so far apart from each other that it has little influence on their movements. As there is no longer any quantum mechanical restriction on such free carriers, either in terms of their energy or inter-particle separation, they relax very fast, non-radiatively, down to their mobility edges and then to their tail states.

Once these carriers are in the tail states, they eventually recombine non-radiatively via dangling bonds, because the energy emitted by the recombination in the tail states is too small, and it gets absorbed in the amorphous lattice structure. We may thus conclude that PL occurs only when the radiative recombination takes place from the excitonic states. This, however, contradicts the radiative tunneling model and supports the non-radiative tunneling model, described in the chapter on the transport of charge carriers in a-solids. The radiative and non-radiative recombination processes are illustrated in Fig. 5.8. A schematic illustration of the formation of an excitonic state in a-solids is shown in Fig. 5.9.

The above answer is based on the following. The energy eigenvalue of an excitonic state calculated from the hydrogenic energy operator given in Eq. (5.42) can have only discrete values and so can the corresponding orbital radii. That means in an excitonic energy state the separation between e and h is fixed and cannot be changed unless a transition occurs to another energy state. Therefore, the center of

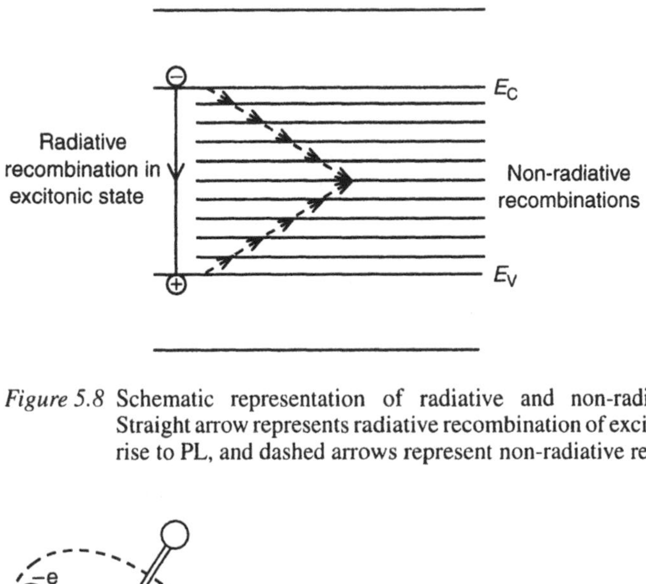

Figure 5.8 Schematic representation of radiative and non-radiative transitions. Straight arrow represents radiative recombination of excitonic states giving rise to PL, and dashed arrows represent non-radiative recombination.

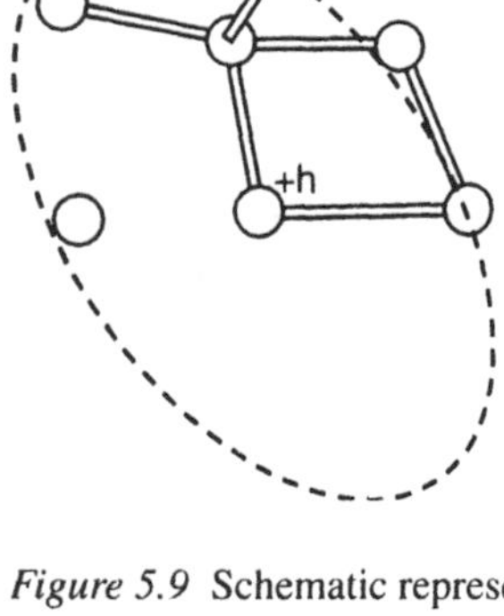

Figure 5.9 Schematic representation of formation of excitonic states in a-solids.

mass of an excited e and h pair bound in an excitonic energy state is free to move around, while the relative distance between e and h does not change much during this motion. Thus there are only two types of transition that can occur when the excited e and h are bound in a hydrogenic state: upward to higher excitonic energy states, including the ionizing state, by absorbing thermal energy, or downward by radiative recombination and subsequent PL. It is expected that most excitonic states will be in their ground state before recombining radiatively. Therefore, no further downward non-radiative transition is possible from the ground excitonic state.

The theory of excitonic states in a-solids, responsible for the radiative recombinations, has been developed in Section 5.3. In the following section, a phenomenological approach will be followed to study the rates of non-radiative transitions under different conditions.

5.5 Non-radiative recombinations

At higher excitation densities, the excited charge carriers cannot relax down to the band tail states because of the increased repulsive interaction between like charge carriers due to high excitation density. The effect of this will be to dissociate excitonic states and pull e and h far apart from each other. These freed e and h relax down non-radiatively to their respective tail states and eventually recombine non-radiatively in the tail states via dangling bonds through quantum tunneling. Therefore, such carriers do not contribute to the PL, but enhance the dissociation of excitonic states resulting in a decrease in the PL intensity, as observed experimentally at higher excitation densities.

In order to understand the role of non-radiative processes, it is more appropriate to define the PL quantum efficiency η as a function of radiative and non-radiative transition rates

$$\eta = K(T)\frac{P_{\mathrm{r}}}{P_{\mathrm{r}} + P_{\mathrm{nr}}}, \tag{5.65}$$

where P_{r} and P_{nr} are the rates of radiative and non-radiative transitions. $K(T)$ is a proportionality factor of the ratio of the number of singlet or triplet excitons to the total number of excitons, and it depends on the temperature according to the Boltzmann distribution. The PL lifetime τ is defined by

$$\frac{1}{\tau} = \frac{1}{\tau_{\mathrm{r}}} + \frac{1}{\tau_{\mathrm{nr}}}, \tag{5.66}$$

where $\tau_{\mathrm{r}} = 1/P_{\mathrm{r}}$ and $\tau_{\mathrm{nr}} = 1/P_{\mathrm{nr}}$ are the radiative and non-radiative lifetimes, respectively. At the low generation rate, the rate of non-radiative transitions, P_{nr}, is expected to be very small and it can be neglected. This gives η from Eqs (5.65) and (5.66) as

$$\eta \approx K(T), \tag{5.67}$$

a constant and

$$\tau \approx \tau_{\mathrm{r}}. \tag{5.68}$$

Assuming that τ_r is independent of the generation rate, we thus find that τ and η are independent of the generation rate, at lower G and temperature. In other words, the region of constant τ and η with respect to G, indicates a region in which the rates of radiative recombinations are high and those of non-radiative recombinations are negligibly small. Accordingly the region of constant τ and η, obtained in both a-Si : H (Fig. 5.6) and a-Ge : H (Fig. 5.7) for both slow and fast components, is the range of G in which the process of radiative recombinations is dominant. A phenomenological equation for this process can be written as

$$\frac{\mathrm{d}n}{\mathrm{d}t} = G - \frac{n}{\tau_r}, \tag{5.69}$$

where n is the concentration of excitonic states. In the steady state, $\mathrm{d}n/\mathrm{d}t = 0$, and then Eq. (5.69) gives $n = \tau_r G$ or $n \propto G$.

As G increases, the density of excitonic states also increases and reaches a value that can dissociate excitonic states into free e and h due to the excessive repulsive interaction between like charge carriers. These free electrons and holes can move around and then recombine non-radiatively. A word of caution is necessary here. The term *free* is used here only to mean that e and h in these pairs are not bound in excitonic states, but they can either be in the extended states or tail (localized) states. As explained above, usually such free e and h will relax down non-radiatively in their tail states and then recombine non-radiatively through the process of distant pair recombination, as it is commonly called. A phenomenological equation for this process can be written as

$$\frac{\mathrm{d}n}{\mathrm{d}t} = G - \frac{n}{\tau_r} - Cn_e n_h, \tag{5.70}$$

where C is a constant but may depend on the defect concentration in the material, and n_e and n_h are the density of thus created free excited electrons and holes, respectively. Using $n_e = n_h = n$, Eq. (5.70) becomes

$$\frac{\mathrm{d}n}{\mathrm{d}t} = G - \frac{n}{\tau_r} - Cn^2. \tag{5.71}$$

In the steady state, Eq. (5.71) gives

$$Cn^2 + \frac{n}{\tau_r} - G = 0. \tag{5.72}$$

At higher G, the non-radiative component is expected to be dominant, that is, $Cn^2 \gg n/\tau_r$, then Eq. (5.72) gives

$$Cn^2 = G, \tag{5.73}$$

which gives

$$n \propto G^{0.5}. \tag{5.74}$$

As $1/\tau_{nr} = Cn$, and $\tau_r \gg \tau_{nr}$, using Eq. (5.74) in Eq. (5.73) we get

$$\tau_{nr} \propto G^{-0.5}. \tag{5.75}$$

Using Eq. (5.75) in Eq. (5.66), we get

$$\tau \approx \tau_{nr} \propto G^{-0.5}. \tag{5.76}$$

The quantum efficiency η defined in Eq. (5.65) can be approximated in this range as

$$\eta = \frac{\tau_{nr}}{\tau_r + \tau_{nr}} \approx \frac{\tau_{nr}}{\tau_r} \propto \tau_{nr}, \tag{5.77}$$

which from Eq. (5.76) gives

$$\eta \propto G^{-0.5}. \tag{5.78}$$

The decrease of τ (Eq. (5.76)) and η (Eq. (5.78)) with G at higher generation rates, agrees with the experimental results shown in Fig. 5.6 for a-Si : H with both slow and fast peaks. The only difference is that the fast peak remains constant for much higher G in comparison with the slow peak. Likewise, results of Eqs (5.76) and (5.78) also agree with the decrease of τ_1 and η_1 observed in a-Ge : H as shown in Fig. 5.7. For τ_2 and η_2 in a-Ge : H, no decrease is observed within the range of measurements.

The starting point of the decrease in τ_2 and η_2 is at a much higher value of G in a-Si : H, and no decrease is observed in a-Ge : H. This may be attributed to the large oscillator strength of the radiative transition from a singlet state. Rates of these transitions are faster than that of dissociation of excitonic states into distant pairs, and hence the radiative recombination from singlets dominates even at higher values of G. Hence the fast peak remains constant up to a much higher value of G in both a-Si : H and a-Ge : H. In a-Ge : H (shown in Fig 5.7), the fast peak remains constant in the whole range of measurements, which means that the value of G at which the decrease may start is much higher and beyond the experimental range of measurements.

There is another non-radiate decay process for which a phenomenological theory gives the same result for the dependence of τ and η on G, as obtained in Eqs (5.76) and (5.78). It is called the Auger type non-radiative recombination process, as described below.

5.5.1 Auger type non-radiative recombinations

An excitonic state recombines due to their (e and h) Coulomb interaction and the excess energy is given to an electron or hole to create a hot carrier. This is called Auger recombination, and it is expected to be efficient only at low temperatures and probably more in doped solids where excess carriers are readily available to absorb the excess energy. For the Auger recombination, the phenomenological equation, for intrinsic solids, can be written as (Ambros *et al.*, 1991)

$$\frac{dn}{dt} = G - \frac{n}{\tau_r} - Knn_e, \tag{5.79}$$

where K is a constant. Using $n = n_e = n_h$, we get

$$\frac{dn}{dt} = G - \frac{n}{\tau_r} - Kn^2, \tag{5.80}$$

which is similar to Eq. (5.71) and hence will produce the same dependence for τ and η on G. Therefore, on the basis of the G-dependence, it is difficult to identify between the non-radiative recombinations due to dissociated excitonic states and excitonic Auger type non-radiative recombinations. At high G, but in the low temperature region, the Auger process may be dominant, and at higher temperatures the dissociation of excitonic states will be dominant. Also, for an Auger process to be effective, a reasonable excess charge density is essential. This process may therefore be more efficient when the excitation energy is higher, so that many e and h pairs are excited at higher energies and will be free before forming excitonic states.

The temperature dependence of non-radiative recombination needs to be considered further, because higher temperatures can induce higher order processes as well. This will be described below.

5.5.2 *Higher order non-radiative processes*

When both temperature and generation rate G are high, another kind of Auger type non-radiative recombination may also occur. At higher temperatures and G, excitonic states will be dissociated and charge carriers will have enough energy to move around freely. Such a situation is not expected to occur in the tail states. All excited charge carriers will be in their extended states. In this case, an e may recombine with an h, and the excess energy is given to a third particle (e or h) to create a hot carrier. The phenomenological equation for this process can be written as

$$\frac{dn}{dt} = G - \frac{n}{\tau_r} - Cn^2 - K'n_e n_h n_e, \tag{5.81}$$

where K' is another constant. The third term on the right-hand side of Eq. (5.81) is due to excitonic dissociation, and the last term is due to the higher order Auger process for which the concentration of all three particles have to be considered. Using again $n = n_e = n_h$, and assuming that the non-radiative part is dominant, in the steady state Eq. (5.42) becomes

$$K'n^3 + Cn^2 - G = 0. \tag{5.82}$$

If the Auger recombination is dominant, that is, $K'n^3 \gg Cn^2$, then neglecting the second term in Eq. (5.39) gives

$$n \propto G^{1/3}. \tag{5.83}$$

In this case the inverse of the non-radiative lifetime can be written as

$$\frac{1}{\tau_{\rm nr}} = K'n^2. \tag{5.84}$$

Using Eq. (5.83) in Eq. (5.84), we get

$$\tau \propto G^{-0.67}, \tag{5.85}$$

and then from Eq. (5.77) we get

$$\eta \propto G^{-0.67}. \tag{5.86}$$

This suggests that if higher order non-radiative processes are not negligible, the dependence of τ and η on G can be like $G^{-\beta}$, with $0.5 \leq \beta \leq 0.67$. In a-Ge : H only $\beta = 0.5$ has been observed for $G \geq 3 \times 10^{20}\,\text{cm}^{-3}\,\text{s}^{-1}$ (Ishii *et al.*, 2001; Aoki, 2001) for the long living PL component. However, in a-Si : H $\beta \approx 0.9$ has been predicted (Ambros *et al.*, 1991) for $G \geq 5 \times 10^{19}\,\text{cm}^{-3}\,\text{s}^{-1}$. There has apparently developed a kind of lack of consensus on the role of radiative and non-radiative recombination, dependence of PL lifetime and quantum efficiency on the generation rate and temperature in a-solids in the last two decades. Some of these issues are addressed in the next section.

5.6 Comparison between theory and experiments

The research activities in PL of a-solids have focussed on the following two problems: (1) The lifetime distribution gives rise to two peaks, slow and fast, and (2) the dependence of PL lifetime and quantum efficiency on the generation rate and temperature. The hydrogenated amorphous silicon is the most studied of all, because of its greater use in the fabrication of opto-electronic devices. The earlier work in these areas are due to Kimberling (1978), Kivelson and Gelatt Jr. (1982), Depina and Dunstan (1984), Boulitrop and Dunstan (1985), Bort *et al.* (1991), Stachowitz *et al.* (1998), Baranovskii *et al.* (1991), Levin *et al.* (1991), Ristein (1991), Kemp and Silver (1992), Oheda (1995) and others cited above in this chapter. Most of these studies, including the recent one in a-Ge : H (Ishii *et al.*, 2000, Aoki, 2001), have observed the double peak PL lifetime distribution. The time of the peak value of one of them is slow at about 1 ms, and that of the other is about 1 μs. The theory presented in Section 4.3, supports that the origin of these peaks is radiative recombination from triplet and singlet excitonic states. One of the important points to note is that excitonic states in a-solids are analogous to excitons in c-solids. Therefore, such a double peak PL structure should also be observed in c-solids. There are no such observations in c-Si to our knowledge, but in porous silicon such a double peak structure has been observed by Fishman *et al.* (1993) and Calcott *et al.* (1993), and the activation energy is estimated to be about $\Delta E_{ex} \approx 10\,\text{meV}$, which is larger than that obtained for a-Si : H both theoretically (see Section 5.3) and experimentally (Aoki, 2001). A larger value of ΔE_{ex} in porous silicon can be attributed to the fact that porous silicon offers more severe confinement than a-Si : H. The exciton binding energy gets enhanced due to the confinement resulting in a larger singlet–triplet splitting. The reason for not observing it in c-Si is probably due to the small binding energy of excitons. Also as Wannier excitons in c-solids are of large radii, the exchange interaction is negligible and hence the energy difference between singlet and triplet states will be nearly zero, giving rise to only one observable peak.

As far as the dependence of τ and η on the generation rate is concerned, it is quite consistent with the phenomenological results presented in Section 5.5. The decrease in τ and η at higher generation rate G occurs due to more efficient dissociation of the excitonic states into e and h and then their subsequent non-radiative recombination. However, it may be clarified here that the dominant contribution to PL is only due to the radiative recombination from excitonic states and once the excitonic states dissociate into e and h they do not recombine radiatively as efficiently as non-radiatively. It is difficult to prove that all dissociated excitonic states can only recombine non-radiatively, but relatively the non-radiative channel is more efficient. The non-radiative recombination would be even further enhanced at higher temperatures, because then the available thermal energy will provide more kinetic energy to the dissociated pairs of e and h to move them far apart. That means, both high temperature and generation rate will enhance the efficiency of non-radiative recombination processes.

There may be a problem in deciding whether the dominant non-radiative process is due to exicitonic dissociation caused by the repulsive energy or thermal energy, due to Auger recombination process, or due to higher order recombination processes. If τ and η depend on the generation rate as $G^{-\beta}$, the value of β can determine the dominant channel of non-radiative recombination, but not unambiguously. For example, when $\beta = 0.5$ one cannot exactly determine whether the dominant non-radiative channel is due to excitonic dissociation or Auger recombination, particularly when the temperature is not very low or high. This is one of the problems with the phenomenological theory, which gives the same solution for different processes involving the same number of particles (channels).

References

Ambros, S., Carius, R. and Wagner, H. (1991). *J. Non-Cryst. Solids* **137–138**, 555.

Aoki, T. (2001). *Proceedings of the International Summer School in Condensed Matter Physics*, Varna, 1–5 September, 2000, Bath, 2001, p. 58.

Aoki, T., Komedoori, S., Kobayashi, S., Fujihashi, C., Ganjoo, A. and Shimakawa, K. (2002a). *J. Non-Cryst. Solids* **299–302**, 642.

Aoki, T., Komedoori, S., Kobayashi, S., Shimizu, T., Ganjoo, A. and Shimakawa (2002b), *Nonlinear Optics* (in press).

Baranovskii, S.D., Saleh, R., Thomas, P. and Vaubel, H. (1991). *J. Non-Cryst. Solids* **137–138**, 567.

Bort, M., Fuhs, W., Liedtke, S. and Stachowitz, R. (1991). *Phil. Mag. Lett.* **64**, 227.

Boulitrop, F. and Dunstan, D.J. (1985). *J. Non-Cryst. Solids* **77–78**, 663.

Calcott, P.D.J., Nash, K.J., Canham, L.T., Kane, M.J. and Brumhead, D. (1993). *J. Phys.: Condens. Matter* **5**, L91.

Cody, G.D. (1984). *Semiconductors and Semimetals*, Vol. **21**, Part B, p. 11.

Conwell, G.A.N. and Pawlik, J.R. (1976). *Phys. Rev. B* **13**, 787.

Davydov, A.S. (1971). *Theory of Molecular Excitons*, Plenum, New York.

Depinna, S.P. and Dunstan, D.J. (1984). *Phil. Mag. B* **50**, 597.

Elliot, R.J. (1962). In: Kuper, K.G. and Whitfield, G.D. (eds), *Polarons and Excitons*. Oliver and Boyd, Edinburgh and London, p. 269.

Fishman, G., Romstain, R. and Vial, J.C. (1993). *J. Lumin.* **57**, 235.

Ishii, S., Kurihara, M., Aoki, T., Shimakawa, K. and Singh, J. (1999). *J. Non-Cryst. Solids* **266–269**, 721.

Kemp, M. and Silver, M. (1992). *J. Non-Cryst. Solids* **141**, 88.

Kimberling, L.C. (1978). *Solid State Electron.* **21**, 1391.

Kivelson, S. and Gelatt, Jr., C.D. (1982). *Phys. Rev. B* **26**, 4646.

Knox, R.S. (1965). Excitons. In: Seitz, F., Turnbull, D. and Ehrenreich, H. (eds), *Solid State Physics*. John-Wiley and Sons, New York.

Levin, E.I., Marianer, S., Shklovskii, B.I. and Fritzsche, H. (1991). *J. Non-Cryst. Solids* **137–138**, 559.

Morigaki, K. (1999a). *Physics of Amorphous Semiconductors*. World Scientific, London.

Morigaki, K. (1999b). Private communication.

Morigaki, K., Hikita, H. and Kondo, M. (1995). *J. Non-Cryst. Solids* **190**, 38.

Mott, N.F. and Davis, E.A. (1979). *Electronic Processes in Non-crystalline Materials*, Clarendon Press, Oxford.

Oheda, H. (1995). *Phys. Rev. B* **52**, 16530.

Ristein, J. (1991). *J. Non-Cryst. Solids* **137–138**, 563.

Schiff, L.I. (1968). *Quantum Mechanics*. McGraw-Hill, Tokyo.

Schubert, M., Stachowitz, R., Saleh, R. and Fuhs, W. (1993). *J. Non-Cryst. Solids* **164–166**, 555.

Searle, T.M., Hopkinson, M., Edmeades, M., Kalem, S., Austin, I.G. and Gibson, R.A. (1987). In: Kastner, M.A., Thomas, G.A. and Ovshinsky, S.R. (eds), *Disordered Semiconductors*. Plenum, New York, p. 357.

Singh, J. (1994). *Excitation Energy Transfer Processes in Condensed Matter*. Plenum, New York.

Singh, J. (2002). *J. Non-Cryst. Solids* **299–302**, 444.

Singh, J. (2003). *J. Mat. Science* (in press).

Singh, J., Aoki, T. and Shimakawa, K. (2002). *Phil. Mag. B* **82**, 855.

Stachowitz, R, Bort, M., Ccarius, R., Fuhs, W. and Liedtke, S. (1991). *J. Non-Cryst. Solids* **137–138**, 551.

Stachowitz, R., Schubert, M. and Fuhs, W. (1998). *J. Non-Cryst. Solids* **227–230**, 190.

Street, R.A. (1991). *Hydrogenated Amorphous Silicon*. Cambridge University Press, Cambridge.

Stutzmann, M. and Brandt, M.S. (1992). *J. Non-Cryst. Solids* **141**, 97–105.

Tsang, C. and Street, R.A. (1979). *Phys. Rev. B* **18**, 3027.

6 Charge carrier–phonon interaction

This chapter presents a brief description of the role of phonons, charge carrier–phonon interaction and associated phenomena observed in amorphous semiconductors (a-semiconductors). All atoms in a solid are always in vibration about their equilibrium positions, including the zero point vibrations occurring at the absolute zero temperature. To solve the problem with all the atoms in a solid as coupled oscillators is very complicated. A simplified way is to consider each atom as harmonic oscillators, and then energy of each normal mode of such coupled oscillators gets quantized. A quantum of such vibrational energy is called a phonon. Thus, although phonons are an intrinsic part of any solid, amorphous or crystalline, they are counted among the defects, particularly in crystalline solids (c-solids), because the vibration of atoms destroys the translational symmetry which is also caused by other defects. In c-solids, atomic displacements due to vibrations can be expanded in terms of waves associated with a wavevector, called a phonon wavevector, but it is not applicable in amorphous solids (a-solids). Only some of the effects of phonons in the extended states of a-solids may possibly be understood in the same way as those in c-solids.

The effect of atomic vibrations on the electronic and optical properties of a solid depends on the strength of charge carrier–phonon interaction, which is usually considered linearly proportional to the atomic displacement coordinates. The theory is developed within the adiabatic approximation. First, the energy eigenvalues of charge carriers are obtained in static equilibrium, and then it is expanded in Taylor's series about positions of static equilibrium. The first-order term of the series is used as charge carrier–phonon interaction Hamiltonian (Taylor, 1970; Singh, 1994). For a-solids, the concept of an interaction coordinate introduced by Holstein (1959) for polarons and Toyozawa (1959, 1962) for self-trapping in molecular crystals, is usually found to be very useful. Accordingly, an atom gets displaced due to electron–phonon interaction in the direction of an interaction coordinate, not necessarily a normal coordinate, and causes a distortion in an otherwise periodic crystal lattice. Thus the electron–phonon interaction introduces a displacement in the equilibrium position of a harmonic oscillator (atom) along the interaction coordinate. Let us first study the problem of a simple one-dimensional (1D) displaced harmonic oscillator to provide an insight into the problem of distortion around a charge carrier caused by atomic vibrations.

6.1 Displaced harmonic oscillator

One-dimensional harmonic oscillator potential with its equilibrium position at the origin is given by

$$V_0(x) = \tfrac{1}{2} m\omega^2 x^2, \tag{6.1}$$

where m is the mass, $\omega = \sqrt{c/m}$ frequency (c being the force constant of vibration), and x position coordinate of the oscillator. Let us introduce a term linear in x in the potential in Eq. (6.1) as

$$V(x) = V_0(x) - \alpha x, \tag{6.2}$$

where α represents the strength of force causing the linear displacement. The potential in Eq. (6.2) can be written as

$$V(x) = \frac{1}{2} m\omega^2 (x - x_0)^2 - \frac{\alpha^2}{2m\omega^2}, \tag{6.3}$$

where $x_0 = \alpha/(m\omega^2)$. The potential in Eq. (6.3) is of a harmonic oscillator, whose position of equilibrium is displaced from $x = 0$ to x_0, and its potential energy lowered by $\alpha^2/(2m\omega^2)$, as illustrated in Fig. 6.1. Thus the effect of introducing a linear term in the harmonic oscillator potential is to displace the oscillator to another position of equilibrium and in the new displaced position the energy of the oscillator gets lowered. The lowering of energy implies that the new position of equilibrium is relatively more stable than the original position of equilibrium. The magnitude of both the displacement coordinate, x_0, and lowering of energy

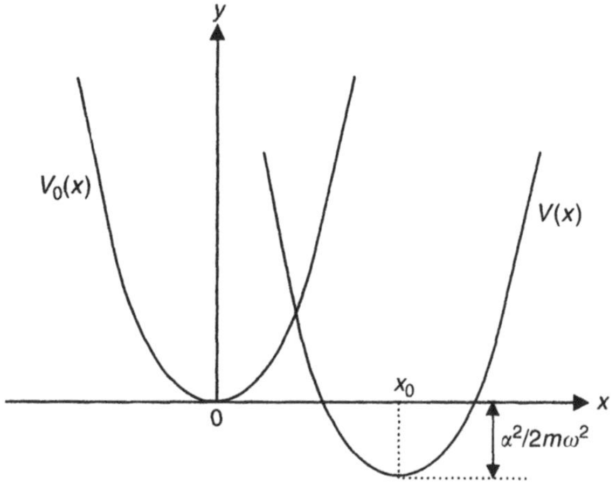

Figure 6.1 Potential energy curves of a 1D displaced harmonic oscillator.

depend on α. These results, although obtained from a very simple theory for a single harmonic oscillator, illustrate very clearly the situation that occurs in solids, as described in the next section.

6.2 Application to solids

The problem of a free charge carrier moving in a solid and subjected to atomic vibrations through a charge carrier–lattice interaction linear in the atomic displacement coordinate becomes analogous to that of displaced harmonic oscillator. Although the electron–phonon interaction is covered in most books of solid state physics, a clear demonstration of its linear dependence on the lattice displacement vector is shown in the theory of polarons developed by Holstein (1959) for a linear chain of diatomic molecules. The Hamiltonian consisting of the energy operator of a free charge carrier and that of vibrating atoms (harmonic oscillators) gets introduced with an interaction term linear in the interaction coordinate q, resulting in a displacement of atom along the interaction coordinate, also called the configurational coordinate (Mott and Davies, 1979), as shown in Fig. 6.2. Such a distortion in the atomic network occurs around the charge carrier, lowering the

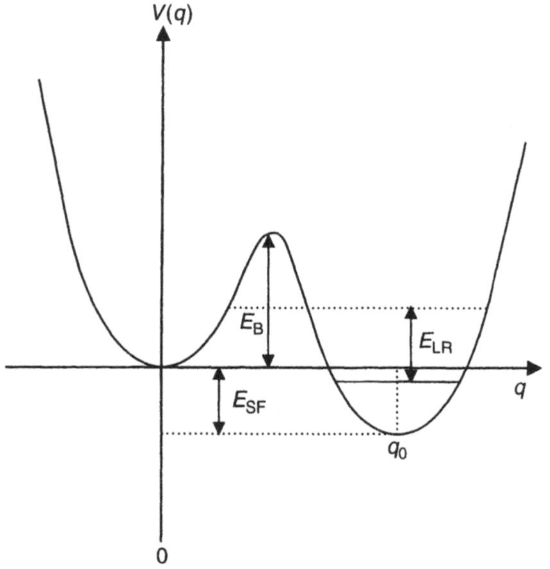

Figure 6.2 Adiabatic potential energy surface (or double well potential) of an electronic state plotted as a function of the interaction coordinate q. E_{SF} is the self-trapped depth, E_B barrier height between the potential energy states of undistorted and distorted lattice, and E_{LR} lattice relaxation energy. q_0 is the measurement of distortion along the interaction coordinate.

energy of the system (charge carrier plus displaced atom), in the same way as in the displaced oscillator. The combined effect of distortion and lowering of energy is such that the charge carrier moves slower than before in a rigid lattice, and the distortion moves with it. This combination of charge carrier and lattice distortion moving together in a lattice is called a polaron.

In solids as the charge carrier–lattice interaction energy operator is derived through adiabatic or Born–Oppenheimer approximation, the potential energy curves thus obtained are called adiabatic potential energy surfaces (APES). One gets a double well potential (see Fig. 6.2), which is frequently used for both c- and a-solids to explain the occurrence of self-trapping and metastable defect states of charge carriers (Shimakawa *et al.*, 1995). The strength of charge carrier–phonon interaction plays the crucial role in creating polarons and self-trapped metastable states. In the weak interaction limit, a charge carrier becomes a polaron, whereas in the strong coupling limit it gets self-trapped. A polaron moves slower than a free charge carrier, because its effective mass increases due to interaction with phonons. A self-trapped charge carrier becomes localized around a center called self-trapping center, and it cannot move away, because its effective mass becomes infinitely large, reducing its kinetic energy to zero due to very strong interaction with phonons (Singh, 1994). The occurrence of a double potential well in solids is different from the potential of a displaced single oscillator. For a single oscillator a double well potential does not arise. In the case of solids, only the atom where the charge carrier is localized will be displaced but others will be vibrating at the original position of equilibrium. Hence the resulting potential is a combination of the two. In Fig. 6.2, the APES with minimum at $q = 0$ represents atomic vibrations of an undistorted lattice, and with a minimum at $q = q_0$, corresponds to vibration of an atom where the charge carrier is localized at any time giving rise to the distortion in the lattice. The latter represents a metastable state, because its lowest energy is lower than that of the undistorted lattice by E_{SF}, which is usually called the self-trapped depth energy. As shown in Fig. 6.2, there also exists a barrier of height E_B between the free charge carrier state and self-trapped state.

There are three forms of charge carrier–phonon interactions known to occur in c-solids. The first one is applicable to polar solids (crystals) where basis atoms have unequal charges so that a moving electron polarizes the electric field around it. Such a polarization can displace an ion along the interaction coordinate q, and then the lattice vibrates with a frequency in resonance with that of the polarization field. The phonons involved in such interactions are optical phonons, and the term "polaron" was initially used for an electron accompanied by its own induced lattice deformation by optical phonons (Fröhlich, 1952) in polar crystals. A polaron has an effective mass larger than the effective mass of an electron in a rigid lattice. A self-trapping caused by charge carrier–optical phonon interaction is not apparently known to occur.

The second type of interaction is due to dilation associated with the acoustic lattice waves, which lowers the energy of band edges and such dilations are expressed in terms of deformation potentials. In this case, the interaction coordinate q

may represent the dilation caused by the deformation potentials (Bardeen and Shokley, 1950). This is the most common form of charge carrier–phonon interaction found in semiconductors (like silicon and germanium) and insulators, and it involves acoustic phonons. In this case, depending on the strength of interaction, an electron becomes a polaron in the weak interaction limit and gets self-trapped in the strong coupling limit. However, the deformation potential interaction vanishes for transverse acoustic (TA) phonons, only the longitudinal acoustic (LA) component is non-zero in cubic semiconductors with isotropic energy bands, as usually is the case with most conduction bands. Thus, the electron–acoustic phonon interaction due to deformation potential has a non-zero contribution only from LA phonons. On the other hand, as the valence band is usually anisotropic, this is not the case with hole–phonon interaction, as recently shown by Oh and Singh (2000). The hole–acoustic phonon interaction is non-zero in the valence band for both LA and TA phonons. It is not obvious, however, whether such a distinction can be applied to a-semiconductors where instead of conduction and valence bands we have conduction and valence extended states, which cannot be associated with any isotropic or anisotropic concepts in the **k**-space.

The third form of charge carrier–phonon interaction occurs in non-centrosymmetric crystals and polar semiconductors, where a strain associated with acoustic phonons can induce a macroscopic electric polarization, called the piezoelectric effect. In this case the interaction coordinate may represent the distortion in the ionic position, due to the electric potential of piezoelectric fields. Here also, depending on the strength of interaction, both polaron and self-trapping states are possible. The charge carrier–acoustic phonon interaction due to piezoelectric coupling has non-zero contribution from both LA and TA phonons. The piezoelectric charge carrier–phonon interaction has not apparently drawn much attention from a-semiconductors.

In addition, there is another kind of deformation that is very well known to occur in alkali halides and it is associated with the interaction of phonons. It gives rise to V_K centers in alkali halides. This occurs due to displacement of an atom or anion containing a hole, when it forms a bond with the neighboring atoms as it occurs in rare gas solids. The interaction coordinate is associated with the bond length in this case. In chalcogenides also the formation of dangling bonds is attributed to such deformation (Shimakawa *et al.*, 1995), which takes place due to strong charge carrier–phonon interaction.

For a-semiconductors, the concept of carrier–phonon interaction is usually used from the c-solids. For example, the derivation of the first three types of charge carrier–phonon interactions; carrier–optical phonon interaction in polar crystals, and carrier–acoustic phonon interaction due to deformation potentials and piezoelectric effects, is strictly applicable only to c-solids. The distinction between acoustic and optical phonons is based on the dispersion relations of phonon energy, which is applicable only to c-solids. In this view, it is only the last carrier–phonon interaction, arising due to bonding between neighboring atoms, that seems to be really appropriate for use in a-solids. However, before addressing the issue of charge carrier–phonon interaction in a-semiconductors, it may be desirable to describe the influence of the strength of electron–phonon interaction that gives rise to polarons and self-trapped electrons in c-solids in more detail.

6.3 Polarons and self-trapped electrons

In c-solids, it is relatively easy to develop a concept of polarons by considering the motion of an electron in the conduction band and interacting with phonons. This will be briefly discussed here. The Hamiltonian consisting of the energy operator of an electron ($\hat{H}_{\rm el}$) and phonons ($\hat{H}_{\rm ph}$) and the interaction between them ($\hat{H}_{\rm I}$) can be written as

$$\hat{H} = \hat{H}_{\rm el} + \hat{H}_{\rm ph} + \hat{H}_{\rm I}, \tag{6.4}$$

where

$$\hat{H}_{\rm el} = \frac{p^2}{2m_{\rm e}^*}, \tag{6.5}$$

$$\hat{H}_{\rm ph} = \sum_{\mathbf{q}} \hbar\omega(q)\left(b_{\mathbf{q}}^+ b_{\mathbf{q}} + \frac{1}{2}\right) \tag{6.6}$$

and

$$\hat{H}_{\rm I} = \sum_{\mathbf{q}} V_q {\rm e}^{{\rm i}\mathbf{q}\cdot\mathbf{r}_{\rm el}} (b_{-\mathbf{q}}^+ + b_{\mathbf{q}}). \tag{6.7}$$

In Eqs (6.5)–(6.7), p, $\mathbf{r}_{\rm el}$ and $m_{\rm e}^*$ are the magnitude of linear momentum, position coordinate and effective mass of an electron, respectively; $\omega(q)$ is the frequency and $b_{\mathbf{q}}^+ (b_{\mathbf{q}})$ is creation (annihilation) operator of a phonon with wavevector $\mathbf{q}$, and V_q represents the strength of interaction between electron and phonons. It should be noted here that the interaction operator in Eq. (6.7) is linear in the atomic displacement coordinates similar to the linear term used in the displaced oscillator in Section 6.1.

For electron and longitudinal optical (LO) phonon interaction in polar crystals, V_q is obtained as (Fröhlich, 1952, 1954; Singh, 1994):

$$V_q = -{\rm i}\hbar\omega\sqrt{\frac{4\pi\alpha}{S}}\frac{1}{q}, \tag{6.8}$$

where

$$\alpha = \frac{e^2}{2\varepsilon\hbar\omega}\sqrt{\frac{2m_{\rm e}^*\omega}{\hbar}}. \tag{6.9}$$

Here, S is a dimensionless parameter and ε is the static dielectric constant of crystal.

For electron–acoustic phonon interaction derived from the deformation potential, V_q is obtained for electron–LA phonon interaction in the conduction band as

$$V_q = \mathrm{i}\sum_{\mathbf{q}} \left(\frac{\hbar}{2NI\nu}\right)^{1/2} E_{\mathrm{D}}^{c} q^{1/2}, \tag{6.10}$$

where N is the number of atoms, I the ionic mass density, ν the velocity of sound in the crystal, and E_{D}^{c} the deformation potential parameter associated with the conduction band. Both electron–transverse optical (TO) and electron–TA phonon interactions are zero for isotropic bands.

As the interaction operator in Eq. (6.7) is not diagonal, one has to diagonalize the Hamiltonian in Eq. (6.4) for determining its energy eigenvalue (Lee and Pines, 1952; Gurari, 1953; Low and Pines, 1953; Lee *et al.*, 1953; Pines, 1963). The diagonalization is carried out using two unitary transformations, which diagonalize the Hamiltonian to a certain extent, with the remaining off-diagonal terms treated as a perturbation. The diagonalized part of the Hamiltonian ($\hat{H}_{\mathrm{D}}$) is then used to solve the following Schrödinger equation

$$\hat{H}_{\mathrm{D}}|\mathbf{k};n\rangle = E_{\mathrm{p}}(\mathbf{k},n)|\mathbf{k};n\rangle, \tag{6.11}$$

where $\mathbf{k}$ is the electronic wavevector, $|n\rangle = |n_1, n_2, \ldots, n_{\mathbf{q}}, \ldots\rangle$ phonon state vector with $n_{\mathbf{q}}$ being the occupation number of phonons with wavevector $\mathbf{q}$, and $E_{\mathrm{p}}(\mathbf{k}, n)$ is the energy eigenvalue of a polaron. The ground state energy ($n_{\mathbf{q}} = 0$) of a polaron, is obtained as (Singh, 1994):

$$E_{\mathrm{p}}(\mathbf{k},0) = \frac{\hbar^2 k^2}{2m_{\mathrm{ep}}^*} - A\hbar\omega, \tag{6.12}$$

where m_{ep}^* represents the effective mass of an electronic polaron and A the strength of coupling. Here "electronic polaron" is used to distinguish it from "hole polaron," whose energy eigenvalue can also be calculated analogously. Using V_q given in Eq. (6.8) for polar crystals, one gets the effective mass of a polaron and strength of coupling due to optical phonons as

$$m_{\mathrm{ep}}^* = \frac{m_{\mathrm{e}}^*}{1-(\alpha/6)} \quad \text{and} \quad A = \alpha. \tag{6.13}$$

For non-polar crystals, using V_q as given in Eq. (6.10), one gets the effective mass of a polaron and strength of coupling due to deformation potential as

$$m_{\mathrm{ep}}^* = \frac{m_{\mathrm{e}}^*}{1-\frac{4}{3}\beta_{\mathrm{ac}}^{\mathrm{D}}\gamma} \quad \text{and} \quad A = \beta_{\mathrm{ac}}^{\mathrm{D}} = \frac{m_{\mathrm{e}}^{*^2} E_{\mathrm{D}}^2}{\pi^2 \delta \nu \hbar^3}. \tag{6.14}$$

Here, γ is the ratio of the Debye phonon energy ($\hbar\omega_{\mathrm{D}}$) to the kinetic energy of an electron with a wavevector equal to q_{D}, which is the wavevector associated with the acoustic phonon of Debye cut-off frequency, that is, $\omega_{\mathrm{D}} = \nu|q_{\mathrm{D}}|$.

It is clear from both Eqs (6.13) and (6.14) that the effective mass of a polaron depends on the strength of coupling, and it increases with the coupling strength as, for example, when α increases from 0 to 6 (Eq. (6.13)) or when $\beta_{ac}^{D}\gamma$ increases from 0 to $\frac{3}{4}$. The kinetic energy of the polaron, first term of Eq. (6.12), decreases with α. For $\alpha < 6$, the kinetic energy of the polaron is non-zero, but it decreases with increasing α, and as long as this kinetic energy is larger than the energy $\alpha\hbar\omega$, the total energy of a polaron will be positive. That means, in this range of coupling strength, according to Eq. (6.12), an electron interacting with phonons, a polaron, moves slower than a free electron. The wavefunction of an electron can still be represented by a free wave given by

$$\psi_p(r) \propto e^{i\mathbf{k}\cdot\mathbf{r}}. \tag{6.15}$$

However, the width of the energy band of a polaron thus obtained will be narrower than that of a free electron.

If the coupling strength increases further, the kinetic energy decreases further and becomes less than $\alpha\hbar\omega$ or reduces to zero. In this situation the total energy of an electron interacting with the lattice becomes negative (see Eq. (6.12)). Considering this negative energy as the kinetic energy of the electron, we find that this can only happen when the momentum becomes imaginary ($\hbar^2k^2/2m_{ep} < 0$, implies k is imaginary). Using an imaginary k in Eq. (6.16) produces an exponentially decreasing wavefunction for the motion of electron. Thus the electron gets self-trapped around a center with an exponentially decreasing wavefunction of the type

$$\psi_p(r) \propto e^{-\rho(|r-s|)}, \quad k = i\rho, \tag{6.16}$$

where s represents the position of the self-trapping center.

This simple theory presented above illustrates very clearly that an electron, a polaron and a self-trapped electron refer to the same particle, but traveling in different types of crystalline media. It is like a ball (electron) rolling from a concrete surface that reaches a sandy beach (becomes a polaron) and then falls into mud and stops (becomes self-trapped electron). Thus, zero coupling between electron and phonon means a very rigid crystalline structure, weak coupling means less rigid and strong coupling means soft crystalline structure. It may be pointed out here that the results derived for polarons in Eqs (6.12)–(6.14) are valid only in the weak coupling limit and therefore to stretch them to discuss the situation of self-trapping occurring in the strong coupling limit is not really valid. However, these results do provide some qualitative insight into the problem of self-trapping.

In Toyozawa's theory of electron self-trapping developed for molecular crystals, it is found that self-trapping occurs when the coupling strength is greater than unity. The coupling strength, g, is given by

$$g = \frac{E_D^2}{2I\nu^2B} = \frac{E_{LR}}{B}, \tag{6.17}$$

where B is the half width of the conduction band and E_{LR} is the lattice relaxation energy as defined in Fig. 6.2. Accordingly it can be written as

$$E_{\mathrm{LR}} = E_{\mathrm{SF}} + B. \tag{6.18}$$

The energy of the barrier height is obtained as

$$E_{\mathrm{B}} = \frac{B^2}{4E_{\mathrm{LR}}}. \tag{6.19}$$

Results of the theory of polarons and self-trapping presented above, although derived for c-solids, are considered applicable for a-solids as well, at least for qualitative interpretations of experimental results. The quantitative features may be difficult to obtain because parts of the theoretical developments, at least for phonons, are done in the reciprocal lattice vector space, which does not apply to a-solids. For a-solids the theory of polarons developed in the real crystal space, first by Holstein (1959) for molecular crystals, appears to be more logical to apply. This will be presented in the next section.

6.4 Tight-binding approach to polarons

Holstein (1959) considered the problem of polarons in a 1D molecular crystal model of identical diatomic molecules and used the tight-binding approach to study the motion of an excess charge carrier in such a system. The diatomic molecules are assumed to be oriented along the chain so the only vibrations molecules have are the intra-molecular ones along the chain. Each term of the Hamiltonian $\hat{H}$ in Eq. (6.4) for the motion of an excess charge carrier (electron or hole) interacting with the lattice vibrations can be written in the real coordinate space as

$$\hat{H}_{\mathrm{el}} = \sum_{l,m} E_{lm} a_l^+ a_m, \tag{6.20}$$

$$\hat{H}_{\mathrm{ph}} = -\sum_{n} \frac{\hbar^2 \nabla_n^2}{2M} + \frac{1}{2} \sum_{m,n} M\, \omega_m\, \omega_n\, x_m\, x_n \tag{6.21}$$

and

$$\hat{H}_{\mathrm{I}} = \sum_{n} A_n x_n a_n^+ a_n. \tag{6.22}$$

Here E_{lm} is the energy transfer matrix element between sites at l and m $(l \neq m)$, and a_l^+ (a_l) is the creation (annihilation) operator of a charge carrier at site l. M is the molecular mass and ω_m the vibrating frequency of a diatomic molecule at m. x_n is the vibrating inter-nuclear separation of molecule at n measured from its equilibrium value, and A_n is the first-order coupling term obtained by expanding the adiabatic carrier energy in Taylor's series about the lattice equilibrium. Thus in Eq. (6.22), A_n is a force of vibration of molecule at n from its position of equilibrium. The interaction term is derived from a general Hamiltonian of the motion of charge carrier in a vibrating molecular system by applying the Born–Oppenheimer and tight-binding approximations (Singh, 1994). The eigenvector of a charge carrier in the tight-binding limit can be written as

$$|i\rangle = \sum_l C_l a_l^+ |0\rangle, \tag{6.23}$$

where i = e or h denotes the state of a charge carrier (electron or hole). As the lattice is vibrating, the translational symmetry cannot be applied and therefore the probability amplitude coefficient is not known in any simple mathematical form. Using Eqs (6.4) and (6.20)–(6.23), a secular equation for determining the energy eigenvalue W_i of the Schrödinger equation

$$\hat{H}|i\rangle = W_i|i\rangle \tag{6.24}$$

is obtained as

$$W_i C_l = \left(-\sum_n \frac{\hbar^2 \nabla_n^2}{2M} + \frac{1}{2}\sum_{m,n} M\, \omega_m\, \omega_n\, x_m\, x_n - A_l x_l + E_0\right) C_l$$
$$- J(C_{l+1} + C_{l-1}), \tag{6.25}$$

where $E_0 = E_{ll}$ is a constant energy term obtained from Eq. (6.20) for $l = m$, and $J = E_{lm}$ is the nearest neighbor carrier transfer energy from site l to $l \pm 1$. Considering only the diagonal terms in the vibrating potential in Eq. (6.25), it reduces to

$$\left[W_i - \left(-\sum_n \frac{\hbar^2 \nabla_n^2}{2M} + \frac{1}{2}\sum_m M\, \omega_m^2\, x_m^2 - A_l x_l + E_0\right)\right] C_l$$
$$+ J(C_{l+1} + C_{l-1}) = 0. \tag{6.26}$$

Although the vibrating frequency ω_m has a subscript m, being the intramolecular vibration frequency of diatomic molecules, it is site independent, because identical diatomic molecules will vibrate with the same frequency. For understanding the static problem the kinetic energy of nuclear motions may be ignored for the time being, and then multiplying Eq. (6.26) by C_l^* and summing over l, we get

$$W_i = -\sum_n \frac{\hbar^2 \nabla_n^2}{2M} + \frac{1}{2}\sum_{m,n} M\omega_m^2 x_m^2 - \sum_l A_l x_l |C_l^2| + E_0$$
$$- J\sum_l (C_{l+1} + C_{l-1}) C_l^*, \tag{6.27}$$

where $\sum_l C_l^* C_l = 1$ is applied. As is obvious from Eq. (6.27), the energy W_i is a function of x_l. For getting information on the effect of carrier–lattice interaction on the charge carrier's energy, it is desirable to minimize the above energy with respect to the vibrating bond length x_l. The easiest way is to consider the probability amplitude coefficient C_l as a continuous function of atomic positions. According to Holstein, this assumption is valid when the linear spread of charge carrier interacting with the lattice is larger than the lattice spacing, and hence it is called the large polaron model. Using this model the probability amplitude coefficient can be expanded as

$$C_{l\pm 1} = C_l \pm \frac{\partial C_l}{\partial l} + \frac{1}{2}\frac{\partial^2 C_l}{\partial l^2}. \tag{6.28}$$

Using Eq. (6.28) we minimize the energy in Eq. (6.27) by setting

$$\frac{\partial W_i}{\partial x_{\mathrm{p}}} = 0. \tag{6.29}$$

This produces $x_{\mathrm{p}}^{(0)}$ corresponding to the minimum energy as

$$x_{\mathrm{p}}^{(0)} = \frac{A_{\mathrm{p}}|C_{\mathrm{p}}^{(0)}|^2}{M\omega_{\mathrm{p}}^2}, \tag{6.30}$$

where superscript (0) denotes the value of quantities at the minimum energy. Substituting Eq. (6.30) into Eq. (6.27) we get the minimum energy denoted by $W_i^{(0)}$ as

$$W_i^{(0)} = E_0 - 2J - E_{\mathrm{p}}, \tag{6.31}$$

where

$$E_{\mathrm{p}} = \frac{(A_{\mathrm{p}}^2/M\omega_{\mathrm{p}}^2)^2}{48J}. \tag{6.32}$$

Thus, the minimum energy of an electron interacting with lattice vibrations is lowered by an energy E_{p} from that of a free electron (not interacting with lattice vibrations), which is $E_0 - 2J$. Also according to Eq. (6.30), there is a finite deviation in the bond length, given by $x_{\mathrm{p}}^{(0)}$, from its value at equilibrium. These two results are similar to those described earlier for a displaced oscillator. That means an electron moving along a chain of diatomic molecules and interacting with their vibrations finds the chain distorted around it as it moves. The interesting point about Holstein's model of diatomic molecules is that it can be applied not only to molecular crystals but to other solids as well. On a microscopic scale, in atomic crystals also a charge carrier moves from one bond between any two atoms to another during the course of its motion regardless of the dimensionality of the system and whether the solid possesses translational symmetry or not. Therefore the model can be applied to a-solids as well. Anderson (1975) has applied the linear charge carrier–phonon interaction in a linear chain of a-solid to study the bonding between a pair of covalent electrons in a-semiconductors. A similar approach has also been followed to present the mechanism of bond breaking in laser sputtering (Singh, 1994; Singh *et al.*, 1994) in non-metallic crystals and polymer ablation (Singh and Itoh, 1990). Anderson's model for the electronic structure of a-semiconductors, commonly known as Anderson's negative-U, will be described in the next section.

6.5 Anderson's model of a-semiconductors

6.5.1 *Like charge carriers*

Assuming that most solid structures are naturally diamagnetic and therefore prefer to have paired electrons in their electronic states, Anderson considered the interaction potential V between a pair of electrons of opposite spins localized on a bond and interacting with the lattice vibrations as

$$V = \sum_l \left[\frac{1}{2}cx_l^2 - \lambda x_l(a_{l\uparrow}^+ a_{l\uparrow} + a_{l\downarrow}^+ a_{l\downarrow})\right], \tag{6.33}$$

where, like in Holstein's model, x_l is the vibrating bond length between a pair of atoms at l. c is a force constant and λ electron–phonon coupling constant like A_l in Holstein's model. The subscript of an arrow used with the creation and annihilation operators represents the direction of spin of electron, for example, $a_{l\uparrow}^+$, is the creation operator of an electron with spin up. The electronic part of the Hamiltonian considered by Anderson is

$$\hat{H}_e = \sum_{l\sigma} E_l n_{l\sigma} + U\sum_l n_{l\uparrow}n_{l\downarrow} + \sum_{lm\sigma} T_{lm} a_{l\sigma}^+ a_{m\sigma}, \tag{6.34}$$

where E_l is the energy of an electron localized at site l, like E_0 in Holstein's model but for a-solids this may not be the same for every site. U represents the repulsive potential between the paired electron on a site, T_{lm} electron transfer energy from site l to m, and $n_{l\sigma} = a_{l\sigma}^+ a_{l\sigma}$ number operator. Thus the total Hamiltonian becomes

$$\hat{H} = \hat{H}_e + \frac{1}{2}\sum_l M\dot{x}_l^2 + V, \tag{6.35}$$

where M is the mass of vibrating pair of atoms on a bond, and $\dot{x}_l$ is the linear velocity of vibrating atoms. Minimizing the potential V with respect to x_l as

$$\frac{\partial V}{\partial x_l} = 0, \tag{6.36}$$

produces x_l^0 at the minimum of V as

$$x_p^0 = \frac{\lambda}{c}(n_{p\uparrow} + n_{p\downarrow}). \tag{6.37}$$

Substituting x_p^0 into Eq. (6.33) gives the minimum potential energy V_{min} as

$$V_{min} = -\frac{\lambda^2}{2c}\sum_l (n_{l\uparrow} + n_{l\downarrow})^2. \tag{6.38}$$

Substituting V_{min} in place of V in Eq. (6.35), we get the Hamiltonian representing the minimum energy of the system as

$$\hat{H}_{min} = \sum_{l\sigma}\left(E_l - \frac{\lambda^2}{2c}\right)n_{l\sigma} + \left(U - \frac{\lambda^2}{c}\right)\sum_l n_{l\uparrow}n_{l\downarrow} + \sum_{lm\sigma} T'_{lm}a^+_{l\sigma}a_{m\sigma}, \tag{6.39}$$

where the term of kinetic energy of vibrating atoms is ignored. This is valid for the static case or for very low frequency processes ($\omega \ll \omega_0$, where $\omega_0 = \sqrt{c/m}$) occurring at low temperatures. The Hamiltonian representing the minimum energy shows the following three features:

1 The first term of Eq. (6.39) shows that the energy of an electron localized on a site gets lowered by $\lambda^2/2c$ due to the interaction with lattice vibrations.

2 The second term shows that the repulsive potential energy U also gets reduced to $U-(\lambda^2/c)$. Denoting it by U_{eff} to represent the effective interaction potential energy, we can write

$$U_{eff} = U - \frac{\lambda^2}{c}. \tag{6.40}$$

3 The transfer energy term, T'_{lm}, also changes from its value T_{lm} in the rigid lattice, because of the phonon overlap functions used in evaluating the transfer energy matrix element. Typically T'_{lm}/T_{lm} is of the order of $\exp(-\lambda^2/2c\omega_0)$.

The lowering of energy, as mentioned in (1), has already been found in Holstein's model. It is also seen from the theory of polarons in Section 6.3 that the kinetic energy of a charge carrier decreases due to its interaction with phonons, which is in effect similar to the reduction in the transfer energy, mentioned in (3). The new result in Anderson's model is the reduction in the repulsive Coulomb potential energy between a pair of electrons due to their interaction with phonons. This means that for sufficiently strong coupling (large λ) between charge carriers and phonons, $U_{eff} = U - (\lambda^2/c)$ can become negative. Hence, this is usually referred to as Anderson's negative-U and then a pair of like charge carriers can be localized on the same bond without being repelled by each other. Accordingly, all covalent bonded materials, involving two electrons on the same bond, may be expected to have strong charge carrier–phonon interaction.

In a-solids, therefore, the existence of charged dangling bonds, D^+ and D^-, is also attributed to the strong charge carrier–phonon interaction, because D^+ carries two holes and D^- two electrons localized on an uncoordinated bond. In a-Si : H, however, D^+ and D^- are not believed to exist. It is interpreted that the coupling between charge carriers and phonons is not very strong, and the effective potential remains positive, so called positive-U in a-Si : H. Although not every detail about the formation of charged dangling bonds is well understood in a-solids, it is anticipated that an uncoordinated bond means a localized softness or weakness in the atomic network providing sufficient condition for strong charge carrier–phonon interaction. It would be difficult otherwise to explain the existence of charged dangling bonds in any a-solid due to the repulsive Coulomb interaction.

Anderson's model explains the pairing of like charge carriers on covalent bonds in semiconductors, regardless of whether these carriers are in ground or excited states. However, the model has not been applied to electron hole pairs excited by photons. In this case, depending on the excitation density, there are three types of Coulomb interactions; between excited electrons (repulsive), between excited holes (repulsive) and between an excited electron and excited hole (attractive). According to the theory developed above, the negative-U can occur in the first two cases of interaction between like charge carriers, but it is not obvious what would happen between unlike charge carriers with attractive Coulomb potential between them. It may therefore be desirable to extend the above approach to unlike charge carriers; this is dealt with in the next subsection.

6.5.2 *Unlike charge carriers (e–h)*

The Hamiltonian for an excited electron in the conduction states and hole in the valence states can be written as

$$\hat{H}_{\text{e-h}} = \sum_{l\sigma} \left(E_l^{\text{e}} n_{l\sigma}^{\text{e}} - E_l^{\text{h}} n_{l\sigma}^{\text{h}}\right) - U \sum_{l} n_{l\uparrow}^{\text{e}} n_{l\uparrow}^{\text{h}} + \sum_{lm\sigma} \left[T_{lm}^{\text{e}} a_{l\sigma}^{+} a_{m\sigma} - T_{lm}^{\text{h}} d_{l\sigma}^{+} d_{m\sigma}\right], \tag{6.41}$$

where $n_{l\sigma}^{\text{h}} = d_{l\sigma}^{+} d_{l\sigma}$, E_l^{e} and E_l^{h} are energies of electron and hole localized on site l, respectively, and T_{lm}^{e} and T_{lm}^{h} are the corresponding transfer energies from site l to m. It should be obvious that in Eq. (6.41) both electron and hole number operators with the Coulomb potential U are considered to have the same spin. This is because we want to get the paired form number operators as in Anderson's Hamiltonian in Eq. (6.34), when hole operators are converted into electron operators.

The potential energy term for electron–phonon and hole–phonon interaction can also be written using Eq. (6.33) as

$$V_{\text{e-h}} = \sum_{l} \left[\frac{1}{2} c x_l^2 - x_l \left(\lambda_{\text{e}} a_{l\uparrow}^{+} a_{l\uparrow} - \lambda_{\text{h}} d_{l\uparrow}^{+} d_{l\uparrow}\right)\right], \tag{6.42}$$

where λ_{e} and λ_{h} are electron–phonon and hole–phonon coupling constants. Now minimizing this with respect to x_{p}, gives $x_{\text{pe-h}}^{(0)}$ at the minimum value as

$$x_{\text{pe-h}}^{(0)} = \frac{1}{c}\left(\lambda_{\text{e}} n_{\text{p}\uparrow}^{\text{e}} - \lambda_{\text{h}} n_{\text{p}\uparrow}^{\text{h}}\right). \tag{6.43}$$

Substituting Eq. (6.43) in (6.42), we get the minimum potential energy as

$$V^{(0)}_{\text{e-h}} = -\frac{1}{2c}\sum_{l}\left(\lambda_{\text{e}} a^{+}_{l\uparrow} a_{l\uparrow} - \lambda_{\text{h}} d^{+}_{l\uparrow} d_{l\uparrow}\right)^{2}. \tag{6.44}$$

Adding this to $\hat{H}_{\text{e-h}}$ in Eq. (6.41), we get the total Hamiltonian in which the effective potential of interaction between a pair of electron and hole, $U^{\text{e-h}}_{\text{eff}}$, appears as

$$U^{\text{e-h}}_{\text{eff}} = -U + \frac{\lambda^{\text{e}}\lambda^{\text{h}}}{c}. \tag{6.45}$$

This suggests that for large values of λ^{e} and λ^{h}, the effective potential of interaction between an electron and a hole can become positive and then they will repel each other. However, we have to be very careful here. It is a misleading conclusion, because we have tried to minimize the Hamiltonian, which is the energy operator, and not an energy eigenvalue. The energy contribution from the minimum potential energy operator in Eq. (6.44) suggests that this contribution will be zero for $\lambda^{\text{e}} = \lambda^{\text{h}}$, because the electron–phonon and hole–phonon coupling constants have opposite signs. This is a very well known result in the theory of exciton–phonon interaction (Rashba, 1982; Singh, 1994). A non-zero contribution will only be found for $\lambda^{\text{e}} \neq \lambda^{\text{h}}$, but then its magnitude will be less than either the electron–phonon or hole–phonon interaction, whichever is bigger in magnitude. Only in extreme situations it may be found that $U^{\text{e-h}}_{\text{eff}} > 0$. Accordingly, it is demonstrated here quite clearly that it is more realistic to reach the limit of having the effective potential of interaction to be attractive between like charge carriers than the limit of having it to be repulsive between unlike charge carriers due to strong carrier–phonon interactions.

Therefore, it is quite conclusive that a solid with strong electron–phonon and strong hole–phonon interaction does not necessarily have strong exciton–phonon interaction. In many solids usually one of the interactions, electron–phonon or hole–phonon, is stronger than the other, resulting in a strong exciton–phonon interaction. In a-Si : H, for instance, it is known (Morigaki *et al.*, 1995) that the hole self-trapping takes place much more easily than electron self-trapping, implying stronger hole–phonon interaction. It may further be extended to conclude that solids with strong exciton–phonon interaction may be expected to exhibit pairing of one type of like charge carriers, electron or hole, more than the other depending on which is stronger, the electron–phonon or hole–phonon interaction.

In the following section, Holstein's diatomic model is extended to a-semiconductors to study the dynamics of excited charge carriers by photons. Then following Anderson's approach the pairing of like excited charge carriers will also be studied.

6.6 Polarons in a-semiconductors

To apply Holstein's approach to a-solid, first we need to consider the differences between a linear crystalline chain of diatomic molecules and random atomic network of a-solids. On a microscopic scale it is the bonding between nearest atoms that dominates the structure of a solid, regardless of whether it is crystalline or amorphous. At a time in any direction an atom will be bonded with another atom and thus the diatomic model can easily be extended to amorphous network. Although finer details may differ depending on the material, it is expected that the results obtained from a linear atomic chain can provide most information useful for the optical and electronic properties of a-solids. The main difference would be in the vibrating frequency, ω, which is the intramolecular frequency of vibration of diatomic molecules and hence it is the same for every identical molecule as considered by Holstein. In a-semiconductors, as there is no translational symmetry, this may be considered as the frequency of the localized mode of vibration along the bond between any two nearest atoms, but it cannot be assumed to be the same for every bond. With this difference in mind, it is assumed here that Holstein's approach can be applied to a-solids.

We consider a linear chain of atoms with an excess electron localized on a bond. Contrary to a crystalline chain, each bond length in the chain is not the same at equilibrium. In this simplified atomic network of linear system, the theory developed in Section 6.4 can easily be applied and the secular equation (Eq. (6.25)) can be written as

$$W_i C_l = \left(-\sum_n \frac{\hbar^2 \nabla_n^2}{2M} + \frac{1}{2} \sum_{m,n} M\, \omega_m\, \omega_n\, x_m\, x_n - A_l x_l + E_l \right) C_l$$
$$- (J_+ C_{l+1} + J_- C_{l-1}), \qquad (6.46)$$

which is obtained from Eq. (6.25) replacing the site independent constant energy E_0 by site dependent energy E_l, and constant J by J_+ for the transfer of energy between sites l and $l+1$ and J_- between sites l and $l-1$. However, in a fully coordinated network, such differences are going to cause little change in the properties of polarons from those obtained in Section 6.4. Thus it may be regarded as a very good approximation to describe polarons created in the extended states by the same theory as presented in Section 6.4. For the tail states of course, it may not be considered as suitable. However, as the theory is developed in the real coordinate space, the relevant and desired information may be obtained for the tail states as well. Assuming then that E_l is site independent, and $J_+ = J_- = J$ for nearest neighbors we will get the same results for polarons in a-semiconductors as obtained in Section 6.4. However, it is important to discuss the expression given in Eq. (6.30) of the bond length at minimum energy of the bond at which electron is localized and that of the polaron's binding energy in Eq. (6.31) for a-semiconductors.

The bond length (Eq. (6.30)) depends linearly on the force constant A_p and inverse square of the frequency of vibration. In a solid with strong charge carrier–lattice interaction, A_p is expected to be larger and hence the magnitude of the bond length is expected to increase more in such solids, particularly for the low frequency processes occurring at low temperatures. At higher temperatures, as the vibrational frequency increases the effect may be reduced. Thus, the deformation around a charge carrier is expected to be enhanced at low temperatures. The argument may be extended further to apply it to the microscopic structure of a-solids. It is known that amorphous structures do not have 100% fully coordinated atoms and there are also many weak bonds. The locations of such defects are expected to provide microscopic regions of softened structure and hence strong charge carrier–phonon interactions. That means, if a charge carrier gets localized on a bond in these regions, the bond length may be enhanced.

The binding energy of a polaron obtained in Eq. (6.31) is very sensitive to the force constant and frequency of vibration, and hence a polaron localized on a weak bond can be expected to have a much larger binding energy, particularly in the low frequency (low temperature) regions. This agrees very well with the common understanding of transport of charge carriers in a-solids. For example, the hydrogenation in a-Si : H is expected to soften the atomic structure network, which gets further softened in the microscopic regions where weak bonds and dangling bonds exist. The bonding and antibonding states of weak bonds as well as dangling bond states contribute to the existence of tail states. Thus, a polaron localized on a weak bond means it is energetically in the tail states, and having a large binding energy means it will not be able to move easily from that state and hence will remain localized. Such a picture of transport of charge carriers has been applied by Emin (1973) as discussed in the next section.

6.7 Small polarons in a-semiconductors

Based on the fact that the observed Hall mobility in non-crystalline materials is found to be very low ($< 0.1\,\mathrm{cm^2\,V^{-1}\,s^{-1}}$), one of the reasons for the slow motion of the charge carriers is attributed to the formation of small polarons (Emin, 1973). As described in Section 6.3, in the strong electron–lattice interaction limit, the deformation induced around an electron tends to localize the electron on a site. Such an electron accompanied by the lattice distortion is called a small polaron or self-trapped electron. If the deformation around the electron is not very severe or weak, then the electron gets localized in a region covering several lattice sites. This is called a large polaron which can be formed in a relatively rigid lattice.

As described earlier, the deformation of lattice around an electron is a measure of the electron–lattice interaction. This is used to classify small and large polarons; a large polaron is formed when the electron–lattice interaction is weak also called the weak coupling limit and a small polaron is formed when the electron–lattice interaction is very strong or the structure is soft. As derived in Section 6.3, the energy of an electron gets lowered due to its interaction with the lattice. Therefore the formation of a polaron state becomes energetically more favorable.

Emin (1973) has demonstrated that the small polaron energy states are the only dynamically stable states possible in non-c-solids. Although the large polaron states are possible in the weak electron–lattice coupling limit, they are not dynamically stable. The weakly coupled electron–lattice situation represents an electron carrying a small lattice distortion with itself as it moves through the lattice.

The radius of such distortion, R_p, is given by

$$R_p \approx [(6J/\hbar\omega_0)(2N+1)]^{1/2}a, \tag{6.47}$$

where J is the nearest neighbor transfer integral as used in Section 6.4 so that $6J$ gives the half bandwidth in a rigid lattice, ω_0 the vibrating frequency of diatomic molecules of which the molecular crystal is modelled to be consisted of, a the lattice constant and N the number of phonons given by the Planck distribution

$$N = [\exp(\hbar\omega_0/kT) - 1]^{-1}. \tag{6.48}$$

Here k is the Boltzmann constant and T temperature. $R_p \gg a$ in Eq. (6.47) because $6J \gg \hbar\omega_0$.

The energy eigenvalue of a large polaron in the weak coupling limit is obtained at $\mathbf{k} = 0$ as

$$E_{\mathbf{k}=0} = -(6J + E_b/R_p^2) + n\hbar\omega_0\left[N + \frac{1}{2}\right], \tag{6.49}$$

where n is the number of molecules in the lattice, E_b the polaron binding energy and $R_p^2 \gg 1$. The energy eigenvalue of a small polaron in the strong coupling limit is obtained as

$$E_{\mathbf{k}=0} = -(6J\mathrm{e}^{-S_0} + E_b) + n\hbar\omega_0\left[N + \frac{1}{2}\right], \tag{6.50}$$

where $S_0 = (E_b/\hbar\omega_0)(2N+1) \geq 1$.

It is obvious from Eq. (6.49), obtained in the weak coupling limit, that the main contribution to the carrier-related energy is the half bandwidth energy, $-6J$, of the rigid lattice. In the strong coupling limit of small polarons (Eq. (6.50)), however, $-6J$ is reduced by the factor e^{-S_0}. Then the major contribution to the electronic energy comes from the polaron binding energy term $E_b = A^2/M\omega_0^2$ obtained from the electron–lattice interaction, where A represents the strength of electron–lattice coupling (see Section 6.4). Thus, if a lattice is sufficiently soft, A will be sufficiently large and $M\omega_0^2$ small, it will produce a large binding energy E_b. In this situation, an electron placed in the lattice will find itself immobile or self-trapped.

In the weak coupling limit, the energy state of a large polaron can become stable only if (Emin, 1973):

$$\frac{\hbar}{6J} < \frac{\left[\hbar\omega_0\left(N+\frac{1}{2}\right)/M\right]^{1/2}}{A/M}. \tag{6.51}$$

According to Eq. (6.51), a large polaron state can be stable only if an electron in a band of half width, $6J$, can be confined on a site for a time of $\hbar/6J$, which is less than the time taken for the surrounding atoms to adjust to its presence. That means the electron will move completely free as in a rigid lattice. Therefore, there are only two possible dynamically stable states, one of the small polaron within the strong electron–lattice coupling and other of the free electron within the weak coupling limit.

This must be noted that the above conclusion is drawn on the basis of a molecular crystal model where the only element that can cause any deviation from the translational symmetry is considered to be the intramolecular vibration of diatomic molecules. In that case the above results appear to be correct. However, the crystalline model is very far from a real sample of a-solids. For example, a soft lattice structure is necessary for satisfying the condition of strong coupling, which does not exist in pure amorphous silicon (a-Si). As the hydrogenation of a-Si softens the a-Si : H network, it may be applicable to a-Si : H but only at a high hydrogen concentration. However, it is well known that the hydrogen concentration more than 10–15% produces poor quality a-Si : H.

One possible way of applying the small polaron theory to a-Si : H would be to consider the hydrogenation being non-uniform. In this situation, some regions will have more contents of hydrogen than others, even in a good quality sample. The theory can then be applied in the regions of higher hydrogen concentration where the strong electron–lattice coupling limit may be applicable. In that case however, a comparison of the results with the experimental ones appears to be problematic, because all measured quantities represent average values of the whole sample not of only specific regions of higher hydrogen concentration. Further, as stated above, the strong coupling limit is also required to observe the effect of negative-U in a solid. However, it is an established concept that a-Si : H does not qualify to have negative-U. These points need to be kept in mind when considering the application of small polaron theory to a-Si : H.

The small polaron theory appears to be more applicable in chalcogenides where the effect of negative-U is established and that means the strong electron–lattice coupling limit exists. This is the basis of the pairing hole model applied to chalcogenides to study the mechanism of photo-induced bond breaking (see Section 6.9 and Chapter 10).

Using the molecular crystal model, where the carrier–lattice interaction is represented by the interaction with a single optical mode, Emin (1973) has calculated the small polaron jump rate through the nearest neighbor interaction in c-solid and then extended it to a disordered system as well. The Hall mobility of small polaron and phonon assisted hopping (Emin, 1975; Gorham-Bergeron and Emin, 1977), involving both optical and acoustic phonons, have also been studied. The topic of small polarons may require a whole volume of its own and therefore interested readers are referred to the original articles.

6.8 Exciton–phonon interaction and Stokes shift

6.8.1 Crystalline solids

In Section 6.3, we have given the form of electron–phonon interaction in c-solids in Eq. (6.7), which is linear in the lattice displacement vector. The similar dependence of the electron–phonon interaction operator on the lattice displacement vector is also considered by Holstein, as described in Section 6.4. The exciton–phonon interaction involves both electron–phonon and hole–phonon interactions, which are added together to get the exciton–phonon interaction. However, as shown in Section 6.5 for unlike charge carriers, electron–phonon and hole–phonon interactions have opposite signs and therefore the exciton–phonon interaction is obtained as electron–phonon minus hole–phonon interaction. In crystalline materials the exciton–phonon interaction Hamiltonian $\hat{H}_{\mathrm{I}}^{\mathrm{ex}}$ is written as (Singh, 1994):

$$\hat{H}_{\mathrm{I}}^{\mathrm{ex}} = \sum_{\mathbf{q}} V_q^{\mathrm{ex}} \mathrm{e}^{\mathrm{i}\mathbf{q}\cdot\mathbf{r}_{\mathrm{ex}}} \left(b_{-\mathbf{q}}^{+} + b_{\mathbf{q}}\right), \tag{6.52}$$

where

$$V_q^{\mathrm{ex}} = V_q^{\mathrm{e}} - V_q^{\mathrm{h}}. \tag{6.53}$$

Here V_q^{e} and V_q^{h} are the coupling constants for electron–phonon and hole–phonon interactions, respectively, and $\mathbf{r}_{\mathrm{ex}}$ is the position coordinate of the excitonic center of mass. It is obvious that for $V_q^{\mathrm{e}} = V_q^{\mathrm{h}}$, the exciton–phonon interaction vanishes, regardless of whatever the strengths of electron–phonon and hole–phonon interactions are. This point was briefly addressed in Section 6.5 as well. In polar crystals, the coupling coefficient of hole–phonon interaction is obtained from Eq. (6.8) replacing the electron effective mass by hole effective mass. Thus, materials with equal effective masses of electron and hole will have zero exciton–phonon interaction. Likewise, in non-polar crystals where the deformation potential is used to calculate the carrier–phonon interaction, the hole–phonon interaction is obtained from Eq. (6.10) replacing the conduction band deformation potential parameter by the valence band deformation potential parameter. Materials which have the same conduction and valence band parameters, the exciton–phonon interaction will become zero.

The effect of exciton–phonon interaction on the transport of excitons, in materials with non-zero exciton–phonon interaction, can be expected to be the same as that of charge carrier–phonon interaction on the transport of charge carriers. This is because the form of the exciton–phonon interaction operator in Eq. (6.52) is the same as that of the electron–phonon interaction in Eq. (6.7). If the strength of coupling is weak, an exciton will behave like excitonic polaron. That means, as described in Section 6.3, the effective mass of exciton increases, kinetic energy decreases and the total energy decreases as well. So an excitonic polaron moves slower than a so-called free exciton (without interacting with phonons). In the strong coupling situation, an exciton can get self-trapped, which means that the lattice gets deformed around it so much that it cannot move out of it, and the energy of a self-trapped state is lower than that of a free exciton state as shown in Fig. 6.2.

Excitons are created in crystalline semiconductors (c-semiconductors) by exciting them with photons of energy equal to or greater than the band gap energy. If a sample is excited with energy bigger than the free exciton state, a free exciton may

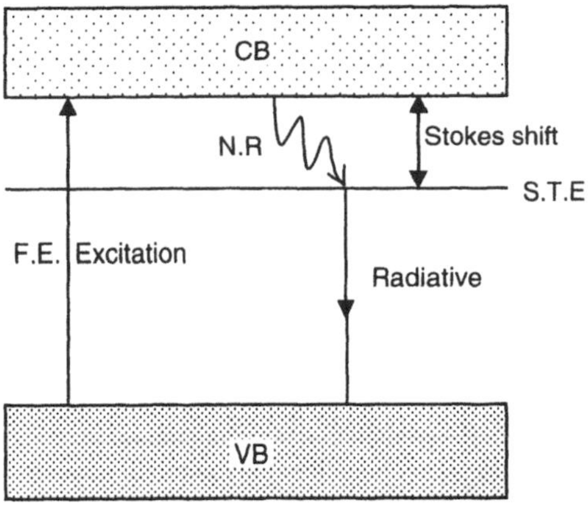

Figure 6.3 Illustration of Stokes shift due to self-trapping of excitons. F.E. represents a free exciton state that is excited. Excitons then non-radiatively move to the self-trapped exciton (S.T.E) state. Radiative transitions down to ground state (VB) then occurs giving rise to exciton photoluminescence from S.T.E. Energy difference between excitation and photoluminescence energies gives the Stokes shift.

relax non-radiatively to its self-trapped state, and then radiatively to the ground state from the self-trapped state. In this case one observes a red shift in the exciton photoluminescence spectra, which is commonly called Stokes shift. On an energy band diagram the mechanism of Stokes shift is illustrated in Fig. 6.3. In materials with strong exciton–phonon interaction large Stokes shifts are observed, for example, in some molecular crystals (Mizuno and Matsui, 1982) and rare gas solids (Fugol, 1978; Zimmerer, 1987).

6.8.2 Amorphous solids

In a-solids, the theory of exciton self-trapping can be developed following Holstein's approach. The exciton–phonon interaction Hamiltonian, $\hat{H}_{\mathrm{IA}}^{\mathrm{ex}}$, for an exciton excited in an a-semiconductor such that the hole is excited in the bonding and electron in the antibonding states formed by any two neighboring atoms in a linear chain, can be written as

$$H_{\mathrm{IA}}^{\mathrm{ex}} = \sum_{n} x_n \left(A_n^{\mathrm{e}} a_n^{+} a_n - A_n^{\mathrm{h}} d_n^{+} d_n \right), \tag{6.54}$$

where A_n^{e} and A_n^{h} are electron–phonon and hole–phonon coupling constants in the real coordinate space along the direction of the chain. This operator is easy to obtain from Eq. (6.23) by considering two energy levels, one for valence states and the other for conduction states, and then converting the operators in the valence state into hole operators. It may be reminded here that Eq. (6.23) involves only a single energy state. There are a few ways in which the theory could be advanced from here onward. One of the easiest ways is to write Eq. (6.54) in terms of exciton operators, replacing electron and hole operators by exciton operators (Singh, 1994). Thus the interaction Hamiltonian reduces to

$$\hat{H}_{\rm IA}^{\rm ex} = \sum_{n} x_n \beta_n B_n^+ B_n, \tag{6.55}$$

where $\beta_n = (A_n^{\rm e} - A_n^{\rm h})$ is the exciton–phonon coupling coefficient along the chain, and B_n^+ is the exciton creation operator defined by

$$B_n^+ = a_n^+ d_n^+. \tag{6.56}$$

The exciton annihilation operator B_n is the complex conjugate of B_n^+. (Spin index is omitted here.) Equation (6.55) is obtained by replacing $a_n^+ a_n$ and $d_n^+ d_n$ by $B_n^+ B_n$, which causes no errors in any calculations.

Likewise writing the electron Hamiltonian in Eq. (6.21) as a sum over two energy states and then going through the above steps, one can write the exciton Hamiltonian as

$$\hat{H}^{\rm ex} = \sum_{l,m} E_{lm}^{\rm ex} B_l^+ B_m, \tag{6.57}$$

where $E_{lm}^{\rm ex} = E_{lm}^{\rm e} - E_{lm}^{\rm h}$ is the energy of exciton. Using now the phonon Hamiltonian given in Eq. (6.22), exciton Hamiltonian in Eq. (6.57) and the exciton–phonon interaction Hamiltonian in Eq. (6.55), we can write the total Hamiltonian as

$$\hat{H}_{\rm tot}^{\rm ex} = \hat{H}^{\rm ex} + \hat{H}_{\rm ph} + \hat{H}_{\rm IA}^{\rm ex}. \tag{6.58}$$

The exciton eigenvector can be written as

$$|{\rm ex}\rangle = \sum_{l} C_l^{\rm ex} B_l^+ |0\rangle, \tag{6.59}$$

which produces a similar secular equation as Eq. (6.26) and hence similar results for the bond length (Eq. (6.30)) of the bond on which exciton is localized and for the lowering of energy (Eq. (6.32)). It may however be remembered here that the exciton operator should be converted into fermion operators according to Eq. (6.56) for deriving the secular equation. Thus we can write the bond length and lowering of energy, respectively, as

$$x_{\rm p}^{\rm ex(0)} = \frac{\beta_{\rm p} |C_{\rm p}^{\rm ex(0)}|^2}{M\omega_{\rm p}^2}, \tag{6.60}$$

and

$$E_{\rm exp} = \frac{\left(\beta_{\rm p}^2 / M\omega_{\rm p}^2\right)^2}{48T}, \tag{6.61}$$

where T represents the energy of transfer of exciton to a nearest neighbor and it is assumed to be constant to get the analytical expression.

Equations (6.60) and (6.61) clearly demonstrate that lattice deformation and lowering of energy similar to those occurring around a polaron also occur around an exciton or excitonic polaron. However, as $\beta_l = A_l^{\mathrm{e}} - A_l^{\mathrm{h}}$, it means that $|\beta_l| < |A_l^{\mathrm{e}}|$ or $|A_l^{\mathrm{h}}|$. These effects may not be as prominent with an exciton as with a polaron, and for $A_l^{\mathrm{e}} = A_l^{\mathrm{h}}$, one may not observe any change at all. Nevertheless, excitonic polaron and exciton self-trapping give rise to Stokes-shift, which is observed in many a-solids such as a-Si : H and chalcogenides. This is because, as pointed out earlier, usually the condition of $A_l^{\mathrm{e}} = A_l^{\mathrm{h}}$ does not get satisfied in most materials, and commonly one finds that $|A_l^{\mathrm{e}}| \ll |A_l^{\mathrm{h}}|$. That means hole self-trapping becomes easier than electron self-trapping, which has been observed in a-Si : H through ODMR studies (Morigaki *et al.*, 1995).

In the next section, we will consider the case of excited charge carriers in which two holes may get paired on the same bond in a-solids. This situation has applications in the creation of light-induced defects in a-semiconductors and chalcogenides.

6.9 Pairing of photo-excited holes

In Section 6.5 we discussed the possibility of creating an attractive potential of interaction between like charge carriers in the ground state under the influence of strong charge carrier–phonon interactions. Here we are going to consider the case of an excited pair of holes in a-solids. When photovoltaic devices are exposed to photons of energy equal to or more than the gap energy, depending on the intensity of radiation, many pairs of electrons and holes are excited. Considering that the hole–phonon coupling is very strong, we want to study the possibility of getting two excited holes localized on a bond.

Like in the previous section, here again we consider a linear chain of an a-solid and assume that two pairs of electrons and holes are excited at some time. The eigenvector of such an excited state can be written as (Singh and Itoh, 1990):

$$|x, x', L\rangle = \sum_l C_l |l, x, x', L\rangle_{\uparrow\downarrow}, \tag{6.62}$$

where C_l is the probability amplitude coefficient, x the separation between one excited pair of electron and hole and x' that between the other excited pair of electron and hole. L is the distance between two excited holes on the chain, and l denotes the center of mass of a bond between neighboring atoms on the chain, similar to the Holstein's model considered in Section 6.4. The only difference is that here we have got two excited electrons and two excited holes. The two arrows as subscripts denote that the like charge carriers are assumed to have opposite spins as considered by Anderson and described in Section 6.5. The ket vector

$|l, x, x', L\rangle_{\uparrow\downarrow}$ is given by

$$|l, x, x', L\rangle_{\uparrow\downarrow} = \frac{1}{2}\sum_{\sigma} a^{+}_{l+x}(\sigma)d^{+}_{l}(-\sigma)a^{+}_{l-L+x'}(-\sigma)d^{+}_{l-L}(\sigma)|0\rangle, \tag{6.63}$$

where σ sums over the two possible spins. As there are two excitations, the secular equation in this case is obtained as (Singh, 1994)

$$\begin{aligned} W(x, x', L)C_l = \Bigg[&(E_{\mathrm{e}}^{l+x} - E_{\mathrm{h}}^{l}) + (E_{\mathrm{e}}^{l-L+x'} - E_{\mathrm{h}}^{l-L}) - (A^{\mathrm{e}}_{l+x}x_{l+x} - A^{\mathrm{h}}_{l}x_l) \\ &- (A^{\mathrm{e}}_{l-L+x'}x_{l-L+x'} - A^{\mathrm{h}}_{l-L}x_{l-L}) + \frac{1}{2}\sum_{m} M\omega_m^2 x_m^2 + U'_{12}\Bigg]C_l \\ &+ J_{\mathrm{e}}(C_{l+x+1} + C_{l+x-1} + C_{l-L+x'+1} + C_{l-L+x'-1}) \\ &- J_{\mathrm{h}}(C_{l+1} + C_{l-1} + C_{l-L+1} + C_{l-L-1}), \end{aligned} \tag{6.64}$$

where E_{e}^{l} and E_{h}^{m} represent the energy of an electron at site l and a hole at m, respectively, U'_{12} is the inter-excitation Coulomb interaction, J_{e} and J_{h} are the energy of transfer of electron and hole, respectively, between nearest neighbors. It is to be noted that E_{e}^{l} and E_{h}^{m} include the total interaction energy due to other unexcited charge carriers present in the solid and therefore these are different from E_l used in Eq. (6.34). It is difficult to solve Eq. (6.64) without some further approximations.

As described above, usually in most solids the coupling constant for one type of charge carrier–phonon interaction is stronger than the other. Accordingly, here we assume that hole–phonon interaction is dominant, as it appears to be the case in most a-semiconductors and chalcogenides. It can be then expected that after the excitation electrons will be moving much faster than holes and hence are far away from each other and also from the two holes. With this assumption, we can neglect the effect of electron–electron interaction from U'_{12}, and then the energy of two electrons can be replaced by the corresponding polaron energy. These two approximations enable us to consider dynamics of only the two holes in the secular Eq. (6.64), which can now be written as

$$\begin{aligned} W(x, x', L)C_l = \Bigg[&2E_{\mathrm{ep}} - E_{\mathrm{h}}^{l} - E_{\mathrm{h}}^{l-L} + A^{\mathrm{h}}_{l}x_l + A^{\mathrm{h}}_{l-L}x_{l-L} \\ &+ \frac{1}{2}\sum_{m} M\omega_m^2 x_m^2 + U_{12}\Bigg]C_l \\ &- J_{\mathrm{h}}(C_{l+1} + C_{l-1} + C_{l-L+1} + C_{l-L-1}), \end{aligned} \tag{6.65}$$

where $E_{\mathrm{ep}} = W_i^{(0)}$ is the energy of an electronic polaron as obtained in Eq. (6.31), U_{12} represents only part of U'_{12} which excludes electron–electron and electron–hole interactions. Now we want to consider the case where both holes are localized on the same bond at l, then Eq. (6.65) can be written as

$$W(x, x', 0)C_l = \left[2E_{\text{ep}} - 2E_{\text{h}}^l + 2A_l^{\text{h}} x_l + \frac{1}{2}\sum_m M\omega_m^2 x_m^2 + U_{12}\right] C_l$$
$$+ 2J_{\text{h}}(C_{l+1} + C_{l-1}). \qquad (6.66)$$

We minimize $W(x, x', 0)$ with respect to x_l such that $\partial W/\partial x_l = 0$, which gives the bond length, at the minimum energy, of the bond at which both the holes are localized as

$$x_{\text{p}}^{\text{hh}} = -\frac{2A_{\text{p}}^{\text{h}} C_{\text{p}}^* C_{\text{p}}}{M\omega_{\text{p}}^2}. \qquad (6.67)$$

The corresponding minimum energy is then obtained as

$$W(x, x', 0) = 2E_{\text{ep}} - 2E_{\text{h}}^l + U_{12} - 2J_{\text{h}} - E_{\text{hh}}, \qquad (6.68)$$

where

$$E_{\text{hh}} = \frac{[(1/M)(A/\omega_{\text{p}})^2]^2}{3J_{\text{h}}}. \qquad (6.69)$$

The energy of two excitations without the charge carrier–phonon interaction can be written as

$$W_0(x, x', 0) = 2(E_{\text{e}} - E_{\text{h}}) - 2(J_{\text{e}} - J_{\text{h}}) + U_{12}. \qquad (6.70)$$

Thus the energy of a pair of excited holes localized on a bond is lower by ΔE given by

$$\Delta E = W_0(x, x', 0) - W(x, x', 0) = 2E_{\text{p}} + E_{\text{hh}}, \qquad (6.71)$$

where $2E_{\text{p}}$ is the polaron binding energy (Eq. (6.32)) associated with the two excited electrons. According to our assumption of weak electron–phonon interaction, it may be expected that $2E_{\text{p}} \ll E_{\text{hh}}$, and that means a dominant contribution to ΔE is expected to come from the hole–hole binding energy E_{hh}.

The above results have the following two features of interest:

1 An excited state with a pair of holes localized on the same bond of a solid is energetically more favorable.
2 At the same time the bond on which the two holes get localized gets enlarged to twice the size as compared to when a single hole is localized. This is easy to see by comparing Eq. (6.67) with (6.30).
3 Reduction in energy due to pairing of holes for a pair of excitations also suggests that the energy gap between conduction and valence states gets reduced by the strong carrier–phonon interaction. This is well known for c-solids and was first demonstrated by Singh (1988) in GaP.

It may be noted that the expression for lowering of energy derived in Eq. (6.61) is very sensitive to the strength of hole–phonon coupling. Depending on the material, the lowering of the energy can be in the range of a fraction of an eV to a few eV. This suggests that the excited holes will have a tendency to get paired on a bond due to strong hole–lattice interaction. The interesting point is that as soon as two excited holes are paired on the same bond, the bond will be broken due to the removal of covalent electrons. Thus the bond will be converted into two dangling bonds and an increase in such bonds can be observed in a sample exposed to radiation. This is the mechanism suggested to be applicable for the creation of light-induced defects in chalcogenides, where no hydrogen exists to be involved in the process. Also the existence of charged dangling bonds, D^+ and D^-, in a-solids can only be explained on the basis of strong carrier–phonon interaction. The theory can be extended further and some estimates for ΔE can be obtained as follows.

It is assumed that a weak bond satisfies the criterion of having strong hole–phonon interaction along the bond. Expressing the vibrational energy associated with a weak bond in terms of the interaction coordinate as

$$E(q) = E_0 + \tfrac{1}{2}(2M)\omega^2(q - q_0)^2, \tag{6.72}$$

where $2M$ represents the mass of two vibrating atoms on a weak bond, and E_0 and q_0 are the energy and the interaction coordinate at the minimum of the vibrational energy, respectively. The vibrational force along the interaction coordinate can be obtained as

$$A = \left(\frac{\partial E}{\partial q}\right)_{q=0} = -M\omega^2 q_0. \tag{6.73}$$

Using this in Eqs (6.32) and (6.69) we can write

$$\frac{E_\mathrm{p}}{J} = \frac{2}{3}\frac{\left[M\omega^2 q_0^2\right]^2}{J^2}, \tag{6.74}$$

and

$$\frac{E_\mathrm{hh}}{J} = \frac{16}{3}\frac{\left[M\omega^2 q_0^2\right]^2}{J^2}. \tag{6.75}$$

Using the phonon energy of 344 cm^{-1} for the symmetric mode of $AsS_{3/2}$ units (Tanaka *et al.*, 1985) and applying Toyozawa's criterion (Toyozawa, 1981; Singh and Itoh, 1990) of strong carrier–phonon interaction as $E_\mathrm{p} \geq J$, we get $J =$ 35 meV from Eq. (6.74) and then using Eqs (6.74) and (6.75) in Eq. (6.71), we get $\Delta E = 0.35$ eV for a-As_2S_3. Application of the above theory for creating light-induced defects in amorphous chalcogenides is presented in Chapter 10.

It may be concluded that under the condition of strong hole–phonon coupling limit in a solid, a pair of excited holes can be localized on a bond, which will break the bond. This is opposite of pairing of electrons on a bond in the ground state, which helps strengthening covalent bonding in a-solids according to Anderson's negative-U. The pairing of like charge carriers on a bond is energetically favorable.

References

Anderson, P. (1975). *Phys. Rev. Lett.* **34**, 953.

Bardeen, J. and Shockley, W. (1950). *Phys. Rev.* **80**, 102.

Emin, D. (1973). In: Le Comber, P.G. and Mort, J. (eds), *Electronic and Structural Properties of Amorphous Semiconductors*. Academic Press, New York, p. 261.

Emin, D. (1975). *Adv. Phys.* **24**, 305.

Feynman, R.P. (1953). *Phys. Rev.* **97**, 660.

Fröhlich, H. (1952). *Proc. Roy. Soc. A* **215**, 219.

Fröhlich, H. (1954). *Adv. Phys.* **3**, 325.

Fröhlich, H. (1963). In: Kuper, C.G. and Whitfield, G.D. (eds), *Polarons and Excitons*. Oliver and Boyd, Edinburgh, p. 2.

Fugol, I.Y.A. (1978). *Adv. Phys.* **27**, 1.

Gorham-Bergeron, E. and Emin, D. (1977). *Phys. Rev. B* **15**, 3667.

Gurari, M. (1953). *Phil. Mag.* **44**, 329.

Holstein, T. (1959). *Ann. Phys.* **8**, 325, 343.

Lee, T.D and Pines, D. (1952). *Phys. Rev.* **90**, 960.

Lee, T.D., Low, F. and Pines, D. (1953). *Phys. Rev.* **90**, 293.

Low, F. and Pines, D. (1953). *Phys. Rev.* **91**, 193.

Mizuno, K. and Matsui, A. (1982). *J. Phys. Soc. Jpn.* **52**, 3260.

Morigaki, K., Hikita, H. and Kondo, M. (1995). *J. Non-Cryst. Solids* **190**, 38.

Mott, N.F. and Davis, E.A. (1979). *Electronic Processes in Non-Crystalline Materials*. Clarendon Press, Oxford.

Oh, I.-K. and Singh, J. (2000). *J. Lumin.* **85**, 233.

Pines, D. (1963). In: Kuper, C.G. and Whitfield, G.D. (eds), *Polarons and Excitons*. Oliver and Boyd, Edinburgh, p. 33.

Rashba, E.I. (1982). In: Rashba, E.I. and Sturge, M.D. (eds), *Excitons*, North-Holland, Amsterdam, p. 543.

Shimakawa, K., Kobolov, A. and Elliott, S.R. (1995). *Adv. Phys.* **44**, 475.

Singh, J. (1988). *Chem. Phys. Lett.* **149**, 447.

Singh, J. (1994). *Excitation Energy Transfer Processes in Condensed Matter*. Plenum, New York.

Singh, J. and Itoh, N. (1990). *Appl. Phys. A* **51**, 427.

Singh, J., Itoh, N. and Truong, V.V. (1989). *Appl. Phys. A* **49**, 631.

Singh, J., Itoh, N., Nakai, Y., Kanasaki, T. and Okano, A. (1994). *Phys. Rev. B* **50**, 11730.

Tanaka, Ke., Gohda, S. and Odajima, A. (1985). *Solid State Commun.* **56**, 899.

Taylor, P.A. (1970). *A Quantum Approach to Solid State*. Prentice, New Jersey.

Toyozawa, Y. (1959). *Prog. Theor. Phys.* **26**, 29.

Toyozawa, Y. (1962). In: Kuper, C.G. and Whitfield, G.D. (eds.), *Polarons and Excitons*. Oliver and Boyd, Edinburgh.

Toyozawa, Y. (1981). *J. Phys. Soc. Jpn.* **50**, 1861.

Zimmerer, G. (1987). Excited state spectroscopy in solids. In: Grassano, U.M. and Terzi, N. (eds), *Proceedings of the International School of Physics*. North-Holland, Amsterdam, p. 37.

7 Defects

Any departure from the perfect *crystalline lattice* of a solid is called a defect. A different definition, however, is required for disordered materials since there is no perfect lattice structure. Hence, we define a defect as a *departure* from the fully coordinated random network. We will discuss the presence of such defects in actual amorphous semiconductors (a-semiconductors), as many electronic and optical properties of such solids can be defect controlled as they can also be for crystalline solids (c-solids).

7.1 Types of defect in a-semiconductors

When an atom has a distinct bonding state different from its full coordination, it is called a coordination defect: In a-Si, for example, a coordination defect is three-fold coordinated, and in a-Ch such defects are due to the presence of atoms with one- and three-fold coordinations. All of these deviate from the normal coordination given by the 8-N rule, where N designates the number of valence electrons in the atom. These coordination defects can be characterized by an electronic correlation energy, either positive or negative, as will be discussed below.

7.1.1 Positive-U defects

In covalently bonded materials with a well-defined local geometry, one type of point defect concerns the atomic coordination, which occurs due to either under- or over-coordination, with respect to the normal coordination. Under-coordinated defects are simply regarded as dangling bonds. There is an important difference between coordination defects in a-Si (or a-Ge) and c-Si (or c-Ge). An isolated dangling bond can be formed only in the non-crystalline state but not in the crystalline state. An example of dangling bonds in an amorphous solid (a-solid) is shown schematically in Fig. 7.1. On the contrary in the diamond–cubic structure of c-Si, for example, four dangling bonds are formed simultaneously, if an atom is removed to create a vacancy.

A neutral dangling bond normally contains one electron. In effect, as it is neutral, it is regarded to contain an electron and a hole (towards the nucleus). However,

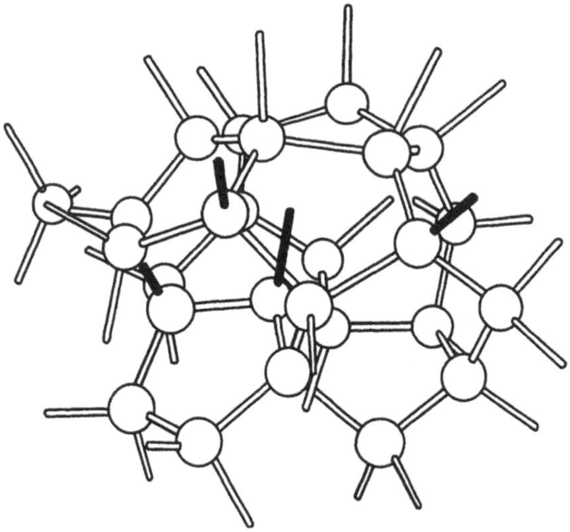

Figure 7.1 Illustration of a dangling bond in tetrahedrally-bonded amorphous semiconductors. Some of them are shown in black.

under certain conditions (e.g. chemical or electronic doping), the electronic occupancy can be changed to create negatively or positively charged dangling bonds. The energy states of neutral, positively and negatively charged dangling bonds, for example, in tetrahedrally bonded materials, are denoted by T_3^0, T_3^+ (two holes localized on a bond) and T_3^- (two electrons localized on a dangling bond), respectively, where the subscript denotes the coordination number and superscript charge state. The two electrons repel each other due to Coulomb interaction and hence the energy level of T_3^- state is raised by the correlation energy (positive correlation energy):

$$U_c = \frac{e^2}{4\pi\varepsilon_r\varepsilon_0 r}, \tag{7.1}$$

where r is the effective separation between the two electrons, which is the localization length of the defect wavefunction (Street, 1991). T_3^- for this case is called a positive-U defect. The energy level of the electronic state of an isolated dangling bond is discussed in terms of a molecular orbital theory described in Chapter 1. Consider the case of tetrahedral semiconductors like a-Si and a-Ge. A neutral dangling bond, that is, non-bonding orbital, containing a single electron will have an energy level lying at the zero energy for the sp^3 hybrids. If the atomic level of sp^3 is set at the zero energy as shown in Fig. 7.2(ii), the structural defects, such as dangling bonds, introduce electron states deep into the forbidden gap, called gap states. A gap state of T_3^0 (or T_3^+) dangling bond lies near the middle of the energy gap at E_d (Fig. 7.2(b)), and T_3^- gets located at $E_d + U_c$. The precise position of the

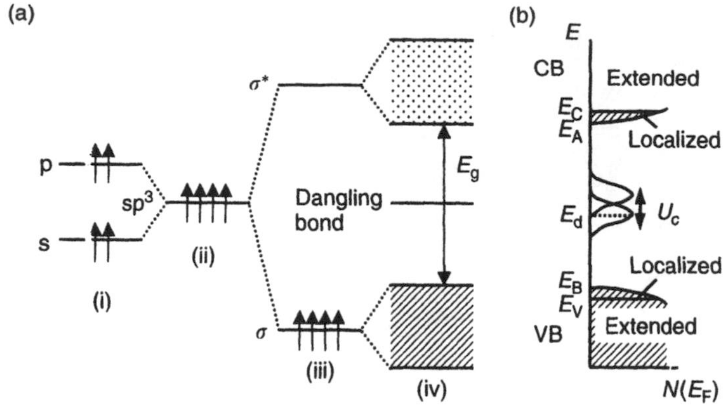

Figure 7.2 (a) Schematic illustration of the origin of valence and conduction band states for a tetrahedrally bonded semiconductor; (i) atomic s- and p-states, (ii) sp^3 hybrid states, (iii) bonding (σ) and antibonding (σ^*) states, and (iv) Condensation leads to broadening of bonding and antibonding states into valence band (VB) and conduction band (CB). (b) Density of states (DOS) for such a band scheme, showing the localized band-tail states. A broadened dangling-bond state appears near midgap. States of negatively charged dangling bonds are shown to be located above E_d by an energy U_c for a positive-U.

energy levels will depend on factors such as atomic (structural) relaxation around the defect. The value of U_c is hard to evaluate theoretically, because one needs to know, r, the extent of localized wavefunction which is difficult to determine theoretically, but it can be estimated experimentally (see Section 7.2).

Over-coordinated defects, on the other hand, are not common in a-Si and a-Ge. When an additional covalent bond is formed between an already fully coordinated atom and another atom by utilizing a non-bonding lone pair electron, over-coordinated defects are formed, for example, T_5^0 is shown in Fig. 7.3, together with T_3^0 configuration. The over-coordinated defects, usually called floating bonds, can also produce gap states and their role in photoinduced degradation in hydrogenated amorphous Silicon (a-Si : H) has been discussed by Pantelides (1986). However, there is no clear evidence of the presence of floating bonds in a-Si : H (Street, 1991).

7.1.2 *Negative-U defects*

The theory of negative-U is presented in Chapter 6. If atomic relaxations around defects are not neglected, both the electronic and local distortion energies should be taken into account in the ground state energy of defect configurations. The electron–phonon coupling causes the network to relax to a new equilibrium state

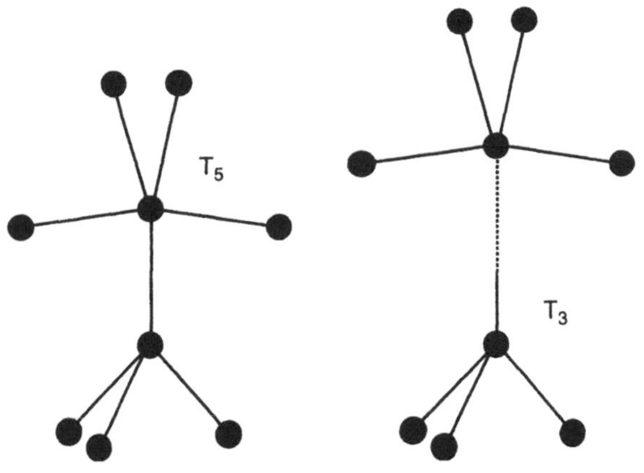

Figure 7.3 Schematic presentation of T_5^0 and T_3^0 configurations.

at a lower total energy. The addition of an electron into a dangling bond, that is, spin pairing of electrons at the defect centers, may cause a change in the bonding, which lowers the electronic energy by the amount U_r. As already discussed in Chapter 6, the effective correlation energy for two spin-paired electrons at the same site is, according to Eq. (6.40), given by

$$U_{\text{eff}} = U_c - U_r = U_c - \frac{\lambda^2}{c}, \tag{7.2}$$

where λ is the electron–phonon coupling constant and the force constant c is related to phonon frequency as $\omega = \sqrt{c/M}$, which is taken to be constant in the Einstein approximation. U_{eff} is negative if $U_c < U_r$, that is, for strong electron-lattice coupling. This type of spin-pairing defect is called the negative-U defect.

This idea proposed by Anderson (1975) was first applied to defects in a-Ch (Street and Mott, 1975) and later by Kastner *et al.* (1976). As such a defect of dangling bond is charged due to pairing of electrons or holes, it is often called a *charged-dangling bond* (CDB). The features of a CDB in a-Se can be presented by a model as shown in Fig. 7.4(a).

The structure of Se is two-fold coordinated and is believed to consist mostly of chains. The dangling bond will contain an unpaired electron if electron–phonon interaction is not strong and is written as C_1^0. However, electron pairing should occur at the C_1^0 center due to the low atomic coordination resulting in a high degree of network flexibility (strong electron–phonon interaction). There are also non-bonding lone-pair $p\pi$ orbitals at chalcogen atoms (Se or S) and hence the transfer of an electron from one C_1^0 center to another produces a C_1^- which is negatively charged (electron spin-paired) and a C_3^+ (hole spin-paired), which is a three-fold-coordinated positively-charged dangling bond. The reaction

$$2C_1^0 = C_1^- + C_3^+ \tag{7.3}$$

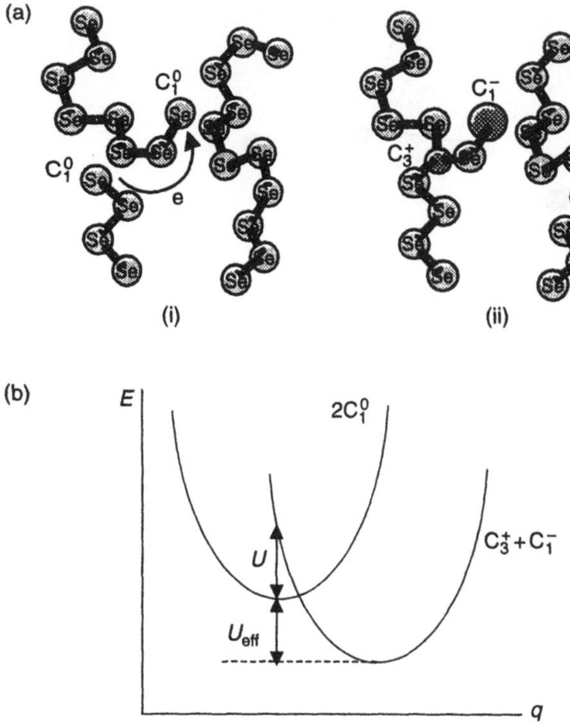

Figure 7.4 (a) Formation of charged defects (valence alternation pairs; VAP) in a-Se. (b) Configuration-coordinate diagram for the formation of a $D^+(C_3^+)$ – $D^-(C_1^-)$ pair. The overall energy is lowered by the effective correlation energy U_{eff}.

is thus exothermic with an effective negative correlation energy. Using a configuration coordinate diagram (CC diagram), this process is illustrated in Fig. 7.4(b). Thus the neutral C_1^0 is unstable and $2C_1^0$ transform into a pair of C_1^- and C_3^+, which is called a *valence-alternation pair* (VAP) (Kastner *et al.*, 1976). A molecular orbital model for different bonding states of Se is shown in Fig. 7.5.

A model for the negative-U dangling bond, similar to a-Ch, has also been proposed in a-Si : H (Adler, 1978; Elliott, 1978): The intimate pair of ${}_{sp}T_3^+ - {}_{p}T_3^-$ is a candidate for negative-U centers, where T_3^+ takes the sp-hybridized configuration and T_3^- the p-orbital. However, in a-Si : H the coordination number (3) does not change when T_3^0 changes into T_3^- or T_3^+. Another type of negative-U center has also been proposed in a specific form of Si-network involving singly-coordinated hydrogen atoms (Bar-Yam and Joannopoulus, 1986). The energy levels for ${}_{sp}T_3^+$ and ${}_{p}T_3^-$ will be located near the conduction and valence extended states.

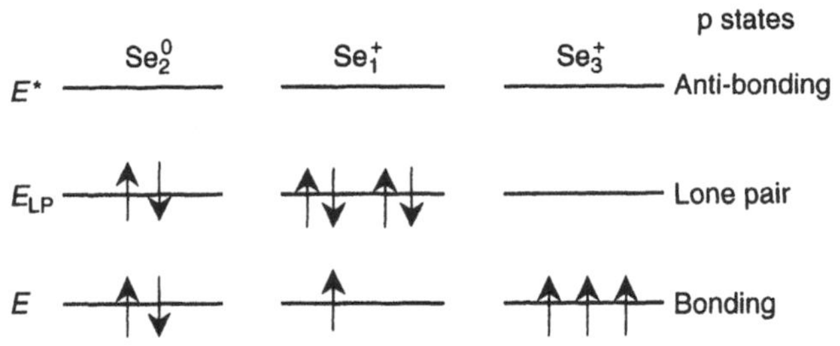

Figure 7.5 A molecular orbital model for the electronic structure of different bonding states of Se, that is, $Se_2^0(C_2^0)$, $Se_1^-(C_1^-)$ and $Se_3^+(C_3^+)$.

7.2 Defect spectroscopy

Defects produce localized states in the band gap of a-solids and hence can be detected by spectroscopic methods. We will discuss here how to estimate the number of defects and their energy levels in a-semiconductors. The nature of defects can be elucidated by optical, electrical and magnetic measurements. The current understanding of the results of defect spectroscopy in a-Si : H and a-Ch will be reviewed in this section.

7.2.1 a-Si : H

We will start with electron spin resonance (ESR), which is one of the major experimental techniques for understanding the microscopic nature of defects.

7.2.1.1 ESR

A quantum state (e.g. a defect state) occupied by a single electron is split by a magnetic field and is called the Zeeman splitting. ESR occurs due to transition between the split energy levels. The transition occurs at microwave frequencies for the conventional magnetic field [~ 0.3 T (3,000 Gauss)] between two Zeeman levels with unpaired electrons whose spin state S is 1/2 (paramagnetic state). This situation is shown in Fig. 7.6. The Hamiltonian describing the electron energy states is

$$H = \mu_B \mathbf{H} \ \ \mathbf{g} \cdot \mathrm{S} + \sum_i \mathbf{I}_i \cdot \mathbf{A}_i \cdot \mathrm{S}, \tag{7.4}$$

where μ_B is the Bohr magnetron, $\mathbf{H}$ external magnetic field, $\mathbf{g}$ gyromagnetic tensor (so-called the g-value or g-factor), and $\mathbf{S}$ electron spin. $\mathbf{I}_i$ and $\mathbf{A}_i$ in the second term are the nuclear spin and hyperfine tensors, respectively, and the sum is over all the contributing nuclei near the electron. The first term on the right hand side of Eq. (7.4) is called the Zeeman interaction term, and the second the

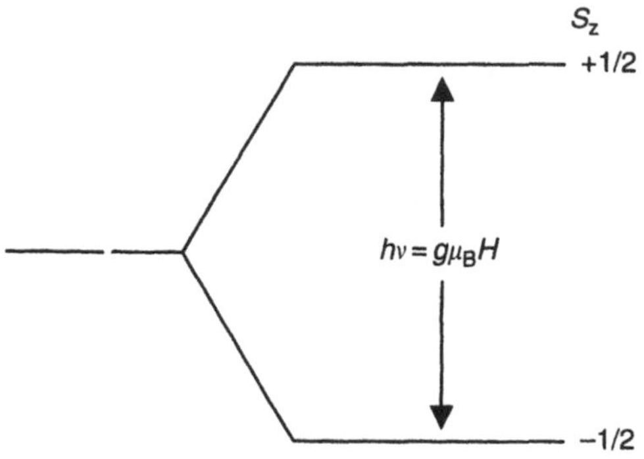

Figure 7.6 Zeeman levels of an unpaired electron (paramagnetic state) with $S = 1/2$ in the presence of a magnetic field.

hyperfine interaction term. The hyperfine interaction occurs when an electron is close to a nucleus with a magnetic moment. This is a magnetic interaction between electron spin and nuclear spin. The ESR spectrum gives information about the local configuration of the paramagnetic states through tensors **g** and **A**. The strength of the microwave absorption produces the density of the paramagnetic states. The free electron g-value ($g_0 = 2.0023$) is shifted due to the spin–orbit interaction to the other electrons, which is called the g-shift. The g-shift, Δg, therefore contains information about the nature of unpaired electrons, for example, on dangling bonds. In amorphous materials, however, most information is lost by the orientational averaging owing to random orientation of the principal axes.

As shown in Fig. 7.7, the wavefunction of a silicon dangling bond has axial symmetry and the g-tensor contains two terms, that is, the direction along the bond $\Delta g_{\parallel}$ and perpendicular to the bond $\Delta g_{\perp}$. It is known for a dangling bond that Δg is small when the bond is oriented along the magnetic field and is a large positive shift when oriented perpendicular (Street, 1991). Any orientation of the defect produces a g-shift which is intermediate between $\Delta g_{\parallel}$ and $\Delta g_{\perp}$. ESR absorption spectra for an axial defect such as the dangling bond therefore should have a peak at the $\Delta g_{\perp}$ and a shoulder at $\Delta g_{\parallel}$. As an additional disorder broadening is added actually, a slightly asymmetrical single peak is observed.

Figure 7.8 shows an example of a typical derivative of ESR spectrum obtained for undoped a-Si : H, showing $g = 2.0055$ with line width 7.5 G at the X-band frequency of 9 GHz (Street, 1991). Note that $g = 2.0055$ for Si-dangling bond is predicted from the values of $\Delta g_{\perp}(= 2.0081)$ and $\Delta g_{\parallel}(= 2.0012)$ (Caplan *et al.*, 1979; Street and Biegelsen, 1984) and the line width is dominated by the g-value distribution, which is the first derivative of the absorption spectrum.

The number of Si-dangling bonds was evaluated using the ESR signal of $g = 2.0055$, which is a measure of the *film quality* and varies with the deposition conditions of a-Si : H. In the optimum quality films, usually obtained from the

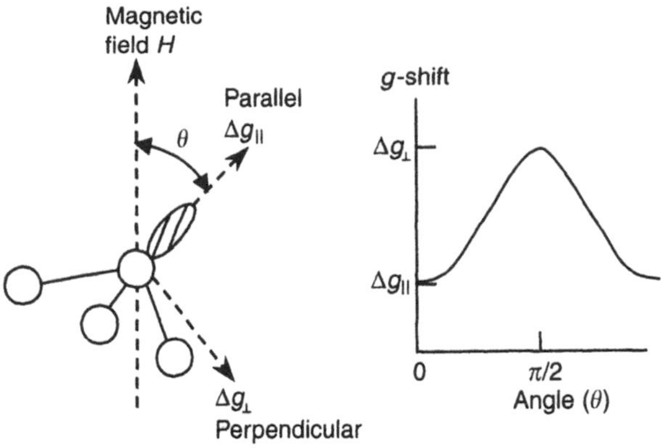

Figure 7.7 Principal components and angular dependence of the g-shift for Si dangling bond (T_3^0) (Street and Biegelsen, 1984).

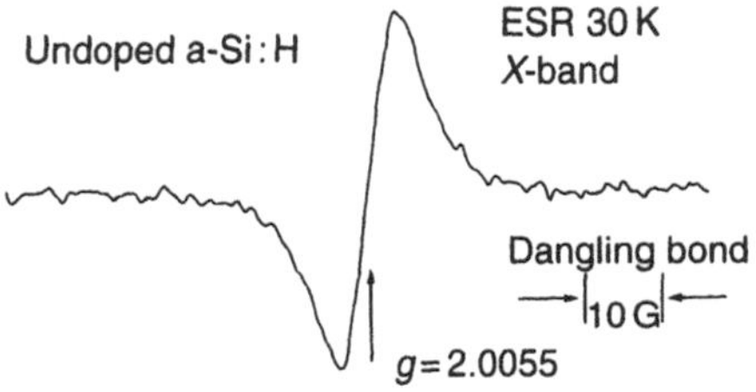

Figure 7.8 Derivative of ESR spectrum for undoped a-Si : H, showing the $g = 2.0055$ dangling bond with line width 7.5 G at the X-band frequency of 9 GHz (Street, 1991).

chemical vapor deposition (CVD) technique, the spin density is below 10^{16} cm^{-3}, at a growth temperature around 250°C. Samples deposited at room temperature have a spin density of about 10^{18} cm^{-3}, which decreases significantly by annealing to 250°C. The $g = 2.0055$ defects decrease in doped a-Si : H. However, it does not mean a decrease in the number of dangling bonds and actually it increases rapidly with doping. The decrease in spin density is due to a change in the charge state of the dangling bonds. The shift of the Fermi level by doping causes T_3^0 to be doubly occupied by electrons (T_3^-) in n-type a-Si : H and empty (T_3^+) in p-type a-Si : H. Neither of these states are paramagnetic and hence no unpaired spin exists there. Note, however, that T_3^+ and T_3^- have positive-U in nature, unlike the intimate pair of $T_3^+ - T_3^-$ as mentioned before and hence the energy level of T_3^- is higher by $+U_c$ (positive correlation) than the T_3^0 state. Thus the equilibrium spin density gives a correct measure of the defect density only in undoped a-Si : H

(Street, 1991). The ESR signal can also originate from paramagnetic surface states and hence a careful method of estimation, separating surface states, is required. Doping induced defects will be discussed in Section 7.3.

When the hyperfine interaction is present, the ESR line is split into its hyperfine components. If a nucleus has nuclear spin I, there is a separate resonance line for each magnetic moment and thus the total number of lines is given by $2I + 1$. For example, there are two hyperfine lines for $I = 1/2$ and three for $I = 1$. In undoped a-Si : H, the atoms of interest are silicon and hydrogen. Hydrogen has a nuclear spin of 1/2 and no obvious hyperfine structure with the $g = 2.0055$ signal suggesting that the overlap of the wavefunctions between unpaired electrons and hydrogen atoms can be ignored. ^{28}Si is the dominant isotope with zero spin but ^{29}Si has spin 1/2 and is naturally present at around 5%. A pair of broad hyperfine structures due to ^{29}Si with a splitting of about 70 G is observed as shown in Fig. 7.9(a); it is enhanced in ^{29}Si-enrich materials (Biegelsen and Stutzmann, 1986) as shown in Fig. 7.9(b). These structures are very weak compared with the main line; only 2.5% of the main line intensity.

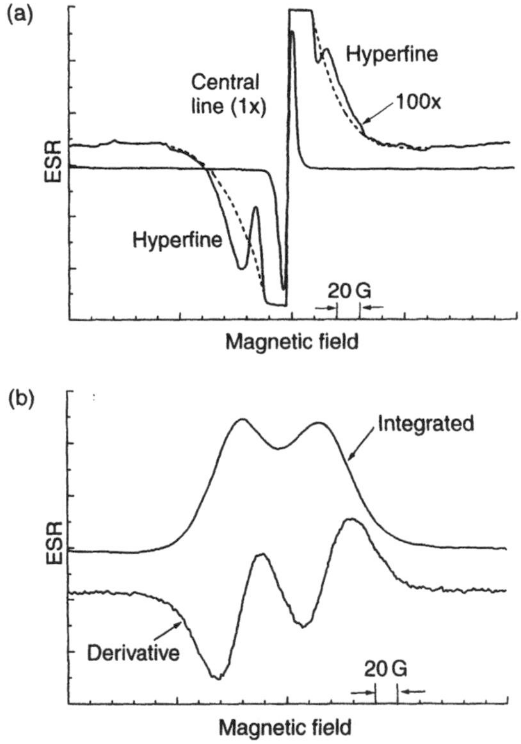

Figure 7.9 Hyperfine interaction in the $g = 2.0055$ defect resonance for (a) normal a-Si : H and (b) ^{29}Si-enriched a-Si : H (Biegelsen and Stutzmann, 1986).

The nature of electron wavefunction at the defect is deduced from the hyperfine structure of ESR. The broad lines imply that the defect is highly localized on a single Si atom. The defect wavefunction for hybridized atomic orbital, s and p, is written as

$$\Psi = c[(\alpha s) + (\beta p)], \quad \alpha^2 + \beta^2 = 1, \tag{7.5}$$

where c is the extent to which the wavefunction is localized. The hyperfine interaction contains the direct overlap of the electron wavefunction with the nucleus, which depends only on the s-component, and a dipole interaction, which depends only on p-component. The total hyperfine splitting is

$$\Delta H = c^2\alpha^2\Delta H(s) + c^2\beta^2(3\cos^2\theta - 1)\Delta H(p), \tag{7.6}$$

where θ is the angle between the magnetic field and p orbital direction, and $\Delta H(s)$ and $\Delta H(p)$ are the atomic hyperfine constants (Street, 1991). Owing to the angular-dependent factor being zero by averaging of different orientations, that is, $\langle 3\cos^2\theta - 1\rangle = 0$, the splitting is given only by the isotropic s-component of interaction. Note, any site-to-site disorders of s and p components produce the line broadening as already described for g-value distribution of the Zeeman term.

A tetrahedral sp^3 orbital has 1/4 s-character and 3/4 p-character and hence α and β should be 0.5 and 0.87, respectively, in the complete tetrahedral form. The values $\alpha = 0.32$ and $\beta = 0.95$ and $c = 0.8$ for a-Si : H are deduced, indicating a smaller s-nature, larger p-nature, and incomplete localization ($c < 1$) (Stutzmann and Biegelsen, 1988; Street, 1991). It should be recognized that the network around T_3^0 should be distorted, since T_3^0 has more p-character which is not complete sp^3. It is not clear why the hydrogen hyperfine interaction is too small to be observed, while hydrogen has a nuclear spin of 1/2 and its content is large. This can be related to stronger localization of electronic wavefunction (Street, 1991). However, the hydrogen hyperfine structure seems to be observed by electron–nuclear double resonance (ENDOR) as will be described below.

7.2.1.2 Electron–nuclear double resonance

ENDOR is more helpful in obtaining the structural information about defects and its surroundings when hyperfine interactions are weak. The ENDOR measurement monitors variation of the ESR with nuclear magnetic resonance (NMR) through hyperfine interaction. The intensity of ESR signal changes when NMR occurs. The details of ENDOR measurements on a-Si : H have been described elsewhere (Yokomichi and Morigaki, 1987, 1993, 1996; Morigaki, 1999) and therefore here we only summarize their results elucidated by the ENDOR measurements. These are: (i) Similar values for α, β and c in Eqs (7.5) and (7.6). (ii) DBs nearby hydrogen (hydrogen-related dangling bond), that is, the distance between the dangling bond site and hydrogen is estimated to be 2 Å.

Pulsed-ESR techniques are also employed to get more precise information about the distance between hydrogen (deuterium) atoms (Isoya *et al.*, 1993; Yamasaki and Isoya, 1993). In pulsed ESR, transient signals produced by coherent excitation using microwave pulses are observed in the time domain. Experimental details are described elsewhere (Yamasaki and Isoya, 1993; Yamasaki *et al.*, 1998) and here we give the principal results obtained by them. DBs become segregated in the hydrogen-depleted regions, which is not consistent with the ENDOR results mentioned above. This controversial issue has not yet been resolved.

7.2.1.3 Electrically detected magnetic resonance (EDMR)

Electronic transport properties of a sample of a-solid can change when it is subjected to ESR (see, e.g. the review by Lips *et al.*, 1998). The sensitivity of a current detection mode is known as 8 orders of magnitude greater than that of a microwave detection mode (i.e. usual ESR technique). First attempts to measure spin-dependent transport were made in c-Si (Honig, 1966; Schmidt and Solomon, 1966). As the photocurrent is usually monitored, this technique is called spin-dependent photoconductivity (Solomon, 1979; Shiff, 1981; Dersh *et al.*, 1983; Brandt and Stutzmann, 1991; Fuhs and Lips, 1991). More generally, however, this is called electrically detected magnetic resonance (EDMR), since the current in the dark condition is often used, for example, space-charge limiting current (Brandt *et al.*, 1993) and diode current (Lips *et al.*, 1998). These conductivities are dominated by the spin selection rule that determines the transition probability between localized states. Because of the conservation of angular momentum in an electron–hole recombination, the trapped electron (hole) can only recombine with a hole (electron) of opposite spin, that is, singlet spin state (see also Chapter 5).

In the zero applied magnetic field, the spin directions of carriers are completely random and hence the probabilities of a singlet or triplet electron–hole spin state are 1/4 and 3/4 (one for singlet and three for triplet). At intense microwave power levels, that is, under the saturation condition of ESR, both spin states will be equally populated, leading to a net increase of the density of singlet pairs. This increases the recombination rate, resulting in a decrease of conductivity. This fractional change in the conductivity can be measured by a lock-in amplifier. If the T_3^0 state participates in non-radiative recombination process, information about T_3^0 state ($N_S \approx 10^{16}\,cm^{-3}$) through a resonance at $g = 2.0055$ is obtained even in thin films, due to its high sensitivity (Lips *et al.*, 1998). Note that a large amount of samples (powder form) should be provided for usual ESR measurements.

Figure 7.10 shows an example of EDMR spectra in a-Si : H (Dersch *et al.*, 1983). Below 200 K, in particular at 125 K, in addition to neutral dangling bonds, ESR of trapped electrons and holes also contributes to the EDMR. These centers at $g = 2.004$ and 2.013 are found under illumination below 200 K and are called light-induced ESR (LESR) centers. The origin of these centers will be discussed later in this chapter.

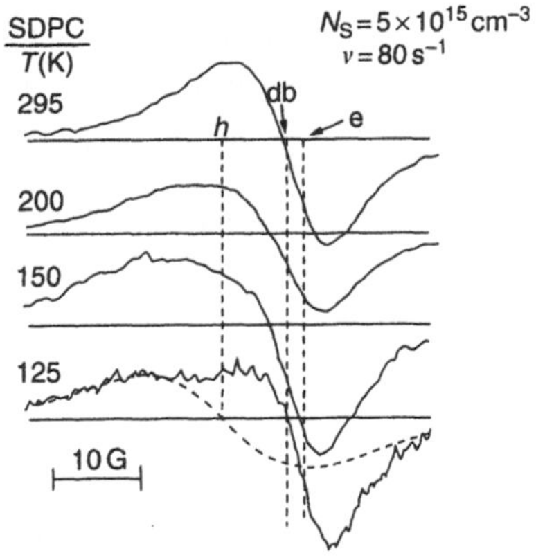

Figure 7.10 EDMR spectra of a-Si : H at various temperatures with the magnetic modulation frequency of 80 Hz, (Dersh *et al.*, 1983).

7.2.1.4 Optically detected magnetic resonance (ODMR)

ESR is also monitored by photoluminescence (PL), that is, detection of a fractional change in PL intensity at ESR (Cavenett, 1981; Morigaki, 1984). The signal is detected either in the total emitted light or monochromatized emission light from a sample in the microwave cavity. This is called ODMR. The principle of ODMR is the same as that of EDMR. The first ODMR measurements on a-Si : H have been carried out by Morigaki *et al.* (1978) and Biegelsen *et al.* (1978) and there is a comprehensive review (Morigaki, 1984) available on the subject. The ODMR signal in a-Si : H has also three components, as described in the EDMR. One at $g = 2.0055$ is related to the dangling bonds and the other two, $g = 2.004$ and 2.013, are related to the LESR centers. Thus the ODMR at $g = 2.0055$ can be useful for obtaining information about T_3^0 state as recombination centers.

One of the non-radiative recombination paths is considered to be the so-called Shockley–Read type recombination, in which recombination occurs at defects (Shockley and Read, 1952). Thus the spin pairing is between electron (hole) and hole (electron) localized on a DB. A net increase of the density of singlet pairs at resonance, as mentioned in the section of EDMR, increases recombination through the T_3^0 states, that is, decrease in PL (quenching signal). Morigaki and Yoshida (1985) observed both the quenching and enhancing signals at 2 K and the quenching signal occurs at $g = 2.0055$, leading to the conclusion that the DBs act as non-radiative recombination centers. In fact, the PL intensity decreases with increasing DB density in a-Si : H.

7.2.1.5 Light-induced ESR

In undoped a-Si : H, the new ESR centers, at $g = 2.004$ and 2.013, with a density of about $10^{16}\,cm^{-3}$ appear under bandgap illumination below around 200 K in addition to the center at $g = 2.0055$ due to DBs (Street and Biegelsen, 1980). An ESR center induced by illumination is called the LESR. These photoinduced centers at $g = 2.004$ and 2.013, are believed to be due to tail electrons and tail holes, respectively (Street and Biegelsen, 1980; Street, 1991; Umeda *et al.*, 1996). However, from the two quenching signals observed in ODMR measurements, Morigaki (1983) has suggested that the center at $g = 2.004$ is attributed to an electron trapped at ${}_{\mathrm{sp}}T_3^+$ and the center at $g = 2.013$ to a hole trapped at ${}_{\mathrm{p}}T_3^-$. The density of negative-U centers is expected to be around $10^{18}\,\mathrm{cm}^{-3}$ (Shimizu *et al.*, 1989; Shimakawa *et al.*, 1995). These defects of relatively higher density may be masked by the band tail states and hence these defects, although they exist, may not dominate electronic and optical properties in a-Si : H.

In doped a-Si : H, LESR at $g = 2.0055$ increases with increasing doping level, which approaches $10^{17}\,\mathrm{cm}^{-3}$, although this signal is very weak in the absence of illumination (Street, 1991). This indicates that the number of T_3^0-defects increases with doping. The LESR data are explained by the presence of charged dangling bonds (positive-U), since the movement of the Fermi level from mid-gap to the band edge by the doping changes the equilibrium electron occupancy of the defects. With n-type doping, DBs capture the excess electrons and become negatively charged in n-type a-Si : H and become positively charged in p-type by capturing excess holes. These configurations are not paramagnetic, so that there is no ESR signal in the equilibrium dark condition. Illumination excites an electron out of a negatively CDB in n-type a-Si : H, leaving it in the paramagnetic state of T_3^0. An electron pushed into a positively charged center by illumination also produces paramagnetic T_3^0 center in p-type. It is also known that only a fraction of these charged defects capture electrons and holes and hence the total number of LESR spin density does not give the total number of these charged centers (Shimizu *et al.*, 1989; Street, 1991).

7.2.1.6 *Optical measurements*

The optical absorption arising from the defect transition under the dark condition is usually weak in a-Si : H owing to low defect density and hence several sensitive techniques have been proposed. In the following, the principal and popular methods will be described.

7.2.1.6.1 PHOTOINDUCED ABSORPTION

Photoinduced absorption (PA), monitors variations in the intensity of probe light absorbed by trapped carriers in defect states after the carriers are created in the extended states by illumination. Owing to the increase in absorption (extra absorption), the transmittance of the probe light T changes with ΔT and expressed as

$$-\Delta T/T \propto (\sigma_{\mathrm{n}} - \sigma_{\mathrm{p}})\Delta n, \tag{7.7}$$

where σ_{n} and σ_{p} are the optical cross-sections corresponding to transitions of trapped electrons into the conduction extended states and transitions of electrons from the valence extended states to empty trapping levels, respectively. Δn is the number of trapped electrons induced by light excitation (Morigaki, 1999). The experimental results for undoped a-Si : H are well explained by the Lucovsky plot (Lucovsky, 1965), which is written as

$$-(\Delta T/T)^{2/3}(\hbar\omega)^2 \propto (\hbar\omega - E_{\text{th}}), \tag{7.8}$$

where E_{th} is the threshold energy of PA corresponding to the depth of the trapping level with respect to the edge of the extended states (Hirabayashi and Morigaki, 1983b). There are two values of E_{th}, one lies at 0.24–0.35 eV above the valence-extended state and the other 0.6–0.75 eV below the conduction extended state. These are identified with the self-trapped hole and negatively CDB levels (T_3^-; positive-U), respectively (Hirabayashi and Morigaki, 1986).

The spin-dependent PA (PA detected ESR) has been also reported in a-Si : H (Hirabayashi and Morigaki, 1983a; Schultz *et al.*, 1997). The principle of this technique is again the same as EDMR and ODMR. As the intensity of PA depends on the number of trapped electrons or holes which is related to their recombination rate, PA should be spin-dependent. In fact, the PA intensity decreases at resonance (quenching of PA) at $g = 2.0055$ (dangling bond).

7.2.1.6.2 PHOTOTHERMAL DEFLECTION SPECTROSCOPY (PDS)

PDS measures the heat absorbed in the sample and is used for measurement of the lower optical absorption coefficient of thin films (Jackson and Amer, 1982). Non-radiative recombination can be a dominant channel at relatively high temperatures (room temperature) and hence generates heat by phonon emission. In this experiment, a sample is immersed in a liquid CCl_4. When heat is generated with an intensity-modulation beam of light (pump beam), the refractive index of CCl_4 changes periodically on the contact surface of the sample and this change is detected by deflection of a laser beam passing through the surface of the sample. A He–Ne laser is usually used as probe light and the periodic deflection is measured with a position sensor. If the wavelength of the pump beam is varied, the deflection of the probe beam becomes a measure of the optical absorption spectrum of the materials. The spectral dependence of absorption coefficient is thus obtained from that of the deflection. Some formulations are required to obtain the optical absorption from the PDS measurements, details of which are given elsewhere (Jackson *et al.*, 1981; Amer and Jackson, 1984). The PDS sensitivity is very high, for example, for a 1-μm thickness, values of α can be determined as low as $0.1\,\text{cm}^{-1}$, which is 100 times greater than photoacoustic spectroscopy (PAS) in the sensitivity.

Examples of the absorption coefficient from PDS measurements are shown for a-Si : H samples deposited at different RF powers (Fig. 7.11). At lower energies (below 1.5 eV) there is a broad absorption band, which is attributed to transitions from deep defect states (dangling bonds). The Urbach tail, that is, $\alpha \propto \exp(\hbar\omega/E_{\text{U}})$, is observed above 1.5 eV. The integrated absorption produces

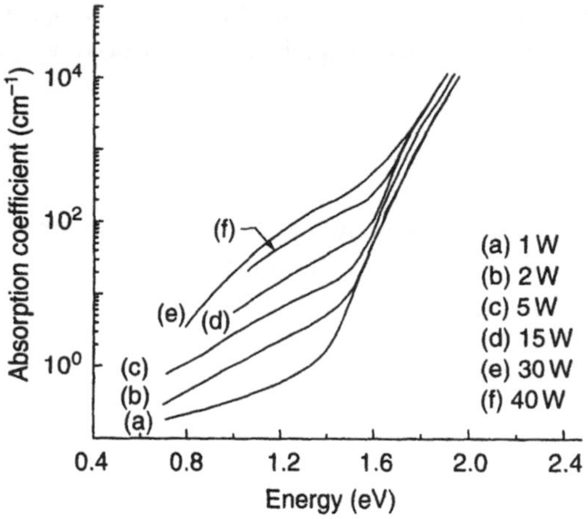

Figure 7.11 PDS spectra for the optical absorption edge and defect transitions of a-Si : H deposited at six different RF powers (Jackson and Amer, 1982).

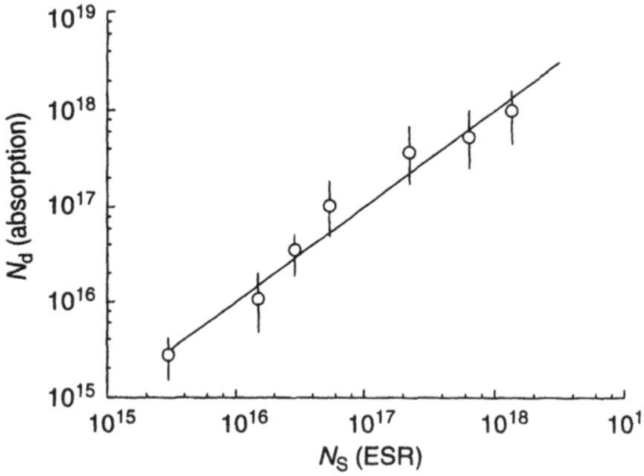

Figure 7.12 Correlation between the spin density N_S at $g = 2.0055$ ESR and defect density N_d deduced from the PDS (Jackson and Amer, 1982).

the defect density N_d as (Jackson and Amer, 1982)

$$N_d = 7.9 \times 10^{15} \int_0^{E_{max}} \alpha(E)\, dE, \tag{7.9}$$

where the upper limit of integration is taken up to an energy E_{max} after subtracting the Urbach tail contribution. When E_{max} is taken to be 1.6 eV, Eq. (7.9) produces an excellent correlation between N_d and the ESR spin density N_S as shown in Fig. 7.12. This agreement shows that the subgap absorption in undoped a-Si : H measured by PDS is due to singly occupied silicon dangling bonds (T_3^0).

In addition to measuring the number of defects, N_d, PDS can be used to deduce the energy level of T_3^0. Since the shape of the defect absorption is given by the product of density of states and the matrix elements M as (Amer and Jackson, 1984, also see Chapter 4)

$$\alpha \hbar\omega \propto M^2 \int N_{\text{defect}}(E) N_c(E + \hbar\omega)\, dE, \tag{7.10}$$

where $N_c(E)$ is the DOS of the conduction extended states, which is usually taken to be parabolic ($\propto E^{1/2}$) and $N_{\text{defect}}(E)$ the DOS of defects. Assuming a constant matrix element, the best fit to the absorption data provide the $N_{\text{defect}}(E)$ and a peak position around 1.25 and 0.9 eV below the conduction band is obtained for undoped and phosphorus-doped a-Si : H, respectively. Examples of optical absorption spectra and DOS for doped and undoped a-Si : H reported by Pierz *et al.* (1987) are shown by solid curves in Fig. 7.13.

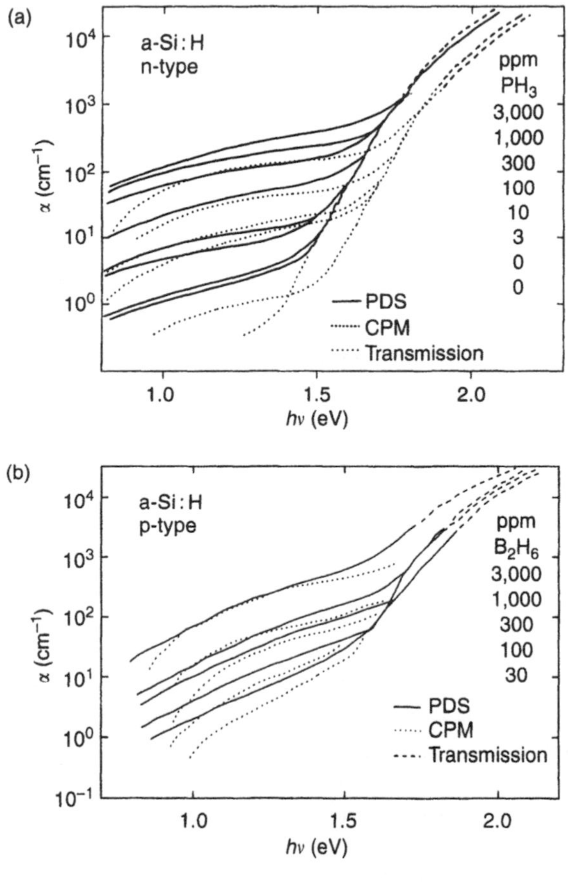

Figure 7.13 Optical absorption spectra for (a) n-type a-Si : H and (b) p-type a-Si : H deduced from both the CPM and the PDS (Pierz *et al.*, 1987).

7.2.1.6.3 CONSTANT PHOTOCURRENT METHOD (CPM)

Another sensitive measurement of optical absorption coefficient α can be made from the spectral dependence of the photoconductivity. The steady state photocurrent I_p is given as

$$I_p \propto \eta G\tau(1 - R)^2[\{1 - \exp(-\alpha d)\}/d], \tag{7.11}$$

where η is the quantum efficiency, G excitation intensity ($\mathrm{W\,cm^{-2}}$), τ recombination time, d thickness of films, and R reflectivity. When αd is very much smaller than unity, so that the non-linear terms can be neglected from the exponential expansion, I_p [Eq. (7.11)] becomes proportional to the product of excitation intensity G and α. Then, by assuming that other parameters are independent of excitation energy, α can be estimated. To ensure a constant value of recombination time, which should be independent of photon energy, a constant (steady) photocurrent is maintained by changing light intensity G. In this method, as the quasi Fermi level is kept the same, recombination path (recombination time in other words) can be unchanged. This method is called the constant photocurrent method (CPM) (Vanecek *et al.*, 1983; Kocka *et al.*, 1988; Hattori *et al.*, 1993) and its sensitivity of α reaches $10^{-1}\,\mathrm{cm^{-1}}$ for a-Si : H films of thickness of 1 μm. The optical absorption spectra obtained from the CPM for doped and undoped a-Si : H are shown by dashed curves in Fig. 7.13, together with the results from the PDS (Pierz *et al.*, 1987). These are almost the same as those from the PDS measurements. The DOS deduced from these measurements are also shown for n- and p-type a-Si : H in Fig. 7.14 (Pierz *et al.*, 1987).

7.2.1.7 *Electrical measurements*

Thermal emission of a trapped electron or hole in defects is basically estimated from the electrical transport measurements. As the thermal emission rate is related to the energy level and the shift of the Fermi level is related to the number of trapped charges, two main important physical parameters, (i) the defect energy levels and (ii) its concentration, can be estimated from the electrical transport measurements. In the following several principal methods for defect spectroscopy are introduced.

7.2.1.7.1 FIELD EFFECT MEASUREMENTS

This technique was first applied to obtain the density of states $N(E)$ in a-Si : H by Madan *et al.* (1976). The principle of this technique is as follows: A voltage is applied across the dielectric thin layer deposited on a-Si : H yields bending band, producing a space charge near the interface between the dielectric and a-Si : H. This gives an excess of carriers in the extended states (excess conductance), which is due to movement of the Fermi level toward the band edge. A low density of defects results in a large shift of the Fermi level and hence a large change in the conductance. The density of states near the Fermi level in this method can be

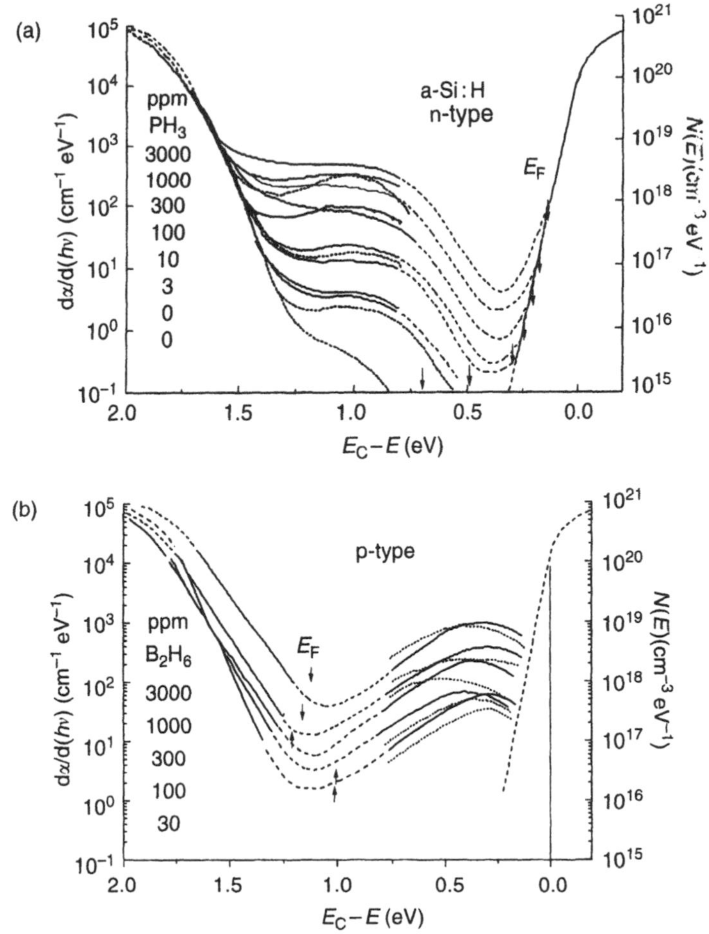

Figure 7.14 DOS, shown on the right *y*-axis, deduced from the CPM and PDS data of Fig. 7.13, (a) n-type and (b) p-type a-Si : H (Pierz *et al.*, 1987).

deduced from voltage-dependent conductivity $\sigma(V)$ and by solving the Poisson equation [see Eq. (7.12)].

The following problems with this technique have been pointed out by Goodman and Fritzsche (1980). First, as the Poisson equation involves the second derivation of the voltage, very precise experimental data are required to obtain the DOS near E_F. An analysis of the experimental data with nearly the same $\sigma(V)$ by Goodman and Fritzshe (1980) produces a different DOS than that deduced by Madan *et al.* (1976). Second, as most of the space charges lie near the interface (within 10 nm) they are strongly affected by the presence of interface states. The DOS in the bandgap obtained from the various techniques are compared later.

7.2.1.7.2 CAPACITANCE MEASUREMENT

The principle of the capacitance measurement is described as follows. Figure 7.15 shows, for example, the depletion layer of a Schottky barrier on n-type semiconductors under zero and reverse bias. The width of the depletion layer is given by the Poisson equation as

$$\frac{d^2V}{dx^2} = -\frac{\rho(x)}{\varepsilon\varepsilon_0}, \tag{7.12}$$

where $\rho(x)$ is the space charge density at x. In a donor-like level with concentration N_D in crystalline semiconductors (c-semiconductors), the charge $\rho(x)$ is constant and thus $V(x)$ takes the well known form of

$$V(x) = \frac{eN_D(W-x)^2}{2\varepsilon\varepsilon_0}. \tag{7.13}$$

Then the depletion width W for the applied voltage V_A is given by

$$W = \left[\frac{2\varepsilon\varepsilon_0(V_A+V_B)}{eN_D}\right]^{1/2}, \tag{7.14}$$

where V_B is the built-in potential. As the capacitance C is defined by $\varepsilon\varepsilon_0 S/W$, where S is the area, one gets the capacitance C from Eq. (7.14) in the popular

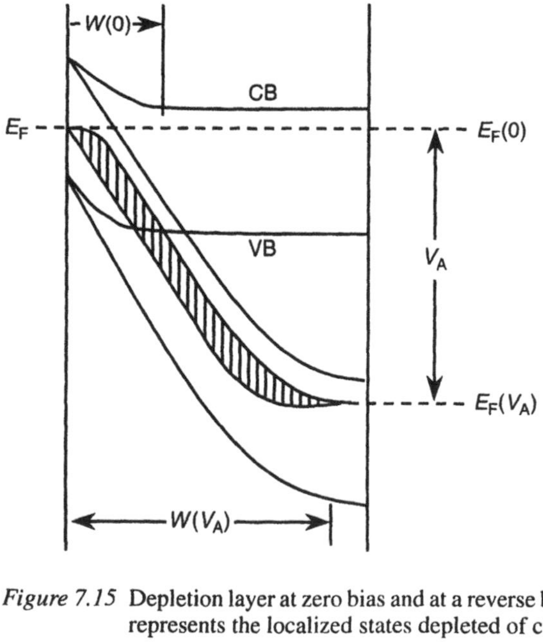

Figure 7.15 Depletion layer at zero bias and at a reverse bias of V_A. The shaded region represents the localized states depleted of charge by the bias voltage.

form of

$$\frac{1}{C^2} = \frac{2(V_A + V_B)}{\varepsilon\varepsilon_0 e N_D S^2}, \tag{7.15}$$

which enables us to deduce N_D from the capacitance measurements.

In a-Si : H, on the other hand, $\rho(x)$ arises from the ionization of distributed localized states in the bandgap, the shaded region in Fig. 7.15 contributes to $\rho(x)$ and is given by

$$\rho(V, x) = \int_{E_F(0)}^{E_F(V(x))} eN(E)\,\mathrm{d}E, \tag{7.16}$$

where a low temperature approximation is adopted. The value of $N(E)$ can be evaluated through the measurement of $C(V_A)$. The capacitance is measured by applying a small applied alternating voltage (AC voltage). The applied frequency ω_A should be smaller than the inverse of the dielectric relaxation time, $\varepsilon\varepsilon_0/\sigma_{dc}$, where σ_{dc} is the bulk DC conductivity, since the depletion layer capacitance is obtained only when the free carriers can respond to an applied AC field. The upper limit of ω_A is around 10 Hz at room temperature in undoped a-Si : H (Street, 1991). Such low frequencies lead to higher impedance of depletion region, $1/\omega_A C$, which may produce leakage current, and hence the precise evaluation of $N(E)$ becomes difficult.

However, the capacitance technique is more easy to apply for doped a-Si : H, since σ_{dc} is considerably higher than that in the undoped a-Si : H (Hack *et al.*, 1987). The $1/C^2$ vs V_A plot does not obey Eq. (7.15), which can be attributed to a broad distribution of localized states in a-Si : H. Therefore, precise information on $N(E)$ cannot be obtained from a simple capacitance measurement (Street, 1991).

The most familiar technique for measuring deep levels in semiconductors can be deep-level transient spectroscopy (DLTS) in which the transient capacitance of a Schottky barrier is measured (Lang, 1974; Lang *et al.*, 1982). As will be discussed below, the DLTS technique may provide more accurate information about $N(E)$ in a-Si : H. In DLTS, first, all the traps are filled by applying a forward bias to the Schottky barrier. Subsequently a reverse bias is applied and then the depletion layer width, which is initially large, decreases with time to its steady state value. Since the carriers are released from traps as mentioned in the capacitance technique, the release time τ_R of carriers from the traps is given by

$$\tau_R = \nu_0^{-1} \exp(E_T/kT), \tag{7.17}$$

where E_T is the trap depth and ν_0 is the attempt-to-escape frequency. This can be of the order of phonon frequency and the change in the depletion layer at time t is therefore related to $E_T(= kTln(\nu_0 t))$. $N(E)$ is thus derived from the transient capacitance. Practically, the capacitance is measured by scanning the temperature T at a fixed time after the application of the reverse bias. This is the principle of DLTS and an example of data for n-type a-Si : H is shown in

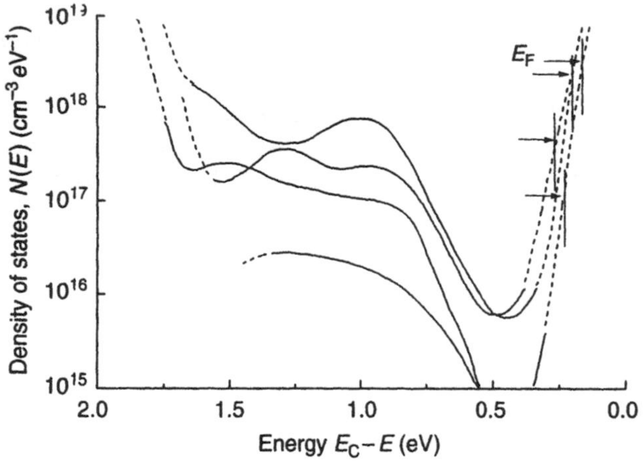

Figure 7.16 DOS of n-type a-Si : H deduced from DLTS. The Fermi energies of different doping concentrations are indicated by arrows (Lang *et al.*, 1982).

Fig. 7.16. The broad peak in $N(E)$ appears at 0.8–0.9 eV below the conduction bands with the magnitude of 10^{16}–$10^{18}\,\mathrm{cm^{-3}\,eV^{-1}}$ which depends on doping levels (Lang *et al.*, 1982). Samples with the largest phosphorous concentration have the largest DOS near midgap.

When the time-dependent capacitance is measured at constant temperature, this technique is called isothermal capacitance transient spectroscopy (ICTS) (Okushi *et al.*, 1981; Johnson, 1983; Okushi, 1985). The $N(E)$ from the ICTS should, in principle, yield the same $N(E)$ as that obtained from the DLTS. However, as shown in Fig. 7.17, the peak in $N(E)$ for n-type a-Si : H appears around 0.5 eV below the conduction band, while the magnitude of $N(E)$ is similar to that from the DLTS (Okushi *et al.*, 1981). These controversial results may originate from the attempt-to-escape frequency ν_0. Lang *et al.* (1982) have estimated $\nu_0 \approx 10^{13}\,\mathrm{s^{-1}}$, which is approximately equal to the phonon frequency, whereas Okushi *et al.* (1981) have used $\nu_0 \approx 10^{8}\,\mathrm{s^{-1}}$. As will be discussed in Chapter 8 (carrier transport), the Meyer–Neldel empirical rule may also dominate the ionization of trapped electrons (pre-exponential term depends on the trap depth) as

$$\tau_R = \nu_0^{-1} \exp(-E_T/kT_0) \exp(E_T/kT), \tag{7.18}$$

where T_0 is a characteristic temperature (or kT_0 is called the Meyer–Neldel energy).

The ESR in combination with DLTS measurements has confirmed that the peak in $N(E)$ observed by DLTS is the same as that of the $g = 2.0055$ ESR center. A 2.0055 spin occurs when an electron is released from the deep traps by applying reverse bias, indicating that the traps observed in DLTS in n-type a-Si : H

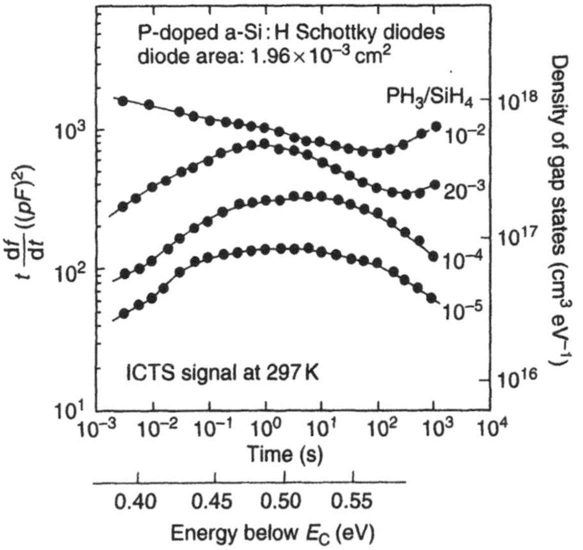

Figure 7.17 DOS of n-type a-Si : H, shown on the right *y*-axis, deduced from ICTS using a Schottky diode cell (Okushi *et al.*, 1981).

is identified with the doubly occupied dangling bonds (T_3^- with positive-U) (Cohen *et al.*, 1982; Johnson and Biegelsen, 1985).

7.2.1.7.3 SPACE-CHARGE-LIMITED CURRENT (SCLC)

Deep traps can be filled by current injection and hence the Fermi level will move up toward the CB if electrons dominate the transport. The current–voltage characteristic, $J(V)$, is then given by

$$J(V) = K \exp\left[-\left(\frac{E_c - E_F^*(V)}{kT}\right)\right] V, \qquad (7.19)$$

where K is a constant, V the applied voltage, and $E_F^*(V)$ quasi-Fermi level, which depends on the applied voltage. The current $J(V)$ is ohmic at low applied voltages, followed by a more rapid voltage dependence which is called the SCLC. The non-linear $J(V)$ depends highly on $E_F^*(V)$. As the change in $E_F^*(V)$ is related to $N(E)$ in the band gap, analysis of $J(V)$ produces $N(E)$ near E_F (Den Boer, 1981). There are several ways of analyzing $J(V)$ and deducing $N(E)$ from such analyses (Mackenzie *et al.*, 1982; Shimakawa and Katsuma, 1986; Shauer, 1994). The most simple analyzing method is the step-by-step (SS) method first adopted by Den Boer (1981), and successively used by Mackenzie *et al.* (1982).

A brief review of the SS method is presented here. Consider points (J_1, V_1), (J_2, V_2), ..., on the J–V curve. As the charge is injected with increasing applied voltage, the quasi-Fermi level moves toward E_c by

$$\Delta E_{f1} = kT \ln[(J_2/V_2)/(J_1/V_1)]. \tag{7.20}$$

The localized charge injected per unit area can be expressed approximately as

$$Q_1 \approx (\kappa\varepsilon_0\varepsilon/d)(V_2 - V_1), \tag{7.21}$$

where ε is the dielectric constant, d electrode separation, and $1 < \kappa < 2$ accounts for the non-uniformity of the internal space-charge field. Q_1 can be related to the average density of states N_1 by

$$Q_1 = ed\Delta E_{f1}N_1, \tag{7.22}$$

where zero-temperature Fermi–Dirac statistics is implicitly assumed. Combining Eqs (7.20)–(7.22), $N(E)$ for the first interval N_1 is obtained as

$$N_1 = (\kappa\varepsilon\varepsilon_0/ed^2\Delta E_{f1})(V_2 - V_1). \tag{7.23}$$

This process can be traced by SS method. The deduced $N(E)$ by Mackenzie *et al.* (1982), for example, is shown in Fig. 7.18, together with the $N(E)$ deduced by other techniques.

To simplify the analysis it is assumed that the injected electrons are uniformly distributed across the sample and the electric field is constant. These two assumptions are contradictory, which may introduce inaccuracy in $N(E)$, particularly near the Fermi level. Shimakawa and Katsuma (1986) have proposed the extended SS method, which may reduce the relative error in determining $N(E)$ in the other analytical methods.

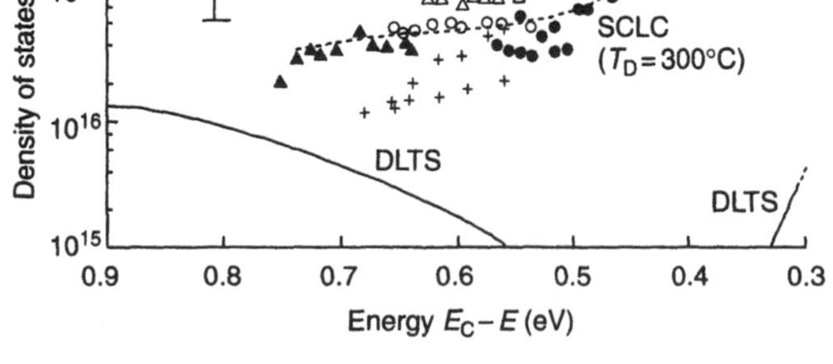

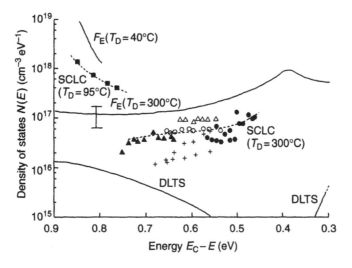

Figure 7.18 DOS for undoped a-Si:H deduced from SCLC, together with that deduced from the other techniques (Mackenzie *et al.*, 1982).

7.2.1.7.4 AC TRANSPORT (AC LOSS)

As will be discussed in Chapter 8, AC conductivity is defined as $\sigma(\omega) = i\omega\varepsilon_0\varepsilon$, where $\varepsilon_0\varepsilon$ is the dielectric constant. As the dielectric constant is a complex number, that is, $\varepsilon = \varepsilon_1 - i\varepsilon_2$, AC conductivity is also a complex quantity. The real part of AC conductivity, $\omega\varepsilon_2$, is usually called AC conductivity or AC loss. There are two types of AC losses. One is defect-related and the other is related to the hopping transport in band tails. The details of AC transport will be discussed in Chapter 8 and therefore we will only briefly mention about the deduced results from the AC loss in a-Si : H in this section.

The defect-related AC loss has been discussed in great details in a-Si : H (Shimakawa *et al.*, 1985, 1986, 1987a,b) (see also Chapter 8). It is now understood that a small temperature-independent loss dominant at low temperatures (2–100 K) is attributed to charge transfer between defects, and a strongly temperature-dependent loss at high temperatures (100–300 K) originates from the macroscopic or mesoscopic Maxwell–Wagner type inhomogeneities. As will be described in Chapter 10 (photoinduced effects), the tunneling of two electrons between $T_3^+ - T_3^-$ pairs is the most probable mechanism for AC loss. Other defects, for example, T_3^0, are not related to the AC loss in a-Si : H (Shimakawa *et al.*, 1995). The number of negative-U pairs ($T_3^+ - T_3^-$) is estimated to be $1 \times 10^{18}\,\mathrm{cm}^{-3}$, which is considerably larger than that ($\sim 1 \times 10^{16}\,\mathrm{cm}^{-3}$) estimated from the LESR signals (Shimizu *et al.*, 1989).

A less stable AC loss is induced by the weak optical excitation ($\mu\mathrm{W\,cm}^{-2}$) (Long *et al.*, 1988). This type of loss decays after removal of the illumination in 500 s at 50 K. Above this temperature, the induced loss decays much more rapidly. This behavior is very similar to that of the LESR signals and photoinduced midgap absorption as already stated. It is expected that this change in AC loss can be attributed to electrons and holes occupying T_3^+ and T_3^- centers (negative-U), respectively. Electronic hopping between neutral and charged states then contributes to the increase in AC loss. The origin of the photoinduced unstable AC loss here must be the same as that of LESR and photoinduced midgap absorption.

7.2.1.8 *Electrophotography (Xerography)*

The electrophotographic or xerographic method is a relatively new candidate to evaluate $N(E)$ in a-Si : H (Imagawa *et al.*, 1984, 1985, 1986). With this technique, one can almost neglect the surface or interface effect which cannot be ignored in the FE and CV methods, which have already been mentioned earlier. This experimental technique is described briefly below.

As shown in Fig. 7.19, charging by the corotron produces a surface potential V_c across a-Si : H films deposited on Al substrates. Schematic configurations of samples with positive corona and negative corona are shown in Figs 7.20(a) and (b), respectively. The light exposure ($\lambda = 450\,\mathrm{nm}$, absorption coefficient $\alpha > 3 \times 10^4\,\mathrm{cm}^{-1}$) creates electron–hole pairs near the free surface. For positive charge, electrons recombine with surface positive charges and holes drift through the film

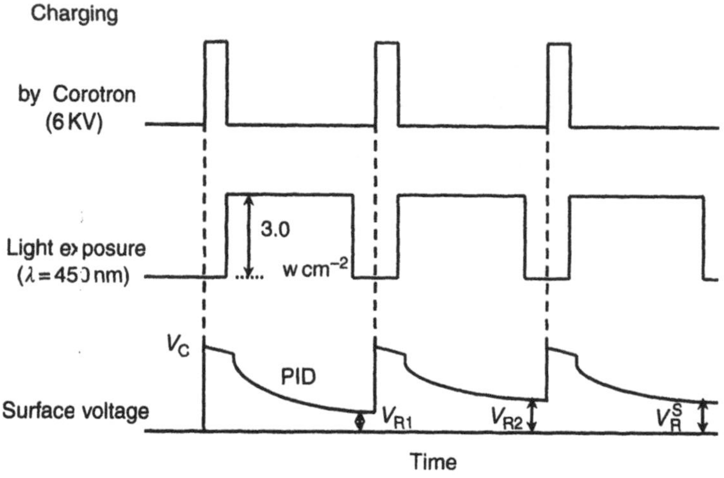

Figure 7.19 Schematic illustration of the basic xerographic process. Here V_C represents the surface voltage, PID the photoinduced decay of the surface voltage and V_R^S the saturated residual voltage.

(see Fig. 7.20(a)). A fraction of drifting holes falls into deep localized states (hole traps), whose release time exceeds the hole transit time, resulting in the decay of the surface potential (photoinduced decay; PID). The trappped holes, which can be uniformly distributed in the film, give rise to the residual surface voltage V_R. After repeating several cycles of these processes, V_R shows a buildup to a certain saturated value V_R^S after the cycling ceases. V_R^S decays in the dark at a temperature-dependent rate as trapped holes thermally get excited to the extended states (VB). The effect of trapping electrons into localized states can be neglected because electrons created near the surface can immediately recombine with the positive surface charges. The above procedure is also applied to a negative corona charge: Holes recombine with surface negative charges and electrons drift through the film (see Fig. 7.20(b)). A fraction of drifting electrons falls into deep electron traps, resulting in the shift of the Fermi level toward the conduction band (CB).

Consider points (t_1, V_1), $(t_2, V_2), \ldots, (t_i, V_i), \ldots$ on the decay curve in Fig. 7.21. The change in charge Q_1 due to thermally emitted holes (electrons) from hole (electron) traps per unit area in the time interval $\Delta t_1 = t_2 - t_1$ can be expressed approximately as given in Eq. (7.21). As the blocking layer is highly conducting (p-type for hole drifting and n-type for electron drifting: see Fig. 7.20), it is not necessary to consider this thin layer as a charge accumulation layer.

The trap depth E_i is related to the emission time by

$$E_i = kT \ln \nu_0 t_i. \tag{7.24}$$

The energy interval $\Delta E_1 = E_2 - E_1$, corresponding to the time interval $\Delta t_1 = t_2 - t_1$ is then obtained from Eq. (7.24) as

$$\Delta E_1 = kT \ln(t_2/t_1). \tag{7.25}$$

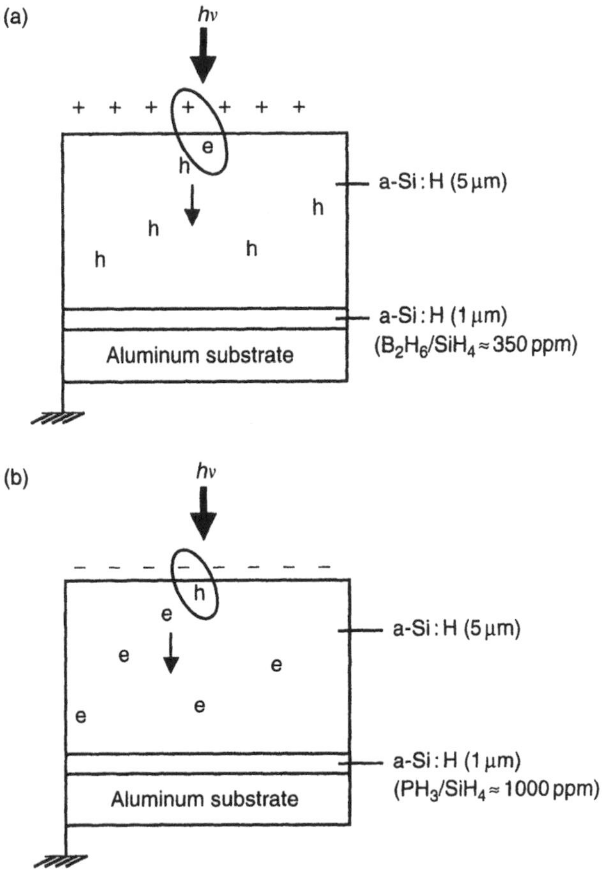

Figure 7.20 Schematic configuration of the sample used for (a) positive corona and (b) negative corona.

The average charge Q_1 from emitted holes (electrons) can be related to the average density-of-states N_1 by

$$Q_1 = ed\Delta E_1 N_1, \tag{7.26}$$

where e is the electronic charge. Following the same procedure as in SCLC analysis mentioned before, and combining Eqs (7.21) and (7.26), produces $N(E)$ for the first interval N_1 as

$$N_1 = (\kappa\varepsilon\varepsilon_0/ed^2\Delta E_1)(V_2 - V_1) = (\kappa\varepsilon\varepsilon_0/ed^2)(V_2 - V_1)/\Delta E_1. \tag{7.27}$$

As indicated in Fig. 6.21(b), $N(E)$ can then be traced out by the SS process.

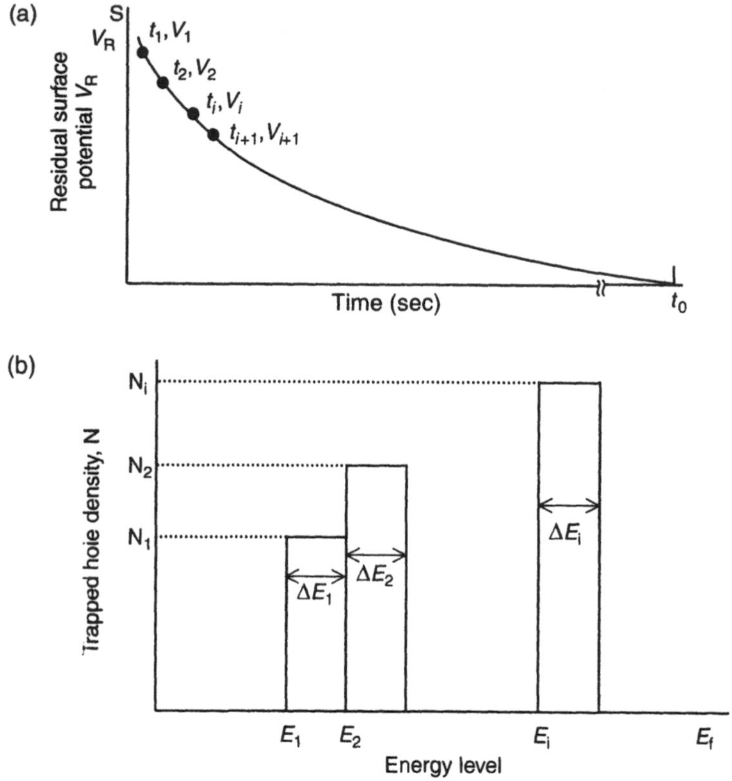

Figure 7.21 (a) Schematic V_R-time curve and (b) trapped-hole (electron) diagrams to illustrate the step-by-step method of analysis.

Figure 7.22 shows $N(E)$ vs $E-E_\nu$ curves for undoped a-Si : H, which are obtained from the positive charge method (see Fig. 7.20(a)). The energy distribution of deep hole traps with reference to the valence-band edge is deduced at temperatures 298, 305 and 309 K, and superimposed to form a master plot. The matching in the superimposition suggests that the analyzing method is valid to evaluate $N(E)$ with attempt-to-escape frequency ν_0 being a constant in the measured temperature range. The value of ν_0 is then evaluated from the intercept on the logarithmic axis of a plot of ln t_m vs $1/T$, where t_m is the time at which $\Delta V_i/\Delta E_i$ [see Eq. (7.27)] becomes maximum at temperature T (Imagawa *et al.*, 1985). Note, therefore, that ν_0 is not a free parameter and ways of determining its numerical value will be discussed later.

Figure 7.23 shows an $N(E)$ vs $E_c - E$ curve for undoped a-Si : H, which is obtained by the negative charge method (see Figure 7.20(b)). Note that the maximum in $N(E)$ for hole traps are two or three times larger than that for electron traps. It is well established that neutral dangling bonds T_3^0 control the deep trapping of

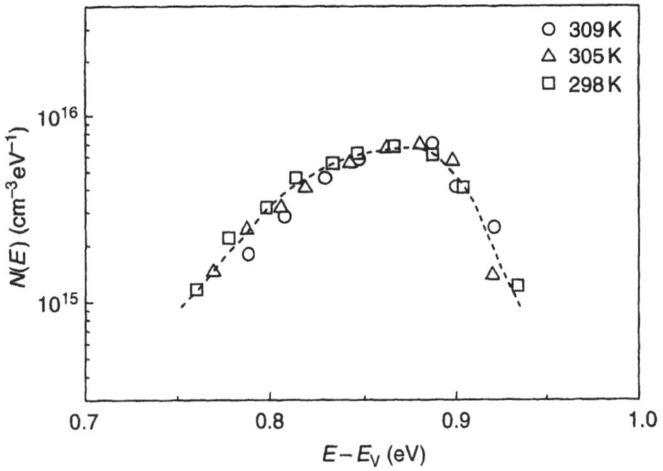

Figure 7.22 The DOS $N(E)$ (hole traps) for undoped a-Si : H calculated from isothermal decay of residual potential (Imagawa *et al.*, 1986).

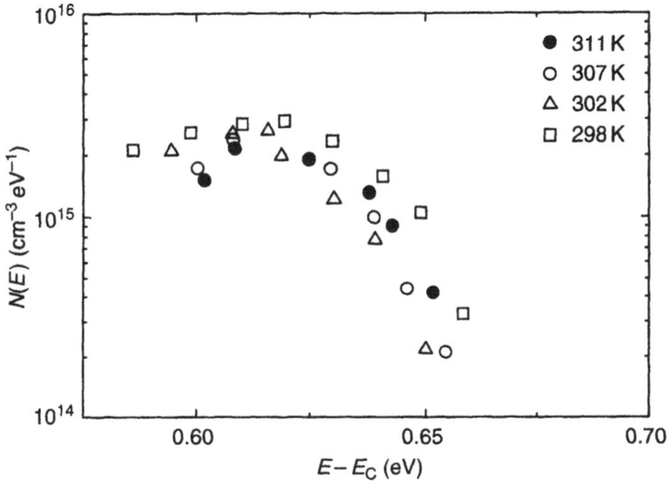

Figure 7.23 The DOS $N(E)$ (electron traps) for undoped a-Si : H calculated from isothermal decay of residual potential (Imagawa *et al.*, 1986).

both electrons and holes for undoped a-Si : H; $T_3^0+e \rightarrow T_3^-$, $T_3^0+h \rightarrow T_3^+$ (Street, 1991). The doubly occupied states T_3^- could be greater in energy by an amount U_c than T_3^0 (or T_3^+) under the positive correlation energy. As the undoped sample is not "intrinsic" but lightly n-type, the doubly occupied state T_3^- may coexist with T_3^0. Then the trapped hole states (THS) can originate from both processes: $T_3^0+h \rightarrow T_3^+$

and $T_3^- + 2h \rightarrow T_3^+$. It is hence understood that the THS density is greater than that of the trapped-electron states (TES) which can only originate from the process $T_3^0 + e \rightarrow T_3^-$.

The density of THS increases with increasing PH_3 concentration (n-type doping) and that of TES increases with increasing B_2H_6 (p-type doping) (Imagawa *et al.*, 1986). These features are qualitatively similar to those deduced from other defect spectroscopic techniques already described. For n-type samples, the doubly occupied T_3^-, which can be introduced by PH_3 doping, (and as discussed later in Section 7.3.2), becomes the dominant defect instead of T_3^0, resulting in $T_3^- + 2h \rightarrow T_3^+$. For p-type samples, on the other hand, the empty T_3^+ is the dominant defect and hence TES ($T_3^+ + 2e$) should increase with doping. Remember that T_3^0 is negligibly small for doped samples and hence no ESR at $g = 2.0055$ is observed.

The correlation energy U_c can be estimated by the difference in peak energy of THS (~0.88 eV from VB) and TES (~0.62 eV from CB), assuming the bandgap $E_g = 1.8$–1.9 eV, $U_c = 0.3$–0.4 eV is obtained. This value is consistent with that ($0.25 < U_c < 0.4$ eV) deduced from the PDS measurement (Jackson, 1982). Energy positions for THS (T_3^+ or T_3^0) and TES (T_3^-) deduced here are very close to their estimates obtained from carrier lifetime measurements (Spear *et al.*, 1984), although the energy for TES (T_3^-) is larger than 0.52 eV in ICTS (Okushi, 1985) and smaller than around 0.9 eV in DLTS (Lang *et al.*, 1982) and PDS (Jackson, 1982; Jackson and Amer, 1982). These will be discussed in next section.

Finally, we should discuss the value of prefactor ν_0. For holes $\nu_{0h} \sim 4 \times 10^{12}\ s^{-1}$ and for electrons, $\nu_{0e} \sim 1 \times 10^{10}\ s^{-1}$ (Imagawa *et al.*, 1986). It is interesting to note that ν_{0e} is smaller than ν_{0h} by two orders of magnitude. The prefactor itself does not seem to have any physical meaning. As it will be discussed in Chapter 8 (electronic transport), according to the Meyer–Neldel rule (compensation rule) a factor ν_0 always appears in thermally activated processes as $\nu_0 = \nu_{00} \exp(\Delta E / E_{MN})$, where ΔE is the activation energy and E_{MN} is called the Meyer–Neldel energy (characteristic energy). Accordingly, smaller ΔE always gives smaller factor ν_0. The origin of this empirical rule is still not clear and it may be a challenging work.

7.2.1.9 Summary

The defect states of T_3^0, T_3^+ and T_3^-, exist in a-Si : H and their densities can be deduced using various techniques. As far as the energy locations of these defects are concerned, results obtained from different techniques are not always in agreement, although some of them do agree. Each technique has a technical limitation to analyze and some aspects of the experimental conditions might be overlooked in the analysis. In spite of inconsistencies in results, gross features of defect structures in a-Si : H have been completely understood. The state of T_3^0 lies near the midgap with a density of about $10^{16}\ cm^{-3}$ and its correlation energy (positive-U) lies between 0.25 and 0.4 eV.

7.2.2 *a-Chalcogenides*

7.2.2.1 ESR

ESR signals in pure a-Ch are generally not observed in the dark conditions (Agarwal, 1973), since unlike a-Si : H only the charged defect states (negative-U) are expected to exist in a-Ch semiconductors. ESR signals are generally observed only after optical excitation (LESR) (Bishop *et al.*, 1975, 1977). A glassy Ge_xS_{1-x} system exhibits exceptionally prominent ESR signals in the absence of optical excitation (Arai and Namikawa, 1973; Cerny and Frumar, 1979; Kordas *et al.*, 1985; Shimizu, 1985; Watanabe *et al.*, 1988).

Figure 7.24 shows LESR signals in a few a-Ch at 4.2 K (Bishop *et al.*, 1977). The arrow shows the location of $g = 2.0023$. At a low illumination intensity about $1\,\mathrm{mW\,cm^{-2}}$, the spin density saturates at about $10^{17}\,\mathrm{cm^{-3}}$ for As_2Se_3 and As_2S_3 and about $10^{16}\,\mathrm{cm^{-3}}$ for Se. It is expected that LESR centers can be attributed to electrons and holes occupying charged dangling bonds (negative-U), that is, $D^+ + e \rightarrow D^0$ and $D^- + h \rightarrow D^0$. The saturation of LESR signals around 10^{16}–$10^{17}\,\mathrm{cm^{-3}}$ suggests that the density of charged defects should also be in this range. The broken curve superimposed on the Se curve is a computer simulation of the line shape. LESR centers are annealed as the temperature increases and are annealed out at around 200 K for all a-Ch. They are also bleached by infra-red (IR) irradiation whose energy is within the optically induced midgap absorption band.

At high excitation intensities above $100\,\mathrm{mW\,cm^{-2}}$, the spin density does not saturate but keeps on increasing and exceeds $10^{20}\,\mathrm{cm^{-3}}$, which is due to creation of defects by prolonged illumination. This issue with more detailed nature of LESR centers will be discussed in Chapter 9 (photoinduced effects).

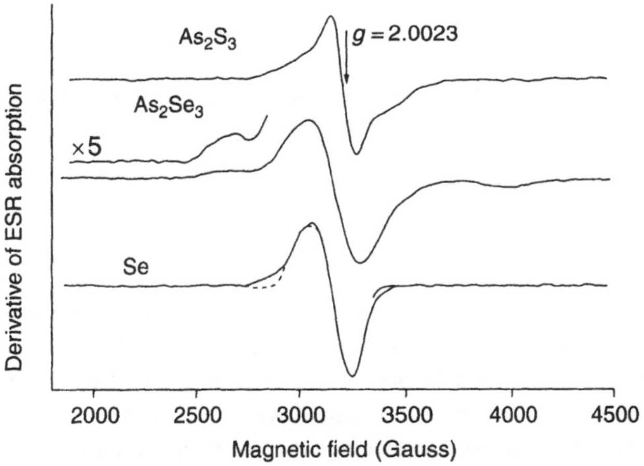

Figure 7.24 Optically induced ESR signals in a-Ch at 4.2 K (Bishop *et al.*, 1977): (- - - -) curve superimposed on the Se curve is a computer simulation.

7.2.2.2 ODMR

ODMR measurements on a-Ch have been done by Suzuki *et al.* (1979), Depinna and Cavenett (1981, 1882a,b), Tada *et al.* (1984), Robins and Kastner (1987) and Ristein *et al.* (1990). The ODMR signal, for example, in a-As_2Se_3 has three components; a narrow enhanced line at $g \approx 2.0$ and two broad enhaned lines at $g \approx 2.3$ and $g \approx 5.2$. The narrow line is attributed to a radiative distant pair recombination of electrons and holes; it is not clearly understood whether or not it involves defect recombination. The broad background line has been interpreted to be due to recombination of excitons (Depinna and Cavenett, 1982a; Robins and Kastner, 1987), which has been examined by comparing it with that in crystalline As_2Se_3 (c-As_2Se_3). Accordingly, the broad line is attributed to the recombination of triplet self-trapped excitons in a-As_2Se_3. Unlike a-Si : H the ODMR cannot provide information about defects in a-Ch.

7.2.2.3 Optical measurements

The weak optical absorption ($\alpha < 10\,\text{cm}^{-1}$) or so-called midgap absorption observed in a-Ch as well as a-Si : H, can be attributed to transitions to states lying between deep localized states (defect states) and extended states. Thus analyzing the weak optical absorption can be useful to get information about defects. It is known that illumination at low temperatures induces a midgap absorption which disappears with low temperature annealing or infra-red illumination (Bishop *et al.*, 1975, 1977). This annealing behavior is very similar to that of LESR signals in a-As_2Se_3. It is therefore expected that the midgap optical absorption occurs due to the existence of the same centers as those in LESR signals. The increase in midgap absorption is observed after prolonged illumination by bandgap light. This is due to creation of new defects by photoirradiation and details will be discussed in Chapter 10 (photoinduced effects).

Recently, CPM measurements have been performed in a-Ch (Kounavis and Mytlineou, 1995, 1996; Mytilineou, 1996; Tanaka *et al.*, 1996; Tanaka and Nakayama, 2000). An example of CPM in As_2S_3 is shown in Fig. 7.25. The DOS in the bandgap, as deduced in a-Si : H, has been calculated from the CPM spectra for some a-Ch but no DOS peak is observed (Kounavis and Mytilineou, 1996).

7.2.2.4 Electrical measurements

Unlike a-Si : H, not much work on electronic transport measurements for the defects spectroscopy has been done in a-Ch. This is due to the fact that a good Schottky barrier cannot be fabricated in a-Ch, because of the relatively lower bulk conductivity and higher defect states compared with a-Si : H. Thus, in principle, the field effect measurements or capacitance measurements are not easy to perform, except for a-As_2Te_3 and related materials (Marshall and Owen, 1976). Higher densities of trapping states, approaching $10^{19}\,\text{cm}^{-3}$, are suggested to exist from the field effect measurements, although the detailed nature, for example,

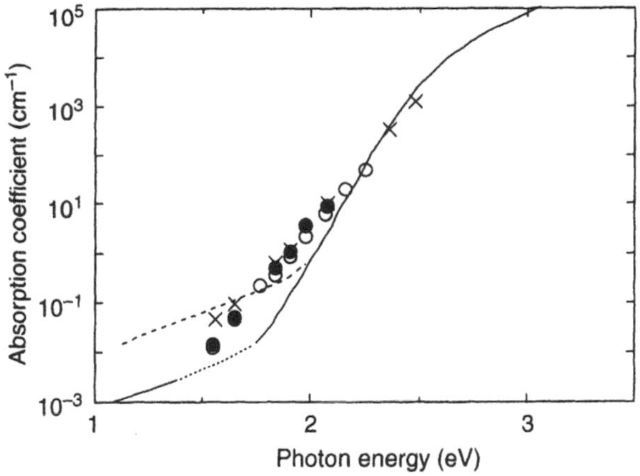

Figure 7.25 CPM spectra (×, ○, ●) for glassy As_2S_3, together with the transmission spectra (solid and dashed curves). Photocurrents are measured with interdigital (×), sandwich cells with front elctrode biased positively (○) and negatively (●) (Tanaka, 1996).

trap distribution, is not clear. Therefore, in a-Ch, electrical techniques for getting information about defects are limited. However, the density and its energetic location of defects can be elucidated from the AC transport measurements. In this section, we discuss about the AC loss technique for a-Ch.

As the AC loss in a-Ch will be discussed later in Chapter 8 in great detail, here we will only briefly mention about the deduced results from the AC loss. Correlated barrier hopping (CBH) of bipolarons, that is, two electrons hopping between charged defects (e.g. C_3^+–C_1^-), has been proposed by Elliott (1977, 1987). This model has been successively extended to the CBH of single polarons (single electron hop between charged and neutral states) (Shimakawa, 1981, 1982). Fitting the experimental data to the CBH model gives fairly good results for the energy position in the bandgap and density of charged defects. The density of defects estimated from the CBH model, 4×10^{18}–$4 \times 10^{19}\,cm^{-3}$ (Shimakawa, 1981, 1982), is always found to be larger than that (10^{16}–$10^{17}\,cm^{-3}$) from other measurements, for example, defect spectroscopy.

The CBH theory is based on pair approximation (PA) in which carriers are confined in a pair of localized states. As the carriers are not confined to a pair of centers at lower frequencies, the PA cannot be valid at lower frequencies. The continuous-time random-walk (CTRW) approximation is an alternative approach that can explain the AC loss in a-Ch (Scher and Lax, 1973; Dyre, 1985, 1988). The CTRW approach has been applied to the CBH mechanism and the fitting to the experimental data produces a reasonable density of charged defects (10^{17}–$10^{18}\,cm^{-3}$) for a-Ch (Ganjoo and Shimakawa, 1994; Ganjoo *et al.*, 1996).

Similar to a-Si : H, a less stable AC loss is induced with optical excitation ($mW\,cm^{-2}$) (Ganjoo *et al.*, 1996) in a-As_2Se_3. This type of loss measured by capacitance method retains 70% of its value during illumination after removal of the illumination at 20 K. This persistent capacitance cannot be annealed out at low temperatures, but it can be annealed out at the room temperature. This behavior is very similar to that of LESR signals and photoinduced midgap absorption as already stated. It is therefore expected that this change in AC loss can be attributed to electrons and holes occupying positively and negatively charged defect centers (negative-U), respectively. Electronic hopping between neutral and charged states then contributes to the increase in AC loss. The origin of the photoinduced unstable AC loss here must be the same as that of LESR and photoinduced midgap absorption. It is emphasized in this section that these unstable losses are commonly observed in both a-Si : H and a-Ch.

7.2.2.5 Electrophotography (Xerography)

An electrographic method to deduce $N(E)$ was applied to a-Se (Abkowitz and Markovics, 1984), and reported almost at the same time by Imagawa *et al.* (1984, 1985, 1986). The principle of the Xerographic method has been described in the previous section in a-Si : H. Abkowitz and Markovics (1984) took the *continuum* analysis, in which $N(E)$ is directly estimated from the time-derivative of residual surface voltage as

$$N(E = kT \ln \nu_0 t) = \frac{\kappa \varepsilon \varepsilon_0}{ed^2} \frac{t}{kT} \frac{\mathrm{d}V}{\mathrm{d}t}, \tag{7.28}$$

where the experimental time scale is mapped onto an energy scale. Note that Eq. (7.27) is equivalent to Eq. (7.28), because $\Delta V_i/\Delta E_i$ in Eq. (7.27) is given as

$$\frac{\Delta V_i}{\Delta E_i} = \frac{\Delta V_i}{\Delta t_i}\frac{\Delta t_i}{\Delta E_i} \equiv \frac{t}{kT}\frac{\mathrm{d}V}{\mathrm{d}t}. \tag{7.29}$$

Figure 7.26 shows the hole trap distribution in a-Se films (Abkowitz and Markovics, 1984). The energy is referred to as the VB edge. The $N(E)$ peak appears around 0.87 eV above the VB and the prefactor $\nu_0 = 1.4 \times 10^{13}\,\mathrm{s}^{-1}$ is reported. The remarkably small integral number of traps near midgap ($\sim 10^{14}\,\mathrm{cm}^{-3}$), however, is inconsistent with the other works already described, although good agreement between the xerographic method and other techniques is obtained in a-Si : H. The reason for this is not clear. We should carefully check the leakage carrier through blocking area (see Fig. 7.20). If blocking of holes is not enough, a small residual voltage and hence small $N(E)$ can be obtained.

7.2.2.6 Summary

Information on DOS, $N(E)$ in the bandgap, for a-Ch is relatively less than for a-Si : H. However, unlike a-Si : H, $N(E)$ deduced from the various techniques are all consistent: The number of charged defects are suggested to exist in the range of 10^{16}–$10^{17}\,\mathrm{cm}^{-3}$ in a-Ch and energy locations of such states are also deduced (Ganjoo and Shimakawa, 1994).

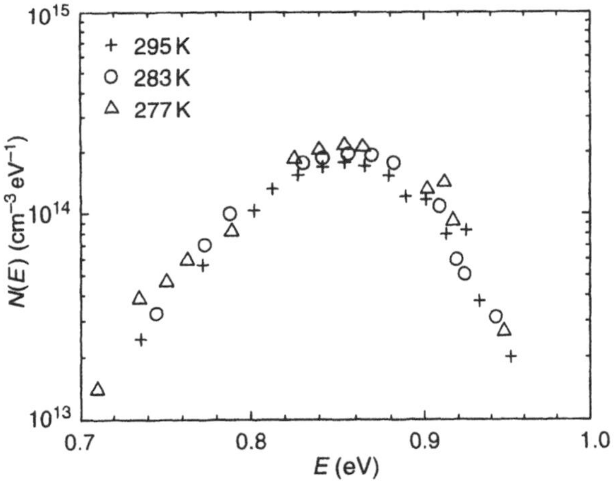

Figure 7.26 Hole trap distribution, $N(E)$, calculated from the isothermal decay of xerographic residual potential. Energy is referred to the valence-band mobility edge.

7.3 Defects in thermal equilibrium

There is a clear evidence of thermal equilibration of defect and dopant states in a-Ch (Mott, 1976; Fritzshe and Kastner, 1978; Uda and Yamada, 1979) and in a-Si : H (e.g. see Street, 1971; Morigaki, 1999). A-Solids are prepared in non-equilibrium state so that defect states, for example, are expected to be frozen in from either preparation temperature or glass transition temperature. It is therefore surprising that the density of native defects and their doping mechanisms are discussed on the basis of thermal equilibrium concepts. It will be concluded from this section that a subset of states, for example, defects, can be in thermal equilibrium, while the whole structural network may not be in the equilibrium phases (lowest energy state).

7.3.1 Native defects in thermal equilibrium

Before proceeding with any discussion on the thermal equilibrium, we should show typical examples for thermal equilibration. Figure 7.27 shows the temperature dependence of the density of thermally-generated recombination centers in a-As_2Se_3 (Thio *et al.*, 1984). The density of defects approaches the thermal equilibrium value above the glass transition temperature $T_g = 450\,K$. Aɔove T_g the density of defects increases with the activation energy (0.8 eV) and the pre-factor is $6 \times 10^{22}\,cm^{-3}$, suggesting that defects are created thermally from normally bonded atoms. Below T_g, the defect concentration is frozen in and the activation energy of 0.35 eV with pre-exponential factor of $4 \times 10^{17}\,cm^{-3}$ is observed. Note that the defects discussed here are considered to be neutral defects C_1^0 (Thio *et al.*,

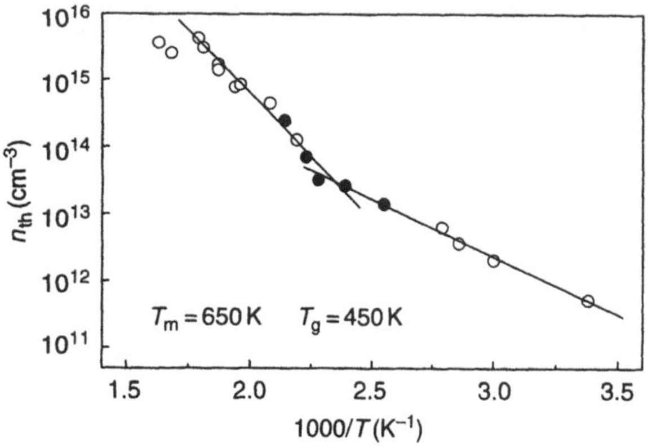

Figure 7.27 Temperature dependence of the density of thermally-generated recombination centers n_{th} for As_2Se_3 in both glassy and liquid states. n_{th} is deduced from transient photoconductivity measurements; open circles (Thio *et al.*, 1984) and closed circles (Orenstein and Kastner, 1981).

1984). However, the number of neutral centers is very small at T_g and hence it is not easy to detect by ESR measurements (Agarwal, 1973). This issue will be addressed further later in this section.

Figure 7.28 shows the temperature dependent DC conductivity in n-type a-Si : H after the thermal treatment of the sample (Street *et al.*, 1988). Similar data for n-type a-Si : H have also been reported by Matsuo *et al.* (1988a). A high conductivity is obtained when the sample is annealed at 250°C and then rapidly quenched, while a lower conductivity is observed when the sample is annealed at 120°C and rapidly quenched. Slow cooling from a higher temperature results in a low conductivity. Below a certain temperature, denoted by T_E (130°C for n-type a-Si : H), the temperature-dependent conductivity behaves differently, depending on their thermal history. The conductivity has a larger activation energy above T_E and no thermal history on electronic transport is observed, as all different curves merge into a single curve at $T > T_E$. T_E is therefore called the thermal equilibration temperature, above which the physical quantity takes a thermal equilibrium value. This means that when a sample is quenched from $T > T_E$, the thermal equilibrium at T_E gets frozen in.

The thermal equilibration effects on the ESR spin density monitored by $g = 2.0055$ are also present in undoped a-Si : H (Smith *et al.*, 1986; Street and Winer, 1989). The ESR measurements were performed by rapidly quenching the material from different temperatures and the signal varies reversibly with temperature, as shown in Fig. 7.29. The defect density increases between 200°C and 400°C, with an activation energy of 0.15–0.2 eV. Note that an endothermic heat capacity in differential scanning calorimetry measurements has been observed in the temperature range of 120–140°C (Matsuo *et al.*, 1988b).

Now, we will discuss the defect formation mechanism under thermal equilibrium. Let us consider that there are N_b broken bonds on N_0 available sites.

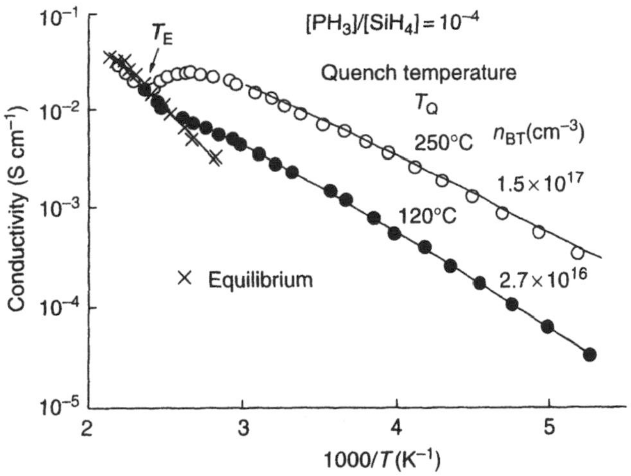

Figure 7.28 Temperature dependence of the DC conductivity of n-type a-Si : H, after annealing and cooling from different temperatures (Street *et al.*, 1988). Open circles are obtained from samples annealed at 250°C and closed circles at 120°C.

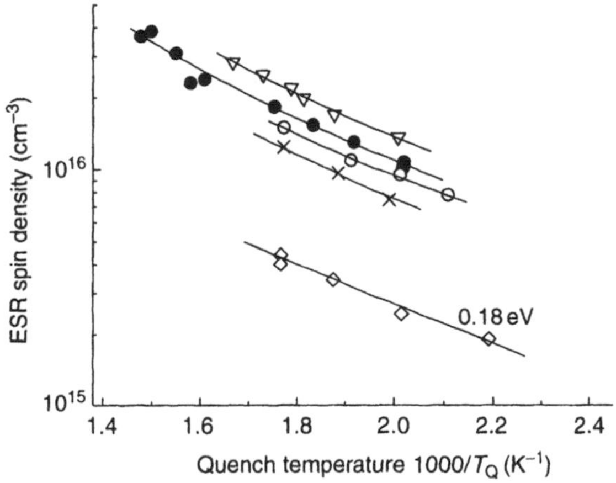

Figure 7.29 Temperature dependence of the equilibrium neutral defect density, measured using ESR in undoped a-Si : H deposited with different conditions (Street and Winer, 1989). Starting from top, curves are obtained for film thickness (RF power) as 50 μm (50 W), 20 μm (30 W), 40 μm (10 W), 90 μm (10 W) and 45 μm (10 W).

The thermodynamic equilibrium state is described by a minimum of the Gibbs free energy G as

$$G = U - TS, \tag{7.30}$$

where U is the formation energy and S entropy. As S is described by

$$S = k \ln\left(\frac{N_0!}{N_b!(N_0 - N_b)!}\right), \tag{7.31}$$

G is therefore

$$G = N_b U - kT \ln\left(\frac{N_0!}{N_b!(N_0 - N_b)!}\right). \tag{7.32}$$

Using Stirling approximation, $\ln n! \approx n \ln n - n$, and then minimizing the free energy with respect to N_b, the equilibrium density of broken bonds, denoted by N_b^0 for $U > kT$, is given as

$$N_b^0 = \frac{N_0 \exp(-U/kT)}{1 + \exp(-U/kT)} \cong N_0 \exp(-U/kT). \tag{7.33}$$

It may also be noted that the formation energy U here is for producing a broken bond but not for producing a dangling bond.

In undoped a-Si:H, N_0 is the total density of weak bonds, if there is thermal equilibrium subset between weak bonds and dangling bonds. The number of dangling bonds $N_{db} = 2N_b^0$. As the energy for breaking a weak bond is U, the formation energy for dangling bonds should be $U/2$. As shown in Fig. 7.29, the activation energy (formation energy) for dangling bonds is around 0.2 eV, then U in Eq. (7.33) should be taken as 0.4 eV. Using $N_b^0 = 0.5 \times 10^{16}\,\text{cm}^{-3}$ at $T = 500$ K and $U = 0.4$ eV, the density of weak bonds N_0 is obtained as $\approx 5 \times 10^{19}\,\text{cm}^{-3}$. This value for N_0 appears to be consistent with the predicted values in the range of 10^{19}–$10^{20}\,\text{cm}^{-3}$ (Street, 1991). Practically one should consider some distributions in the formation energy, U, as some calculations have indeed been done with such considerations (Street, 1991).

The law of mass action can be also applied to the above equilibration situation. One weak bond produces one broken bond (two dangling bonds) and hence the chemical reaction equation will be given as $N_0 + U \leftrightarrow N_b^0$. We hence obtain

$$[N_b^0] = [N_0] \exp(-U/kT), \tag{7.34}$$

where $[N_b^0]$ and $[N_0]$ are the densities of broken and weak bonds, respectively, in a notation of the law of mass action. Note that Eq. (7.34) is equivalent to Eq. (7.33). When hydrogen is involved in the defect creation process (Street and Winer, 1989), the reaction can be described as

$$\text{Si-H } (N_H) + \text{weak bond } (N_0) + U \leftrightarrow 2 \text{ dangling bonds } (N_{db}). \tag{7.35}$$

We then obtain

$$[N_{db}]^2 = [N_0][N_H] \exp(-U/kT). \tag{7.36}$$

As already mentioned, the reaction, $2C_2^0 + E_{VAP} \rightarrow C_1^- + C_3^+$, is suggested to occur in a-Ch, where C_2^0 is a normal bond and E_{VAP} is the creation energy of charged defects (VAP centers). The law of mass action therefore predicts the number of defects frozen-in on quenching through T_g as

$$[C_1^-] = [C_3^+] = N_0 \exp(-E_{VAP}/2kT_g), \tag{7.37}$$

where N_0 is the total atom number. Since both the VAP centers, C_1^- and C_3^+, have equal and opposite charges, they may pair up due to their mutual Coulomb interaction (intimate VAP; IVAP). The creation energy E_{IVAP} becomes lower by some amount ($E_{IVAP} < E_{VAP}$) and hence the concentration for IVAP would be higher than randomly distributed charged defects (Kastner *et al.*, 1976).

As stated already, the neutral center is not easy to detect by ESR measurement. However, the neutral defects, C_1^0, can be detected by ESR at higher temperatures. This is due to the reverse reaction; $C_1^- + C_3^+ + U_{eff} \rightarrow 2C_1^0$ (Shimakawa, 1981, 1982). The concentration of C_1^0 is therefore expected to be

$$[C_1^0] = \sqrt{[C_1^-][C_3^+]}\exp(-U_{eff}/2kT_g) = N_0 \exp\{-(E_{VAP} + U_{eff})/2kT_g\}. \tag{7.38}$$

In fact, Kawazoe *et al.* (1988) have observed ESR signals in liquid chalcogenides that do not agree with the existence of VAP centers. For temperatures below T_g, the total defect concentration is frozen in, but the conversion of C_1^0 from C_1^- and C_3^+ centers continues to be thermally activated with an activation energy of 0.35 eV (see Fig. 7.17). Hence, we suggest that U_{eff} is 0.7 eV for a-As_2Se_3, which is consistent with the other experimental studies, for example, AC conduction in a-Ch (Shimakawa, 1982). The pre-exponential factor ($4 \times 10^{17}\,cm^{-3}$) can be the number of frozen-in charged defects, which is also consistent with other experiments (Ganjoo and Shimakawa, 1994).

7.3.2 *Defects induced by doping*

The discovery of substitutional doping into a-Si : H by Spear and LeComber (1975) created a big stir in the field of disordered condensed matter and opened a new stage for the subsequent development. In a-semiconductors, the constituent atoms had been thought to obey the so-called 8-N rule to satisfy the covalent bonding with their neighboring atoms, where N is the number of valence electrons (Mott, 1969). However, the pioneering work of Spear and LeComber (1975) made it clear that n-type and p-type doping in a-Si : H by using phosphine (PH_3) and diborane (B_2H_6), respectively, are possible in plasma CVD. This historically important result is shown in Fig. 7.30. With doping, the room temperature conductivity increases by more than 8 orders. The activation energy decreases from ~0.7 eV to ~0.15 eV with phosphrous (P) doping and ~0.3 eV for boron (B). The details of electronic transport will be discussed in Chapter 8.

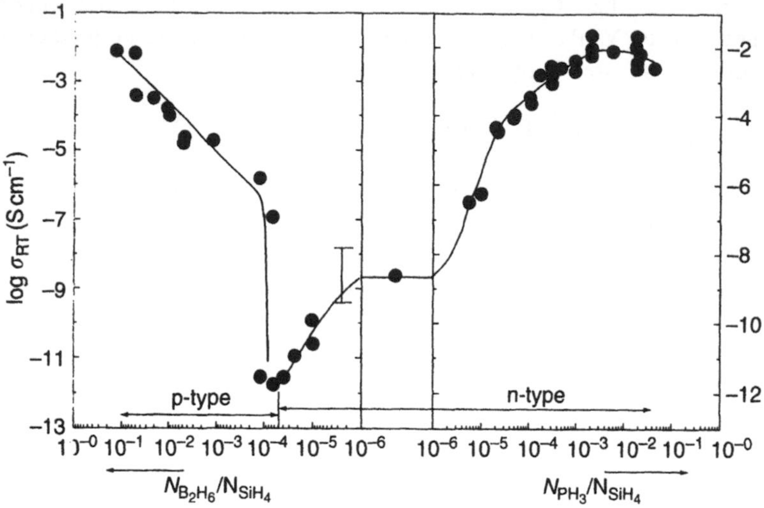

Figure 7.30 Variation of the room temperature DC conductivity of a-Si : H doped with PH_3 and B_2H_6 gases (Spear and LeComber, 1975).

As shown in Fig. 7.31, the doping efficiency is low and is inversely proportional to the square root of the gas-phase dopant concentration, C_{dopant}. Note that the doping efficiency in gas-phase, for example, for n-type, is defined as N_{donor}/C_{dopant}. The reason for this is discussed as follows. Most of P and B are expected to be three-fold coordinated followed by the 8-N rule. Following reaction for doping into a-Si : H was proposed (Street, 1982):

$$\mathrm{Si}_4^0 + \mathrm{P}_3^0 + U \leftrightarrow \mathrm{P}_4^+ + \mathrm{Si}_3^-, \tag{7.39}$$

where P_4^+ and Si_3^- are the positively charged four-fold P and negatively charged three-fold Si and U is the formation energy for isolated P_4^+ and Si_3^-. Note that Si_3^- is the doping induced dangling bond. This reaction may occur under thermal equilibrium conditions during deposition. The densities of P_4^+ and Si_3^- can be given by the law of mass action as

$$[\mathrm{P}_4^+] = [\mathrm{Si}_3^-] = \sqrt{[\mathrm{Si}_4^0][\mathrm{P}_3^0]}\exp(-U/2kT), \tag{7.40}$$

where $[P_4^+]$ is the density of N_{donor} and $[P_3^0]$ that of C_{dopant}. This predicts that the density of four-fold P, acting as donors, and charged dangling bonds are proportional to the square root of the incorporated phosphorus density. The doping efficiency is therefore proportional to $[P_3^0]^{1/2}/[P_3^0] = [P_3^0]^{-1/2}$. Experimentally, such a dopant dependence has been observed by Street (1982) and Stutzmann *et al.* (1987). As the density of both P_4^+ and Si_3^- increases with that of the dopant P_3^0,

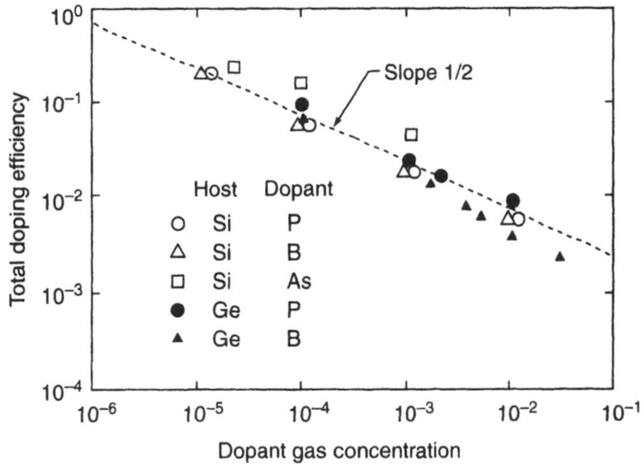

Figure 7.31 Doping efficiency as a function of dopant gas concentration (Stutzmann *et al.*, 1987).

the doping efficiency becomes low. The Fermi level should be pinned at the mid-point between the P_4^+ and Si_3^- states, since the Fermi level is given by the charge neutral conditions as shown in Fig. 7.32. It is expected, therefore, that the Fermi level cannot cross the respective mobility edge and consequently we cannot obtain degenerate or metallic a-Si : H.

Unlike a-Si : H, the effect of doping is not as drastic in a-Ch. This can be due to the fact that the Fermi level is pinned midway between positive and negative charged defect states (e.g. C_1^- and C_3^+) (Adler and Yoffa, 1976). As a result, it is difficult to dope a-Ch. Note that the electronic transport in undoped a-Ch is slightly p-type. Nevertheless, the doping under certain conditions is possible. For example, the addition of more than 10 at.% Bi to Ge–S and Ge–Se glasses increases the DC conductivity by seven orders of magnitude (Tohge *et al.*, 1979, 1980; Tichy *et al.*, 1985). The overall features of electronic transport are not dominated by substitutional (electronic) doping, but by the alloying effect (Elliott and Steel, 1986). Cu in As–Se system also shows electronic doping effects (p-type) (Hautala *et al.*, 1991) and the conductivity at room temperature approaches 1 S cm^{-1} at an activation energy of 0.06 eV.

There are a few doping models on a-Ch (Mott, 1976; Okamoto and Hamakawa, 1977; Fritzsche and Kastner, 1978; Uda and Yamada, 1979). From the law of mass action and the charge neutrality, the total VAP defect induced by doping is obtained as follows (Fritzsche and Kastner, 1987):

1 The reaction, $2C_2 + E_{VAP} \rightarrow C_3^+ + C_1^-$, leads to

$$[C_3^+][C_1^-] = [N_{ch}]^2 \exp(-E_{VAP}/kT_g) \equiv N_0^2, \qquad (7.41)$$

where N_{ch} is the number of chalcogen atoms.

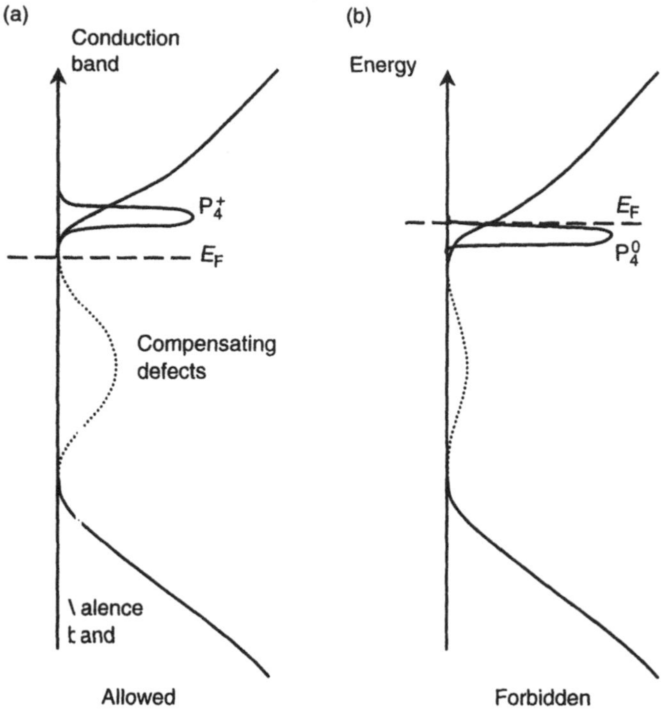

Figure ̓.32 Schematic presentation of DOS for a-Si : H: (a) E_F is located between P_4^+ and compensated defects $D^-(T_3^-)$, and (b) forbidden fourfold coordination of P_4^0 (Street, 1982).

2 The well-known np *product* is described as:

$$np = N_c N_\nu \exp(-E_g/kT) \equiv n_1^2, \tag{7.42}$$

where n and p are number of free electrons and holes, respectively, N_c and N_ν are the effective density of states for the conduction and valence bands, respectively, and E_g is the bandgap.

3 The reaction, $C_3^+ + 2e \rightarrow C_1^-$, is written as

$$[C_3^+] + 2n = [C_1^-] + 2N_c \exp\{-(E_c - E_F)/kT\}. \tag{7.43}$$

This leads to

$$[C_3^+][n]^2 = [C_1^-] N_c^2 \exp\{-2(E_c - E_F)/kT\}, \tag{7.44}$$

and hence

$$\left[\frac{C_3^+}{C_1^-}\right] n^2 \approx N_c^2 \exp(-E_g/kT) \equiv n_0^2. \tag{7.45}$$

4 The charge neutrality requires

$$[C_3^+] + [A^+] + p = [C_1^-] + n, \tag{7.46}$$

for n-type doping or

$$[C_3^+] + p = [C_1^-] + [A^-] + n, \tag{7.47}$$

for p-type doping, where $[A^+]$ and $[A^-]$ are the densities of positively and negatively charged doping additives, respectively. Let us discuss n-type doping in a-Ch. Note that $n_0(\approx n_i)$, n and p are negligible compared with N_0 and thus Eq. (7.46) reduces to

$$[C_1^-] = [C_3^+] + [A^+] \tag{7.48}$$

Substituting Eq. (7.48) into Eq. (7.41) produces

$$[C_3^+] = -\tfrac{1}{2}[A^+] + \sqrt{N_0^2 + \tfrac{1}{4}[A^+]^2} \tag{7.49}$$

and

$$[C_1^-] = \tfrac{1}{2}[A^+] + \sqrt{N_0^2 + \tfrac{1}{4}[A^+]^2}, \tag{7.50}$$

The total number of VAPs, N, introduced by doping is therefore

$$N = [C_3^+] + [C_1^-] = 2\sqrt{N_0^2 + \tfrac{1}{4}[A^+]^2}. \tag{7.51}$$

For n-type doping in a-Ch, Bi, for example in Bi–Ge–S(Se) system, should be positively charged when the doping model mentioned above is applied. For n-type doping, using $[C_1^-] \gg [C_3^+]$ [see Eqs. (7.49) and (7.50)], produces $[Bi^+] \approx [C_1^-]$, which should be satisfied and hence the Fermi level shifts upward. Thus the increase in dopant results in an increase in charged defects, indicating the doping induced defects. This self-compensation effect makes it difficult to produce heavily doped materials, similar to a-Si : H. Recent discovery of n-type conduction in a Cd-In-S system with 10^{-2} S cm^{-1} is of interest (Hosono *et al.*, 1998). Note, however, that this system does not belong to usual a-Ch family, because its valence band is not formed from the chalcogen's lone pair orbitals.

References

Abkowitz, M. and Markovics, J.M. (1984). *Philos. Mag. B* **49**, L31.

Adler, D. (1978). *Phys. Rev. Lett.* **27**, 1755.

Adler, D. and Yoffa, E.J. (1976). *Phys. Rev. Lett.* **36**, 1197.

Agarwal, S.C. (1973). *Phys. Rev. B* **7**, 685.

Amer, N.M. and Jackson, W.B. (1984). In: Pankove, J. (ed.), *Semiconductors and Semimetals*, Vol. 21, Part B. Academic Press, Orland, p. 83.

Anderson, P.W. (1975). *Phys. Rev. Lett.* **34**, 953.

Arai, K. and Namikawa, H. (1973). *Solid State Commun.* **13**, 1167.
Bar-Yam, Y. and Joannopoulus, J.D. (1986). *Phys. Rev. Lett.* **56**, 2203.
Biegelsen, D.K. and Stutzmann, M. (1986). *Phys. Rev. B* **33**, 3006.
Biegelsen, D.K., Knights, J.C., Street, R.A., Tsang, C. and White, R.M. (1978). *Philos. Mag. B* **37**, 477.
Bishop, S.G., Strom, U. and Taylor, P.C. (1975). *Phys. Rev. Lett.* **36**, 134.
Bishop, S.G., Strom, U. and Taylor, P.C. (1977). *Phys. Rev. B* **15**, 2278.
Brandt, M.S. and Stutzmann, M. (1991). *Phys. Rev. B* **43**, 5184.
Brandt, M.S., Stutzmann, M. and Kocka, J. (1993). *J. Non-Cryst. Solids* **164–166**, 693.
Caplan, P.J., Poindexter, E.H., Deal, B.E. and Razouk, R.R. (1979). *J. Appl. Phys.* **50**, 5847.
Cavenett, B.C. (1981). *Adv. Phys.* **30**, 475.
Cerny, V. and Frumar, M. (1979). *J. Non-Cryst. Solids* **33**, 23.
Cohen, J.D., Harbison, J.P. and Wecht, (1982). *Phys. Rev. Lett.* **48**, 109.
Den Boer, W. (1981). *J. Phys.* (Paris) **42**, C-4, 451.
Depina, S.P. and Cavenett, B.C. (1981). *Solid State Commun.* **40**, 813.
Depina, S.P. and Cavenett, B.C. (1982a). *Phys. Rev. Lett.* **48**, 556.
Depina, S.P. and Cavenett, B.C. (1982b). *Philos. Mag. B* **46**, 71.
Dersh, H., Schweizer, L. and Stuke, J. (1983). *Phys. Rev. B* **28**, 4678.
Dyre, J.C. (1985). *Phys. Lett. A* **108**, 457.
Dyre, J.C. (1988). *J. Appl. Phys.* **64**, 2456.
Elliott, S.R. (1977). *Philos. Mag.* **36**, 1291.
Elliott, S.R. (1978). *Philos. Mag. B* **38**, 325.
Elliott, S.R. (1987). *Adv. Phys.* **36**, 135.
Elliott, S.R. and Steel, A.T. (1986). *Phys. Rev. Lett.* **57**, 1316.
Fritzshe, H. and Kastner, M.A. (1978). *Philos. Mag.* **37**, 285.
Fuhs, W. and Lips, K. (1993). *J. Non-Cryst. Solids* **164–166**, 541.
Ganjoo, A. and Shimakawa, K. (1994). *Philos. Mag. Lett.* **70**, 287.
Ganjoo, A., Yoshida, A. and Shimakawa, K. (1996). *J. Non-Cryst. Solids* **198–200**, 313.
Goodman, N. and Fritzsche, F. (1980). *Philos. Mag. B* **42**, 149.
Hack, M., Street, R.A. and Shur, M. (1987). *J. Non-Cryst. Solids* **97&98**, 803.
Hattori, K. (1993). In: Sakurai, Y. *et al.* (eds), *Current Topics in Amorphous Materials*, North-Holland, Amsterdam, p. 351.
Hautala, J., Moosman, B. and Taylor, P.C. (1991). *J. Non-Cryst. Solids* **137&138**, 1043.
Hirabayashi, I. and Morigaki, K. (1983a). *J. Non-Cryst. Solids* **59&60**, 133.
Hirabayashi, I. and Morigaki, K. (1983b). *J. Non-Cryst. Solids* **59&60**, 433.
Hirabayashi, I. and Morigaki, K. (1986). *Philos. Mag. B* **54**, L119.
Honig, A. (1966). *Phys. Rev. Lett.* **17**, 186.
Hosono, H., Maeda, H., Kameshima, Y. and Kawazoe, H. (1998). *J. Non-Cryst. Solids* **227–230**, 304.
Imagawa, O., Akiyama, T. and Shimakawa, K. (1984). *Appl. Phys. Lett.* **45**, 438.
Imagawa, O., Iwanishi, M., Yokoyama, S. and Shimakawa, K. (1985). *J. Non-Cryst. Solids* **77&78**, 359.
Imagawa, O., Iwanishi, M., Yokoyama, S. and Shimakawa, K. (1986). *J. Appl. Phys.* **60**, 3176.
Isoya, J., Yamasaki, S., Ohkushi, H., Matsuda, A. and Tanaka, K. (1993). *Phys. Rev. B* **47**, 7013.
Jackson, W.B. (1982). *Solid State Commun.* **44**, 477.
Jackson, W.B. and Amer, N.M. (1982). *Phys. Rev. B* **25**, 5559.
Jackson, W.B., Amer, N.M., Boccara, A.C. and Fournier, D. (1981). *Appl. Opt.* **20**, 1333.

Johnson, N.M. (1983). *Appl. Phys. Lett.* **42**, 981.
Johnson, N.M. and Biegelsen, D.K. (1985). *Phys. Rev. B* **31**, 4066.
Kastner, M., Adler, D. and Fritzsche, F. (1976). *Phys. Rev. Lett.* **37**, 1504.
Kawazoe, H., Yanagita, H., Watanabe, Y. and Yamane, Y. (1988). *Phys. Rev. B* **38**, 5661.
Kocka, J., Vanecek, M. and Triska, A. (1988). In: Fritzshe, H. (ed.), *Amorphous Silicon and Related Materials.* World Scientific, Singapore, p. 297.
Kordas, G., Weeks, R.A. and Kinser, D.L. (1985). *J. Non-Cryst. Solids* **71**, 157.
Kounavis, P. and Mytilineou, E. (1995). *Philos. Mag. Lett.* **72**, 117.
Kounavis, P. and Mytilineou, E. (1996). *J. Non-Cryst. Solids* **201**, 119.
Lang, D.V. (1974). *J. Appl. Phys.* **45**, 3023.
Lang, D.V., Cohen, J.D. and Harbison, J.P. (1982). *Phys. Rev. B* **25**, 5285.
Lips, K., Lerner, C. and Fuhs, W. (1998). In: Marshall, J.M., Kirov, N., Vavrek, A. and Maud, J.M. (eds), *Thin Film Materials and Devices-Development in Science and Technology.* World Scientific, Singapore, p. 141.
Long, A.R., Anderson, M.J., Shimakawa, K. and Imagawa, O. (1988). *J. Phys. C* **21**, L1199.
Lucovsky, G. (1965). *Solid State Commun.* **3**, 299.
Mackenzie, K.D., LeComber, P.G. and Spear, W.E. (1982). *Philos. Mag. B* **46**, 377.
Madan, A., LeComber, P.G. and Spear, W.E. (1976). *J. Non-Cryst. Solids* **20**, 239.
Marshall, J.M. and Owen, A.E. (1976). *Philos. Mag.* **33**, 457.
Matsuo, S., Nasu, H., Akamatsu, C., Hayashi, R., Imura, T. and Osaka, Y. (1988a). *Jpn. J. Appl. Phys.* **27**, L132.
Matsuo, S., Nasu, H., Akamatsu, C., Hayashi, R., Imura, T. and Osaka, Y. (1988b). *Mat. Res. Soc. Symp. Proc.* **118**, 297.
Morigaki, K. (1984). In: Pankove, J.I. (ed.), *Semiconductors and Semimetals*, Vol. 23, Part C. Academic Press, Orlando, p. 155.
Morigaki, K. (1999). *Physics of Amorphous Semiconductors.* World Scientific & Imperial College Press, London.
Morigaki, K. and Yoshida, M. (1985). *Philos. Mag. B* **52**, 289.
Morigaki, K., Dunstun, D.J., Cavenett, B.C., Dawson, P., Nicholls, J.E., Nitta, S. and Shimakawa, K. (1978). *Solid State Commun.* **26**, 981.
Mott, N.F. (1969). *Philos. Mag.* **19**, 835.
Mott, N.F. (1976). *Philos. Mag.* **34**, 1101.
Okamoto, H. and Hamakawa, Y. (1977). *Solid State Commun.* **24**, 23.
Okushi, H. (1985). *Philos. Mag. B* **52**, 33.
Okushi, H., Tokumaru, Y., Yamasaki, S., Oheda, H. and Tanaka, K. (1981). *J. Phys.* (Paris), **42**, C4-613.
Orenstein, J. and Kastner, M.A. (1981). *Phys. Rev. Lett.* **46**, 1421.
Pantelides, S.T. (1986). *Phys. Rev. Lett.* **57**, 2979.
Pierz, K., Hilgenberg, B., Mell, H. and Weiser, G. (1987). *J. Non-Cryst. Solids* **97&98**, 63.
Ristein, J., Taylor, P.C., Ohlsen, W.D. and Weiser, G. (1990). *Phys. Rev. B* **42**, 11845.
Robins, L.H. and Kastner, M.A. (1987). *Phys. Rev. B* **35**, 2867.
Scher, H. and Lax, M. (1973). *Phys. Rev. B* **7**, 4491.
Schiff, E.A. (1981). *AIP Conf. Proc.* **73**, 233.
Schmidt, J. and Solomon, I. (1966). *Comt. Rend. Acad. Sci.* Paris **263**, 169.
Shauer, F. (1995). In: Marshall, J.M., Kirov, N. and Vavrek, A. (eds), *Electronic, Optoelectronic and Magnetic Thin Films.* John Wiley & Sons Inc, New York, p. 26.
Shimakawa, K. (1981). *J. Phys.* Paris, **42**, C4-167.
Shimakawa, K. (1982). *Philos. Mag. B* **46**, 123.
Shimakawa, K. and Katsuma, Y. (1986). *J. Appl. Phys.* **60**, 1417.

Shimakawa, K., Watanabe, A. and Imagawa, O. (1985). *Solid State Commun.* **55**, 245.
Shimakawa, K., Watanabe, Hattori, K., Imagawa, O. (1986). *Philos. Mag. B* **54**, 391.
Shimakawa, K., Long, A.R. and Imagawa, O. (1987a). *Philos. Mag. Lett.* **56**, 79.
Shimakawa, K., Long, A.R., Anderson, M.J. and Imagawa, O. (1987b). *J. Non-Cryst. Solids* **97&98**, 623.
Shimakawa, K., Kolobov, A. and Elliott, S.R. (1995). *Adv. Phys.* **44**, 475.
Shimizu, T. (1985). *J. Non-Cryst. Solids* **77–78**, 1363.
Shimizu, T., Kidoh, H., Matsumoto, M., Morimoto, A. and Kumeda, M. (1989). *J. Non-Cryst. Solids* **114**, 630.
Shockley, W. and Read, W.T. (1952). *Phys. Rev.* **87**, 148.
Smith, Z.E., Aljish, S., Slobodin, D., Chu, V., Wagner, S., Lenahan, P.M., Arya, P.R. and Bennett, M.S. (1986). *Phys. Rev. Lett.* **57**, 2450.
Solomon, I. (1979). In: Brodsky (ed.), *Amorphous Semiconductors*. Springer-Verlag, Berlin, p. 189.
Spear, W.E. and LeComber, P.G. (1975). *Solid State Commun.* **17**, 1193.
Spear, W.E., Steemers, H.L., LeComber, P.G. and Gibson, R.A. (1984). *Philos. Mag. B* **50**, L33.
Street, R.A. (1982). *Phys. Rev. Lett.* **49**, 1187.
Street, R.A. (1991). *Hydrogenated amorphous silicon*. Cambridge University Press, Cambridge.
Street, R.A. and Biegelsen, D.K. (1980). *Solid State Commun.* **33**, 1159.
Street, R.A. and Biegelsen, D.K. (1984). In: Joannopoulos, J.D. and Lucovsky, G. (eds), *The Physics of Hydrogenated Amorphous Silicon II*. Springer-Verlag, Berlin, Chapter 5.
Street, R.A. and Mott, N.F. (1975). *Phys. Rev. Lett.* **35**, 1293.
Street, R.A. and Winer, K. (1989). *Phys. Rev. B* **40**, 6236.
Street, R.A., Kakalios, J. and Hack, M. (1988). *Phys. Rev. B* **38**, 5603.
Stutzmann, M. and Biegelsen, D.K. (1988). *Phys. Rev. Lett.* **60**, 1682.
Stutzmann, M., Biegelsen, D.K. and Street, R.A. (1987). *Phys. Rev. B* **35**, 5666.
Suzuki, H., Murayama, K. and Ninomia, T. (1979). *J. Phys. Soc. Japan* **46**, 693.
Tada, T., Suzuki, H., Murayama, K., Ninomiya, T. (1984). In: Taylor, P. C. and Bishop, S. G. (eds), *AIP Conf. Proc.* No. 120, American Inst. Physics, New York, p. 326.
Tanaka, Ke. and Nakayama, S. (2000). *J. Optoelectron. Adv. Mater.* **2**, 5.
Tanaka, Ke. Nakayama, S. and Toyosawa, N. (1996). *Philos. Mag. Lett.* **74**, 281.
Thio, T., Monroe, D. and Kastner, M.A. (1984). *Phys. Rev. Lett.* **52**, 667.
Tichy, L., Ticha, H., Triska, A. and Nagel, P. (1985). *Solid State Commun.* **53**, 399.
Tohge, N., Yamamoto, Y., Minami, T. and Tanaka, M. (1979). *Appl. Phys. Lett.* **34**, 640.
Tohge, N., Minami, T., Yamamoto, Y. and Tanaka, M. (1980). *J. Appl. Phys.* **51**, 1048.
Uda, T. and Yamada, E. (1979). *J. Phys. Soc. Jpn.* **46**, 515.
Umeda, T., Yamasaki, S., Isoya, J., Matsuda, A. and Tanaka, K. (1996). *Phys. Rev. Lett.* **77**, 4600.
Vanecek, M., Kocka, J., Stuchlik, J., Kozisek, Z., Stika, O. and Triska, A. (1983). *Solar Energy Mater.* **8**, 411.
Watanabe, Y., Kawazoe, H. and Yamane, M. (1988). *Phys. Rev. B* **38**, 5688.
Yamasaki, S. and Isoya, J. (1993). *J. Non-Cryst. Solids* **164–166**, 169.
Yamasaki, S., Umeda, T., Isoya, J., Zhou, J.H. and Tanaka, K. (1998). *J. Non-Cryst. Solids* **227–230**, 332.

8 Electronic transport

The electronic transport is presented in this chapter. First the transport of charge carriers in a rigid lattice will be considered and then the effect of lattice vibrations (polaronic transport) will be presented.

8.1 Carrier transport in a rigid lattice

In a rigid network of atoms in an amorphous solid (a-solid), electrons (holes) are assumed to move through the conduction extended-states (valence extended-states) and/or through the localized states without being subjected to the lattice vibrations. Therefore, the electron–phonon coupling can be ignored in this case. In this chapter, we will summarize the theory that can be applied to the mechanism of electronic transport in a rigid structure in dark conditions, and then we will examine these theories in the light of experimental results obtained for hydrogenerated amorphous silicon (a-Si : H), amorphous Chalcogenides (a-Ch), and some other materials.

The electronic configuration of individual atoms in a solid remains the same in both c- and a-solids. However the atomic configurations in a-solids are different from c-solids, because of the absence of long-range orders in the former, as stated above. Nevertheless, a fully coordinated atomic network of a-solids, regardless of the lack of long-range orders, is expected to offer crystalline-like behavior to the transport of charge carriers. This is easy to understand on the basis of the tight-binding approach, described in Chapter 1.

An electron spends an equal amount of time on each of the two-bonded atoms. Thus, as mentioned earlier, such networks give rise to the extended states and therefore the transport of charge carriers in the conduction and valence extended states of a-solids is basically the same as that of charge carriers in the conduction and valence bands of c-solids. However, one can expect the effective mass of a charge carrier to be different in an a-solid from in a c-solid, as presented in Chapter 3. The density of states in the region of extended states deviates little from that in the band regions of c-solids and is given by the density of the free electron states. In which case it is proportional to the square root of the energy, as described in Section 4.1. It is the presence of tail states and dangling bond states in a-solids that makes an a-solid behave different from its c-form. Both tail states and dangling bond states are localized states and transport of charge carriers in these states at low temperatures can only be described by quantum tunneling from one site to another. The concept of quantum tunneling is described in Sections 1.6 and 8.1.1.3.

The border between the extended states and localized states is called the mobility edge. Mott (1970) claimed that the zero-temperature electronic transport should vary discontinuously with energy at the mobility edge, leading to the famous term, *minimum metallic conductivity*. The concept of the minimum metallic conductivity, however, is not consistent with the scaling theory of the Anderson localization. As there are many excellent reviews and monographs on this matter (e.g. Kamimura and Aoki, 1989; Mott, 1993; Morigaki, 1999), we will not discuss this issue here. In Section 8.1.2, interestingly, the readers will notice that the Boltzmann conductivity gets altered at the mobility edge.

8.1.1 Electrical conduction

8.1.1.1 Band conduction in non-degenerate state

As discussed in Chapter 3, the nature of the transport of charge carriers gets altered when a charge carrier crosses the mobility edges E_c and E_v. The transport above E_c is the band conduction type for electrons and transport below E_v is band conduction type for holes. Transport through localized states is called the hopping conduction, and this will be discussed in Section 8.1.1.3.

First, the band conduction in non-degenerate cases will be discussed. The electronic transport at relatively high temperatures (near room temperature) both in a-Si : H and a-Ch is believed to occur in the extended states: electrons in the conduction extended states for undoped and n-type a-Si : H and holes in the valence extended states for a-Ch (Mott and Davis, 1979). For electrons this yields an activation-type temperature dependence for the conductivity as

$$\sigma = \sigma_0 \exp(-\Delta E/kT), \tag{8.1}$$

where $\Delta E = E_c - E_F$ is called the activation energy and is the separation of E_F from the mobility edge E_c. The pre-exponential correlates empirically with the activation energy ΔE as

$$\sigma_0 = \sigma_{00} \exp(\Delta E/E_{MN}), \tag{8.2}$$

where σ_{00} is a constant and E_{MN} is often called the Meyer–Neldel characteristic energy.

The relation given in Eq. (8.2), called the Meyer–Neldel rule (MNR) or compensation law, is known to be applicable to various thermally activated phenomena, for example, kinetics (hopping) and thermodynamics (number of carriers in the band states) in crystalline and liquid semiconductors (see, e.g. Meyer and Neldel, 1937; Overhof and Thomas, 1989; Yelon and Movaghar, 1990; Fortner *et al.*, 1995). The relations in Eqs (8.1) and (8.2) are also interpreted as the conductivity σ approaches σ_{00} at a focal point of temperature given by $T_c = E_{MN}/k$. This is

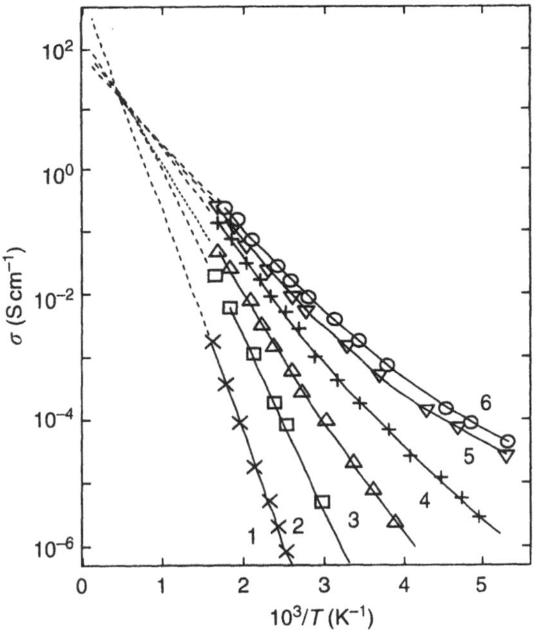

Figure 8.1 Temperature dependence of DC conductivity in p-type a-Si : H: (1) $[B_2H_6]/[SiH_4] = 10^{-5}$, (2) 10^{-4}, (3) 10^{-3}, (4) 10^{-2}, (5) 5×10^{-2} and (6) 2×10^{-1} (Beyer and Mell, 1977).

shown in Fig. 8.1 for p-type a-Si : H at six different doping concentrations (Beyer and Mell, 1977).

Typical examples of obeying MNR in a-Si : H and a-Ch are shown in Figs 8.2 and 8.3, respectively. In a-Si : H, ΔE can be changed by impurity doping (n-type and p-type) and a linear relation between $\ln \sigma_0$ and ΔE is observed as shown in Fig. 8.2. The fitting of the experimental data produces $E_{MN} = 67$ meV and $\sigma_{00} \approx 1\,S\,cm^{-1}$. Note also that ΔE can be changed by photoillumination due to the photodegradation (this will be discussed in Chapter 11), producing $E_{MN} = 43$ meV and $\sigma_{00} \approx 0.3\,S\,cm^{-1}$ (Overhof and Thomas, 1989). The relationship between $\ln \sigma_0$ and ΔE for one a-Ch system, As–Se–S, is shown in Fig. 8.3. This also produces $E_{MN} = 28$ meV and $\sigma_{00} \approx 1 \times 10^{-15}\,S\,cm^{-1}$. It may be noted here that for a-Ch, E_{MN} lies in the range 25–60 meV and σ_{00} in the range 10^{-5}–$10^{-15}\,S\,cm^{-1}$ (Shimakawa and Abdel-Wahab, 1997). Although E_{MN} lies in almost the same range for both a-Si : H and a-Ch, σ_{00} for a-Ch is very much smaller than that for a-Si : H. σ_{00} in a-Si : H is close in value to the microscopic conductivity, $e\mu_0 N_c$, for the standard band-transport (Mott and Davis, 1979). The validity of MNR is also found in organic semiconductors (Eley, 1967), hydrogenated microcrystalline silicon (μc-Si : H) (Fluckiger *et al.*, 1995) and insulating state in high-T_c superconducting material of $YBa_2Cu_3O_y$ (YBCO) films (Abdel-Wahab *et al.*, 1998). Here similar small values of $\sigma_{00}(10^{-3} - 10^{-15}\,S\,cm^{-1}$ for

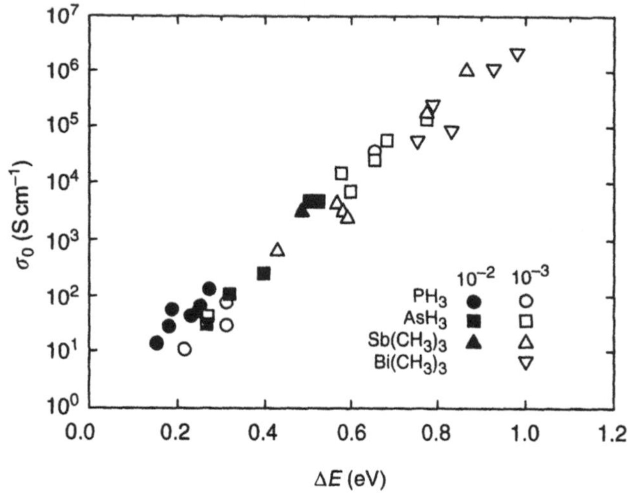

Figure 8.2 The pre-exponential factor σ_0 plotted as a function of ΔE for a-Si : H doped with various dopants (Carlson and Wronski, 1979).

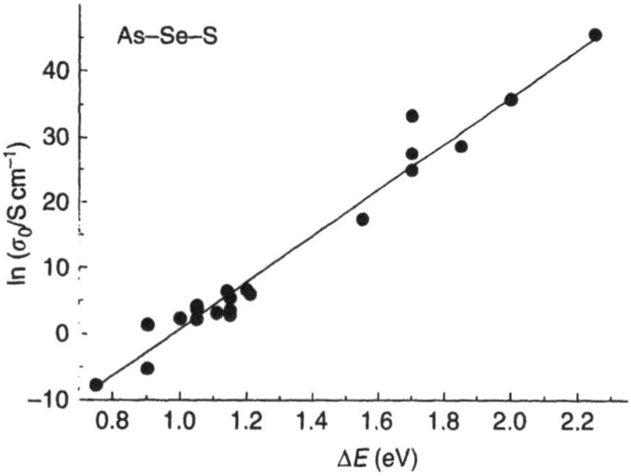

Figure 8.3 The pre-exponential factor σ_0 plotted as a function of ΔE for As–Se–S system (Shimakawa and Abdel-Wahab, 1997).

organic semiconductors, 10^{-6} S cm^{-1} for μc-Si : H, 10^{-3} S cm^{-1} for YBCO) have been found. The common features for small σ_{00} among these materials can be attributed to quantum tunneling through barriers which may exist in organic semiconductors μc-Si : H, and a-Ch, as it has been discussed elsewhere (Shimakawa and Abdel-Wahab, 1997).

There are many suggestions put forward on the origin of the MNR (see Yelon and Movaghar, 1990; Yelon *et al.*, 1992). The most accepted cause of the origin of MNR is apparently a significant shift of the Fermi level with temperature if the density of localized states is highly asymmetric with respect to the Fermi energy. The shift in the Fermi energy seems to be successfully applicable to a-Si : H, as the model calculations produce reasonable values of the physical parameter, such as the density of localized states in the bandgap (Overhof and Thomas, 1989).

To calculate the shift in the Fermi level with temperature, the following charge conservation law has to be satisfied at all temperatures, i.e., the following integral has the same value at all temperatures

$$\int N(E) f(E, E_F(T), T)\, dE, \tag{8.3}$$

where f is the Fermi–Dirac distribution function and $N(E)$ is the density of states (DOS) in the bandgap. Assuming an appropriate form of DOS, for three different $N(E)$, shown in Fig. 8.4, $E_F(T)$ is calculated numerically using Eq. (8.3). The shifts thus obtained for two different $N(E)$, 10^{15} and $10^{17}\,\text{cm}^{-3}\text{eV}^{-1}$ are shown in Fig. 8.5. In a certain temperature range, as shown in Fig. 8.6, $E_F(T)$ can be approximated to vary linearly with the temperature as

$$E_F(T) = E_F^*(0) + \gamma_F^* T, \tag{8.4}$$

where $E_F^*(0)$ and γ_F^* can be deduced from the intercept and slope of the relation between $E_F(T)$ and T (Fig. 8.6). Combining Eqs (8.4) and (8.1), we get

$$\sigma = \sigma_{00} \exp\left(\frac{\gamma_F^*}{k}\right) \exp\left(-\frac{E_c - E_F^*(0)}{kT}\right). \tag{8.5}$$

Comparing Eq. (8.1) together with Eq. (8.2), we find that $E_c - E_F^*(0)$ and γ_F^*/k correspond to ΔE and $\Delta E/E_{MN}$, respectively. As $E_c - E_F^*(0)$ is independent

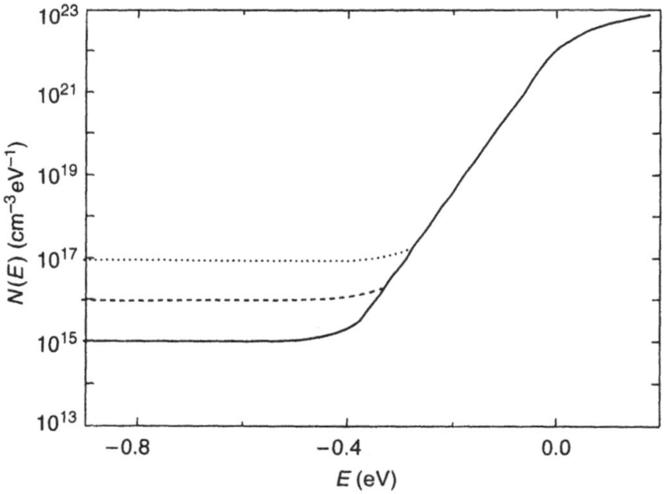

Figure 8.4 Model of DOS for calculating a statistical shift of the Fermi level in a-Si : H at 3 different assumed values of $N(E)$ (Overhof and Thomas, 1989).

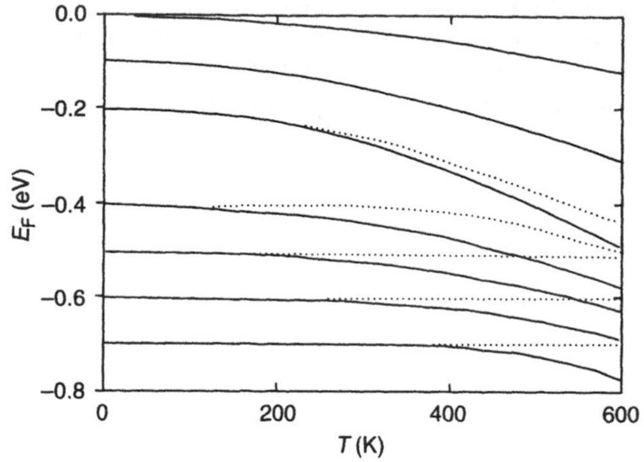

Figure 8.5 Statistical shift of the Fermi level for two different DOS described in Fig. 8.4. The location of $E_F(T = 0)$ is taken as a free parameter. (The solid and dotted lines correspond to the solid and dotted DOS in Fig. 8.4.)

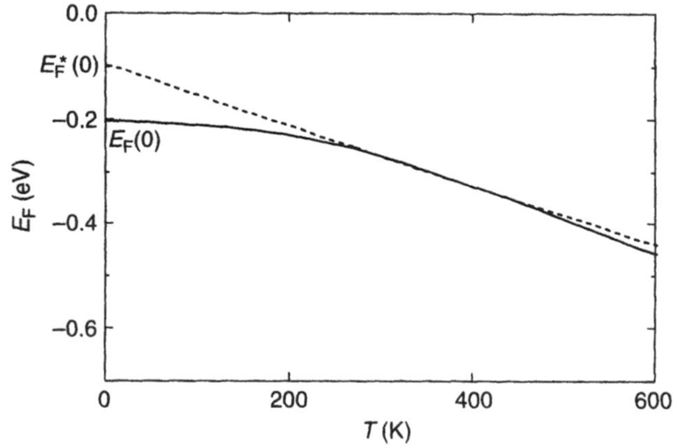

Figure 8.6 A typical statistical shift of E_F and its linear approximation (dashed line).

of temperature, the ΔE observed experimentally does not produce the "true" location of the Fermi (i.e. $E_c - E_F(T)$) at a-temperature. However, γ_F^*/k calculated for the forms of $N(E)$ shown in Fig. 8.4 (also shown in Fig. 8.7) replicates the experimental results well (Overhof and Thomas, 1989). In these cases we find that $\gamma_F < 0$.

Interestingly, the model calculation stated above predicts the inverse of MNR, since at lower activation energies the pre-exponential factor behaves opposite to how it behaves at higher activation energies. Therefore, it is called the inverted MNR and its effect has been observed in a-Si : H using a thin-film transistor (TFT)

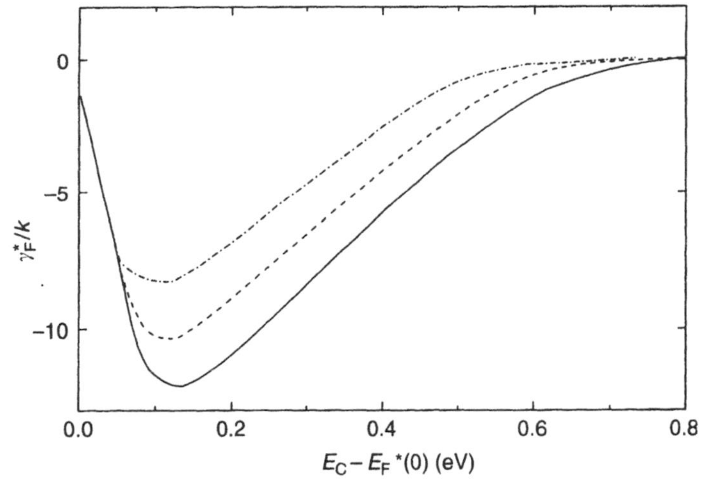

Figure 8.7 Calculated curves of γ_F^*/k vs $E_c - E_F^*(0)$.

(Kondo *et al.*, 1996), in doped μc-Si (Lucovsky *et al.*, 1993), and in high-T_c superconductors (Abdel-Wahab and Shimakawa, 1998).

However, MNR cannot be easily applied to a-Ch. Instead, as a general explanation, multiphonon excitations, for example, small polarons, have been used (Yelon and Movaghar, 1990). In this model, the prefactor σ_0 in conductivity is proportional to the number of ways of assembling these excitations (entropy effect) and hence E_{MN} corresponds to the optical phonon energy. In fact, in a-Si : H the optical phonon energy is 40 meV, which is close to E_{MN}. It is not clear whether or not small polarons (electron–phonon coupled states) dominate electronic transport in a-Si : H (and a-Ch). This will be discussed in Chapter 9.

Finally, a surprising correlation is observed between σ_{00} and E_{MN} in several a-Ch and a-Si : H as illustrated in Fig. 8.8. As the correlation between σ_{00} and E_{MN} appears to be good, the following empirical relation can be predicted

$$\sigma_{00} = \sigma_{00}' \exp(E_{MN}/\varepsilon), \tag{8.6}$$

where ε is a constant (1.7 meV).

Therefore, the origin of MNR is still open to question and it is difficult to obtain information about the microscopic properties of free carriers, such as microscopic mobility μ_0 from the activated-type conductivity only. Study of the electronic transport in degenerate states can be useful, and this will be discussed in the next section.

8.1.1.2 Band conduction in degenerate states

The free-carrier properties are still not known in disordered materials, because some of the fundamental physical parameters such as the effective mass, scattering

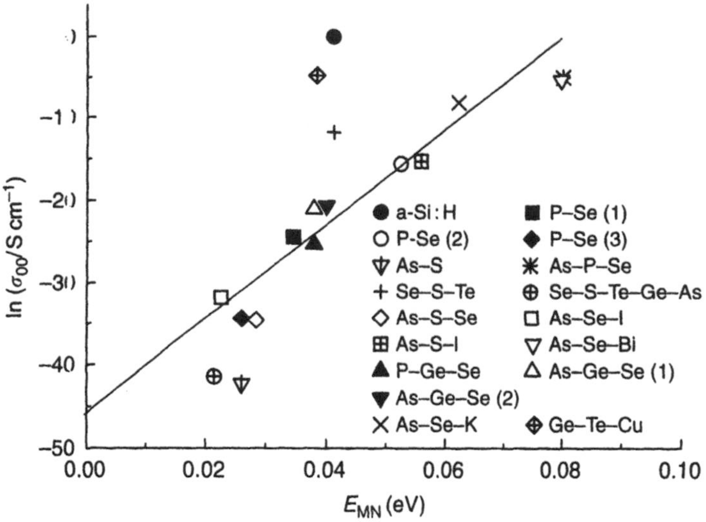

Figure 8.8 Correlation between σ_{00} and E_{MN} for some a-Ch and a-Si : H (Shimakawa and Abdel-Wahab, 1997).

time, and the mobility are not easy to obtain directly from experiments. One of the primary reasons for this difficulty is the existence of a sign anomaly in the Hall effect in many materials, which prevents us from getting a proper estimation of the Hall mobility μ_H and number of carriers. However, films of certain degenerate n-type a-oxides, for example, InO_x (Ovadyahu, 1986; Bellingham *et al.*, 1990; Pashmakov *et al.*, 1993) and Cd_2GeO_4 and Cd_2PbO_4 (Hosono *et al.*, 1996), exhibit normal sign of the Hall effect and may therefore be amenable to analysis. Unfortunately, as degenerate a-Si : H and a-Ch have not yet been prepared, we cannot discuss the free-carrier properties of such a-semiconductors. Therefore, in this section, we will focus on the electronic properties of other degenerate semiconductors like a-oxide semiconductors.

An example of the optical absorption coefficient $\alpha(\omega)$ of a-Cd_2PbO_4 measured at room temperature is shown in Fig. 8.9. Reasons for developing this oxide have been described elsewhere (Hosono *et al.*, 1996). The fundamental absorption rises at photon energies beyond about 1.8 eV and free-carrier absorption is observed below around 1.2 eV. A blue shift observed in the fundamental edge (broken curve), together in the free-carrier absorption, on annealing the films, suggests a Burstein effect (Burstein, 1954) for which the Fermi level E_F lies in the conduction band, which makes this oxide a degenerate semiconductor.

The frequency (energy)-dependent optical conductivity $\sigma(\omega)$ obtained from the absorption coefficient $\alpha(\omega)$ for annealed (250°C) a-Cd_2PbO_4 is shown in Fig. 8.10, where solid circles represent the experimental data. The full curve is fitted curves using the Drude formula. The fact that $\sigma(\omega)$ is almost proportional to ω^{-2} indicates that the system may be described by a single relaxation time τ (Shimakawa *et al.*,

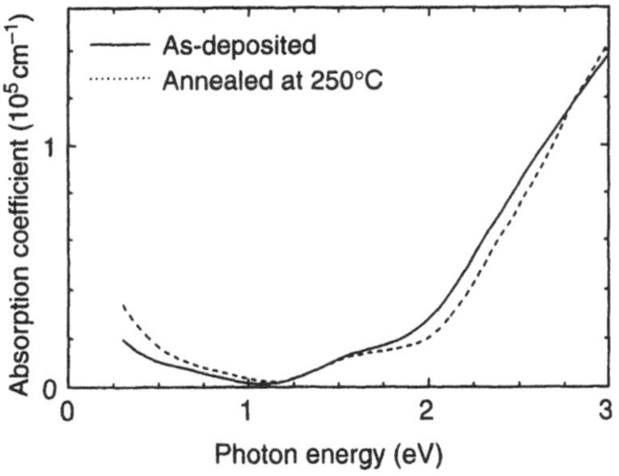

Figure 8.9 Optical absorption spectra of as-deposited (solid line) and annealed (dashed line) a-2CdO·PbO_2 films measured at room temperature (Shimakawa *et al.*, 1999).

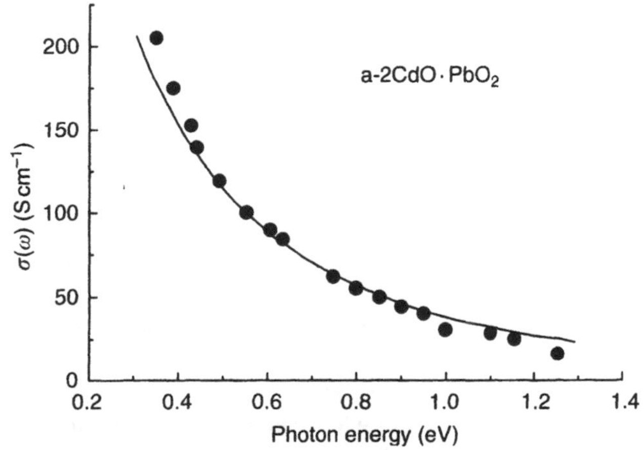

Figure 8.10 Optical conductivity obtained from the absorption coefficient (Fig. 8.9) for annealed a-2CdO·PbO_2 films (Shimakawa *et al.*, 1999). The solid curve indicates the calculation.

1999). Note that $\sigma(\omega) \propto \omega^{-p}(1.5 < p < 3.5)$ is well known in c-semiconductors (Pankove, 1971). $\sigma(0)$ is taken to be the experimentally obtained DC conductivity $\sigma_{DC} \approx 380\,\mathrm{S\,cm^{-1}}$ at room temperature. The fit produces $\tau = 2.0 \times 10^{-15}$ s. The number, n_f, of free electrons and Hall mobility $\mu_H(=\sigma_{DC}/en_f)$ are estimated from the Hall effect and the DC conductivity measurements, respectively, at room temperature. The values of n_f and μ_H, together with other physical parameters, are listed in Table 8.1 for both Cd_2GeO_4 and Cd_2PbO_4 (Shimakawa *et al.*, 1999).

A value of $\mu_H = 10$–$12\,\mathrm{m^2V^{-1}\,s^{-1}}$ seems to be quite reasonable for free electrons and thus is expected to be the same as the microscopic mobility μ_0, which

Table 8.1 Physical parameters deduced from the Hall effect, DC conductivity and optical absorption measurements, for amorphous oxides

	a-CdO·PbO$_2$	a-2CdO·1GeO$_2$
Tauc gap E_0 (eV)	1.8	3.4
σ_{DC} at 300 K (S cm^{-1})	380	210
Hall mobility μ_H (cm^2 V^{-1} s^{-1})	10	12
Number n_f of free electrons (cm^{-3})	2×10^{20}	1×10^{20}
Fermi energy E_F (eV) relative to the conduction-band edge	0.41	0.23
Relaxation time (s)	2.0×10^{-15}	2.7×10^{-15}
Effective mass	0.30 m_0	0.33 m_0
Mean free path l (nm)	1.4	1.4

is very close to the predicted value from a theory (Mott and Davis, 1979). The microscopic mobility μ_0 for a-Si : H and a-Ch can therefore be expected to have values similar to those in those a-oxides. The effective mass may remain a valid concept, even if **k** space is not well defined in disordered matter (Kivelson and Gelatt, 1979; Singh *et al.*, 2002), which is discussed in Chapter 3. From the values of τ and μ_H $(=\mu_0)$, the electron effective mass m_e^* in Cd_2GeO_4 and Cd_2PbO_4 can be estimated to be 0.3 m_e and 0.33 m_e, respectively.

The Fermi energy, which lies above the conduction band edge, is estimated from the standard three-dimensional model, $E_F = h^2(3\pi^2 n_f)^{2/3}/8\pi^2 m_e^*$. These are also listed in Table 8.1 and are larger than those estimated from the Burstein shift (0.1–0.2 eV). The mean free path, $l = (2E_F/m_e^*)^{1/2}\tau$, is estimated to be 1.4 nm for both materials, which is close to those found in liquid metals (Mott and Davis, 1979), and is several times larger than the inter-atomic distance. This value is, of course, very much smaller than that (~50 nm) for normal metals, suggesting that the degenerate electrons are still affected by a disorder potential. Note that the values of the carrier concentrations, $n_f = 1 \times 10^{20}$ and 2×10^{20} cm^{-3} are also considerably smaller than those for normal metals.

Next, let us discuss DC conductivity. Figure 8.11 shows the temperature dependence of the DC conductivity σ_{DC} in Cd_2PbO_4. In contrast to normal metals, in which σ_{DC} is commonly proportional to T^{-1}, an approximately linear dependence ($\sigma \propto T$) on T is observed. This can be interpreted in terms of the Kawabata equation (Kawabata, 1981) extended to a weak-localization regime, in which quantum interference between scattered waves is taken into consideration (see Mott, 1993). The equation takes the form of

$$\sigma_{DC} = \sigma_B \left[1 - \frac{C}{(kl)^2}\left(1 - \frac{l}{L_i}\right)\right] \equiv A + B\frac{l}{L_i}, \tag{8.7}$$

where σ_B is the Boltzmann conductivity, C is of the order of unity, l and L_i are the elastic and inelastic diffusion lengths, respectively. Note that Eq. (8.7) is valid under the condition of $L_i > l$.

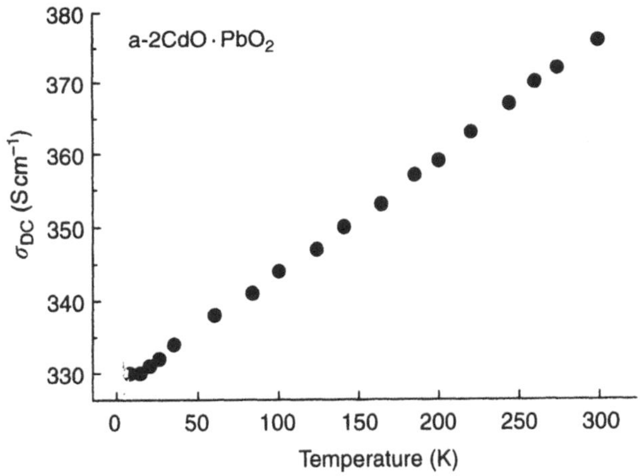

Figure 8.11 Temperature dependence of DC conductivity for a-$2CdO{\cdot}PbO_2$ films (Shimakawa *et al.*, 1999).

For $L_i < l$, the transport should follow the classical behavior, which has been observed in the optical absorption and Hall effect measured at *room temperature*. It is therefore suggested that the electronic transport at lower temperatures does not follow the classical theory. If electron–electron collisions dominate, the scattering time τ_i is proportional to $(kT/E_F)^{-2}$ and hence $L_i(\propto \tau_i^{1/2})$ in Eq. (8.7) should be proportional to T^{-1} (Mott, 1993). Present data correspond to this formulation. The coefficient, $\Delta\sigma_{DC}/\Delta T \approx 0.1$ (S cm^{-1} K^{-1}), is obtained for both the materials. It may be noted that the linear dependence of σ_{DC} on T has also been reported for a-InO_x films with the coefficient $\Delta\sigma_{DC}/\Delta T \approx 0.1$ (Ovadyahu, 1986). Detailed arguments concerning inelastic scattering have been given by Ovadyahu (1986).

Finally, we should discuss the Heisenberg uncertainty in the electron's energy $\Delta E(\sim h/2\pi\tau)$, which is expected to be large for short relaxation times. Using the values of τ in Table 8.1, $\Delta E = 0.33$ and 0.24 eV are obtained for Cd_2GeO_4 and Cd_2PbO_4, respectively. These values are comparable with the Fermi energies already estimated (see Table 8.1), which means that the Fermi statistics breaks down and the Fermi surface is expected to be considerably "blurred." Similar and more severe conditions have been reported in a-$Si_{1-y}Ni_y$: H (Davis *et al.*, 1989) and a-metals (Theye and Fisson, 1983) in interpreting the imaginary part of the optical conductivity (reflectance) due to free carriers. No fundamental theory has yet been developed for such a strong scattering regime.

In summary, the scattering time, mean free path, effective mass and microscopic mobility for free electrons have been estimated from the Hall effect, free-carrier absorption, and DC-conductivity measurements in a new class of degenerate oxide glasses. No sign anomaly was found in the Hall effect and the optical conductivity is found to obey the classical Drude formula. A linear temperature dependence of

DC conductivity found below room temperature suggests that the degenerate electrons are in a weak localization regime and inelastic electron–electron scattering dominates their transport. Although this kind of work is not possible for a-Si : H and a-Ch, the behavior of degenerate electron gas is expected to be common in all disordered semiconductors. The uncertainty in the electrons' energy cannot be ignored and thus a fundamentally important problem still remains unsolved.

8.1.1.3 Hopping conduction

The term hopping conduction means that localized electrons jump (diffuse) quantum mechanically from site to site. The mechanism of hopping conduction was first discussed in doped semiconductors (Kasuya and Koide, 1958; Miller and Abraham, 1960) and was then extended to a-semiconductors (Mott and Davis, 1979). The hopping may be assisted by phonons and hence phonon-assisted hopping between sites will be treated in this section. Two types of hopping can be possible as will be described in the following section.

8.1.1.3.1 NEAREST-NEIGHBOR HOPPING (NNH)

As briefly described in Chapter 1, the quantum tunneling of charge carriers can be easily modeled as follows. Let us first consider the transport of an electron in the conduction tail states, localized on a weak bond. The energy of an electron in this case can be considered to be equal to that of the anti-bonding orbital of the weak bond. The energy of the electron mobility edge, E_c, can be regarded as the height of a barrier in front of the electron that it has to tunnel in order to be able to move to another weak bond with the same or lower energy of its anti-bonding orbital. Following on from Chapter 3, the eigenfunction of the electronic state with an electron in the conduction tail states can be written as

$$|1\rangle = \sum_{\mathbf{l}} C_{1\mathbf{l}}(E_{\mathbf{l}}) a_{1\mathbf{l}}^{+}(\sigma)|0\rangle, \tag{8.8}$$

where $E_{\mathbf{l}}$ is the energy of an electron in the anti-bonding orbital of a bond of site $\mathbf{l}$, and the probability amplitude coefficient, $C_{1\mathbf{l}}(E_{\mathbf{l}})$, is given by Eq. (3.30).

The choice of the probability amplitude coefficient (Eq. (3.30)) suggests that the electron in the conduction states moves as a wave in the extended states because in this situation $E_{\mathbf{l}} > E_c$, and hence ρ_e is real. During the course of its motion, when an electron reaches a weak bond with its anti-bonding orbital energy $E_{\mathbf{l}} < E_c$, it gets trapped in the corresponding tail state with energy $E_{\mathbf{l}}$. In this case the corresponding probability amplitude coefficient becomes

$$C_{1\mathbf{l}}(E_{\mathbf{l}}) = N^{-1/2} \exp(-\rho_t \cdot \mathbf{l}), \tag{8.9}$$

where ρ_t is given by

$$\rho_t = \frac{\sqrt{2m_t^*(E_c - E_{\mathbf{l}})}}{\hbar}. \tag{8.10}$$

Here m_t^* is the effective mass of electron in the tail states. Let us consider a situation that another tail state due to another weak bond's anti-bonding orbital exists at a distance R from the weak bond of site at **l**. Then the probability of tunneling the barrier by a distance R is given by (see Eq. (1.17))

$$p \approx \exp(-2\rho_t R). \tag{8.11}$$

The probability of tunneling the barrier given by Eq. (8.11) from one tail state to another is only valid for tunneling to an energy state equal to or lower than the initial tail state from which the tunneling is taking place. This tunneling can contribute to the transport of electrons localized in the conduction tail states and hence to the photoconductivity even at a very low temperature. However, tunneling from a lower to a higher energy state within the conduction tail states will have to be thermally activated and hence phonon assisted. The rate Γ of such hopping or tunneling from a state of energy E_i to E_f, such that $E_i < E_f$ is given as (Mott and Davis, 1979):

$$\Gamma = \nu_{ph} \exp(-2R/a - \delta E/kT), \tag{8.12}$$

where ν_{ph} is usually taken to be the phonon frequency, $1/a = \rho_t$ and $\delta E = E_f - E_i$. When the tunneling occurs between the nearest neighbor sites, we call it the NNH. The conduction in tail states at relatively low temperatures in a-semiconductors can be dominated by the NNH mechanism. This means that the conduction path moves from the extended states to localized tail states with decreasing temperature, because the number of carriers in tail states increases with decreasing temperature. The NNH mechanism also dominates the photoconductivity at lower temperatures; this will be discussed in Chapter 10.

8.1.1.3.2 VARIABLE-RANGE HOPPING (VRH)

At low temperatures transport near E_F becomes dominant, since the number of carriers in the extended and tail states decrease significantly. Instead, hopping transport near E_F may occur. Since there are $(4\pi/3)R^3N(E_F)$ states available in a spherical region of radius R, where $N(E_F)$ is the DOS at E_F, the average separation of these energy levels can be expressed as

$$\delta E = \left[\tfrac{4}{3}\pi R^3 N(E_F)\right]^{-1}. \tag{8.13}$$

This suggests that the hopping distance R is correlated with the energy difference δE. Using Eq. (8.13), an optimal distance R_{opt} (Mott, 1969) may be defined, at which the hopping rate will be maximum. Thus hopping rate Γ at R_{opt} becomes larger than that in the case of NNH. The former is called the VRH and has been applied to electronic transport in many disordered materials (Mott, 1993).

Let us briefly introduce the VRH mechanism. Figure 8.12 shows the schematic illustration of localized DOS near E_F. Mott (1969) has suggested the most simple case in which DOS around E_F is constant with respect to energy E. Using Eq. (8.13) in Eq. (8.12), and then optimizing the hopping rate with respect to R, we obtain R_{opt} as

$$R_{opt} = \left(\frac{9a}{8\pi N(E_F)kT} \right)^{1/4}. \tag{8.14}$$

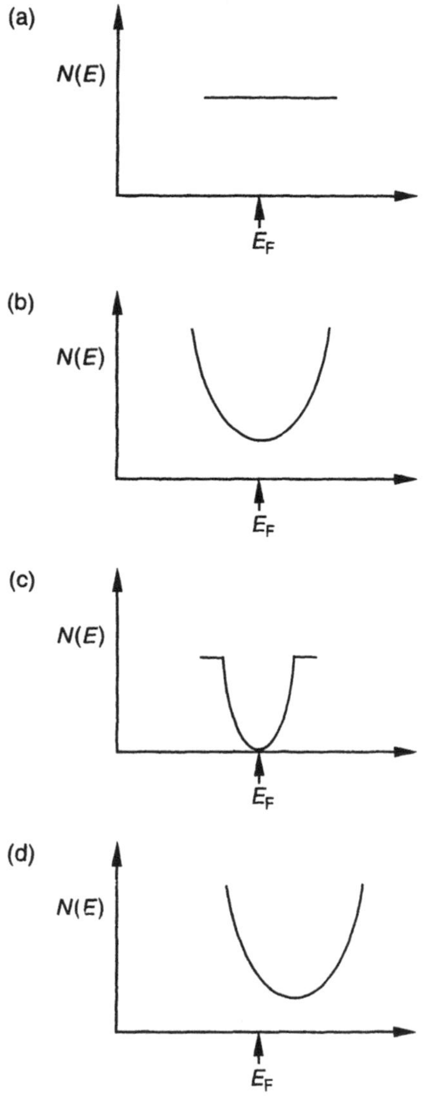

Figure 8.12 Schematic illustration of the localized DOS near E_F: (a) a constant DOS, (b) a parabolic DOS, (c) a Efros–Shklovskii type Coulomb gap and (d) a general case of DOS.

Using Eq. (8.14) in (8.13) gives

$$\delta E_{\text{opt}} = 0.20k(T_0T^3)^{1/4}, \tag{8.15}$$

where

$$T_0 = 18.1/kN(E_F)a^3. \tag{8.16}$$

Using Eqs (8.14) and (8.15) in Eq. (8.12), the optimal hopping rate is then obtained as

$$\Gamma_{\text{opt}} = \nu_{\text{ph}} \exp[-(T_0/T)^{1/4}]. \tag{8.17}$$

The DC hopping conductivity can be expressed as (Mott and Davis, 1979)

$$\sigma_{\text{DC}} = \frac{N_c(eR)^2\Gamma}{6kT}, \tag{8.18}$$

where N_c is the number of carriers. Substituting Eq. (8.17) in Eq. (8.18) and using $N_c = N(E_F)kT$, we get

$$\sigma_{\text{DC}} = \sigma_0 \exp[-(T_0/T)^{1/4}], \tag{8.19}$$

where

$$\sigma_0 = e^2\nu_{\text{ph}}[N(E_F)a/32\pi kT]^{1/2}. \tag{8.20}$$

Equation (8.20) has been applied to many a-semiconductors (Mott and Davis, 1979), doped c-semiconductors (Shklovskii and Efros, 1984), and even high-T_c superconducting materials (Abdel-Wahab and Shimakawa, 1995). Note also that a percolation approach supports the VRH concept (Ambegaokar *et al.*, 1971). One example is of a-Si–Au systems as shown in Fig. 8.13 (Kishimoto and Morigaki, 1979). In a certain temperature range, the fitting to the $T^{-1/4}$ law (Eq. (8.16)) is very good. However, the temperature exponent is not always found to be 1/4. In practice, the DC conductivity is found to follow

$$\sigma_{\text{DC}} = \sigma_0' \exp[-(T_0'/T)^{\gamma}], \tag{8.21}$$

with $1/4 \leq \gamma \leq 1.0$, and hence the Mott VRH theory may be generalized as follows.

For the case considered in Fig. 8.12(b), DOS can be expressed as: $N(E) = N_0|E - E_F|^p + C$, where C is a constant term. The total number of states in an

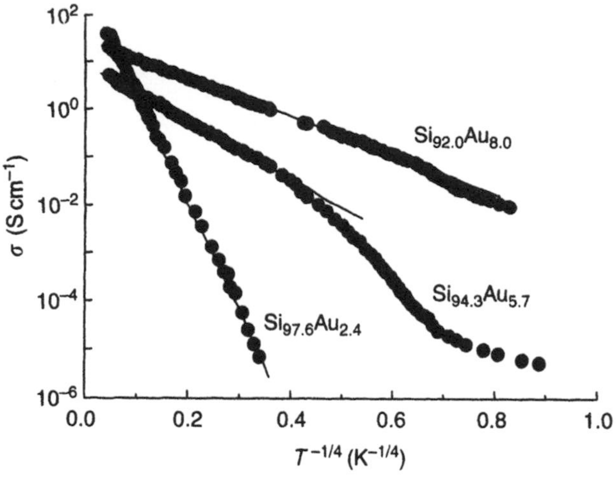

Figure 8.13 Temperature dependence of DC conductivity for amorphous Si–Au alloys (Kishimoto and Morigaki, 1979).

energy range E_F and $E_F + \delta E$ is obtained as

$$\int_{E_F}^{E_F+\delta E} N(E)dE = \frac{N_0(\delta E)^{p+1}}{p+1} + C\delta E.$$

Multiplying this by the spherical volume $\frac{4}{3}\pi R^3$and setting

$$\frac{4}{3}\pi R^3 \left(\frac{N_0(\delta E)^{p+1}}{p+1} + C\delta E \right) = 1,$$

gives

$$\delta E = \left[\frac{4\pi N_0 R^3}{3(p+1)} \right]^{-(p+1)} . \tag{8.22}$$

where $C\delta E$ is assumed to be negligible.

In this case we get $\gamma = (p+1)/(p+4)$ (Hamilton, 1973). If Coulombic correlation is important, the DOS vanishes at E_F and is called *Coulomb gap* (Shklovskii and Efros, 1984). The Shklovskii–Efros type Coulomb gap is shown in Fig. 8.12(c). In this case, DOS near E_F is given as $N(E) = N_0|E - E_F|^{d-1}$, where d is the space dimension (Shklovskii and Efros, 1984). In three-dimensional space, $d = 3$, which gives $p = 2$ in Eq. (8.22) and hence $\gamma = 1/2$ is obtained for the special case of constant density. However, in general, as shown in Fig. 8.12(d), E_F can be located in any position in convex or concave form of DOS. VRH mechanisms for such cases have so far not been discussed.

Any value of γ less than unity may also originate from a modified VRH by taking into account the fractal nature of the system (van der Putten *et al.*, 1992; Hayashi *et al.*, 1994). In this model, a *superlocalization concept* on fractals is assumed, in which the localized wavefunction at E_F decays with distance r as $\exp[(-r/a)^{\zeta}]$; $\zeta > 1.0$, for example, $\zeta = 1.9$ for $d = 3$, and 1.43 for $d = 2$ (Deutscher *et al.*, 1987; Levy and Souillard, 1987). Assuming again a constant DOS near E_F, $\gamma = \zeta/(D + \zeta)$ is obtained, where D is the fractal dimension (van der Putten *et al.*, 1992). Note that $D = 3$ gives Mott's $\gamma = 1/4$. It is actually not easy to choose which model to apply.

Finally, the pre-exponential factor σ_0 appearing in Eq. (8.19) should be addressed. A large discrepancy is found between experimental and theoretical values of σ_0 in most a-semiconductors such as a-Ge, a-Si and a-C etc. (Brodsky and Gambino, 1972; Shimakawa *et al.*, 1994). As $N(E_F)$ estimated from T_0 (Eq. (8.19)) is known to produce reasonable values, VRH is applied for many material systems. However, $N(E_F)$ estimated from σ_0 (Eq. (8.16)) always produces unreasonably high values, for example, $10^{27}\,\mathrm{cm^{-3}\,eV^{-1}}$. This difficulty in Mott's VRH has not been considered to be so serious and it appears also in the modified VRH (Shimakawa *et al.*, 1994). We may, hence, conclude that VRH theory is still incomplete. An alternative model that seems to overcome this problem will be discussed in Chapter 9.

8.1.1.4 AC conduction

AC conductivity is defined as $\sigma(\omega) = \mathrm{i}\omega\varepsilon_0\varepsilon$, where $\varepsilon_0\varepsilon$ is the dielectric constant. As dielectric constant is a complex number, that is, $\varepsilon = \varepsilon_1 - \mathrm{i}\varepsilon_2$, AC conductivity is also a complex quantity. The real part of AC conductivity, $\omega\varepsilon_2$, is usually called AC conductivity or AC loss. An AC conductivity σ_{AC} in the audio-frequency range, which depends on frequency ω and temperature T as

$$\sigma(\omega, T) = A\omega^{s}, \tag{8.23}$$

where A and $s(\leq 1.0)$ are temperature-dependent parameters, has been commonly observed in a-semiconductors (Long, 1982; Elliott, 1987). This simple behavior of AC conductivity is often called dispersive AC loss. In the next section, we outline some of the models or theories used to explain the above empirical law and then describe how these models are applied to a-Ch, a-Si : H and other amorphous systems.

8.1.1.4.1 THEORY

When atomic or molecular dipole relaxation occurs, AC loss is induced and AC conductivity is given in the general form as

$$\sigma(\omega) = N_{\mathrm{p}} \int \alpha(\tau) \frac{\omega^2 \tau}{1 + \omega^2 \tau^2} P(\tau)\,\mathrm{d}\tau, \tag{8.24}$$

where N_p is the number of dipole, $\alpha(\tau)$ polarizability and $P(\tau)$ the distribution of relaxation time (see, e.g. Böttcher and Bordewijk, 1978). Note that the response has a Debye form, that is, $1/(1+i\omega\tau)$. It is known that $\sigma(\omega)$ is nearly proportional to ω when $P(\tau)$ is proportional to $1/\tau$ (Long, 1982; Elliott, 1987). Equation (8.24) is valid in the case of an isolated pair such as electric dipoles. However, within some restrictions, the above equation can also be applied to the electronic hopping case. The AC impurity conduction in compensated Si, $\sigma(\omega, T) = A\omega^s (s = 0.8)$, was first explained by this equation (Pollak and Geballe, 1961). Austin and Mott (1969) then applied the equation to hopping of electrons near E_F in a-semiconductors. For the hopping case, N_p is used for the number of carriers, and the polarizability is expressed as (Pollak and Geballe, 1961):

$$\alpha(\tau) = \frac{(eR)^2}{12kT\cosh^2(\delta E/2kT)}. \tag{8.25}$$

We discuss first the case of quantum tunneling of electrons near E_F, which is apparently applicable to a-Si, a-Ge etc. Defining the relaxation time as $\tau = 1/\Gamma$, and then using Eq. (8.12) for Γ, calculate the derivative of τ with respect to R to get

$$d\tau = \frac{2\nu_{ph}^{-1}}{a}\exp(2R/a)\,dR = \frac{2\tau}{a}\,dR, \tag{8.26}$$

where the Boltzmann factor is ignored for simplification.

For a random distribution of hopping sites N, $P(\tau)$ in Eq. (8.24) can be written as

$$P(\tau)\,d\tau = P(R)\,dR = 4\pi NR^2\,dR. \tag{8.27}$$

Substituting Eqs (8.25)–(8.27) into Eq. (8.24), we get

$$\sigma(\omega) = \frac{\pi NN_p e^2 a\omega}{6kT}\int R^4 \frac{\omega\tau}{1+\omega^2\tau^2}\frac{d\tau}{\tau}. \tag{8.28}$$

For hopping conduction near E_F, the number of carriers is given by $N(E_F)kT$, which is also equal to the number of unoccupied sites. This gives $NN_p = [N(E_F)kT]^2$, which can be substituted in Eq. (8.25). Since $\omega\tau/(1+\omega^2\tau^2)$ is sharply peaked at $\omega\tau = 1$, like a δ-function, R can be taken out of the integral. The expression for R at $\omega\tau = 1$, denoted by R_ω can be obtained from Eq. (8.12) as

$$R_\omega = \frac{a}{2}\ln(\nu_{ph}/\omega), \tag{8.29}$$

where the Boltzmann factor has been neglected. Then the integral in Eq. (8.28) can be evaluated as

$$\int_0^\infty \frac{\omega\tau}{1+\omega^2\tau^2}\frac{d\tau}{\tau} \approx \frac{\pi}{2}. \tag{8.30}$$

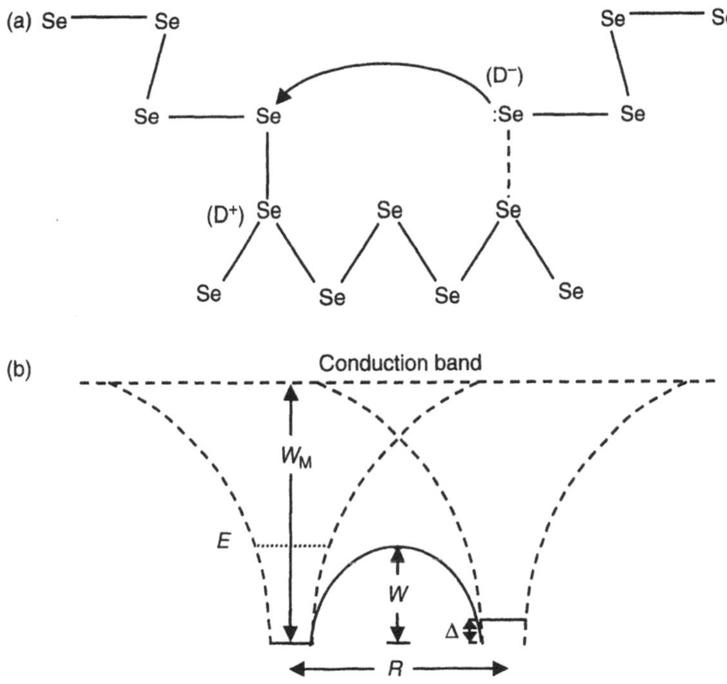

Figure 8.14 (a) Schematic illustration of two-electrons hopping transport in a-Ch. Two electrons transfer from a $D^-(C_1^-)$ to a $D^+(C_3^+)$ center by exchanging their places and a new bond (dotted) is formed. (b) Potentials diagram experienced by carriers at a distance R apart. W_M is the maximum energy required to move two electrons from a $D^-(C_1^-)$ to form a $D^+(C_3^+)$, W is the height of the barrier over which the carriers must hop and Δ is the disorder energy (Elliott, 1977).

Using Eqs (8.29) and (8.30) in Eq. (8.28), we get

$$\sigma(\omega) = Ce^2a^5N(E_F)^2kT\omega\{\ln(\nu_{ph}/\omega)\}^4, \tag{8.31}$$

where C is a constant (numerical value). This is called the *quantum mechanical tunneling* (QMT) model. In the early stages of research in a-semiconductors, AC transport data have been analyzed using Eq. (8.31) (Mott and Davis, 1979). However, due to some difficulties in applying the QMT model to a-Ch, another model of *correlated barrier hopping* (CBH) has been proposed for a-Ch (Elliott, 1977). CBH model is based on over barrier hopping (classical hopping) developed by Pike (1972).

As already discussed in Chapter 7, principal defects in a-Ch are assumed to be negatively and positively charged. Figure 8.14(a) presents a schematic illustration of two-electron hopping transport (for a simple example, in a-Se). Two electrons, as a bipolaron, can hop between oppositely charged coordination defect sites (e.g. $C_3^+ - C_1^-$). These defects are classified as negative-U systems, which may appear in deformed lattice and hence it will be discussed in Section 8.2. However, DC conductivity can occur in the extended states and hence we discuss this type of bipolarons here.

As shown in Fig. 8.14(b), the hopping rate for such processes involves an activation energy W, which is correlated with intersite separation R through a Coulombic interaction between charge carriers and defects. Therefore, for the case of a bipolaron we can write

$$W = W_M - 2e^2/\pi\varepsilon_0\varepsilon R, \tag{8.32}$$

where W_M is the maximum barrier height (for $R = \infty$) and has a value comparable with the bandgap ($W_M = B - W_1 + W_2$; see Fig. 8.14) (Elliott, 1977). $\varepsilon_0\varepsilon$ is the dielectric constant of background. The hopping time is given as

$$\tau = \nu_{ph}^{-1} \exp(W/kT). \tag{8.33}$$

Using Eqs (8.32) and (8.33) and assuming W_M to be a constant we get

$$d\tau = \frac{2e^2\tau}{\pi\varepsilon_0\varepsilon kT R^2} dR. \tag{8.34}$$

Again, for random distribution of hopping site (Eq. (8.27)), the AC conductivity can be written as

$$\sigma(\omega) = \frac{\pi\varepsilon_0\varepsilon N N_p e^2\omega}{12kT} \int R^6 \frac{\omega\tau}{1+\omega^2\tau^2} \frac{d\tau}{\tau}. \tag{8.35}$$

Applying the same arguments as used in deriving Eq. (8.29) for R_ω (R at $\omega\tau = 1$), we get it as

$$R_\omega = \frac{2e^2}{\pi\varepsilon_0\varepsilon[W_M - kT\ln(\nu_{ph}/\omega)]}, \tag{8.36}$$

and then taking $NN_p = N_T^2/2$ (Elliott, 1977) in Eq. (8.35), where N_T is the total number of charged defects, we get the AC conductivity as

$$\sigma(\omega) = \frac{\pi^3}{6} N_T^2 \varepsilon_0\varepsilon\omega R_\omega^6. \tag{8.37}$$

Polarons are assumed to hop between C_3^+, C_1^0 or C_1^- or C_1^0 pairs of sites, and this process is expected to dominate the AC conductivity at higher temperatures, since the number of neutral defects is expected to increase with temperature due to the reverse reaction of $2C_1^0 \rightarrow C_3^+ + C_1^-$ (Shimakawa, 1982). Shimakawa (1983) has confirmed that the microwave absorption in a-Ch can also be interpreted in terms of the bipolaron hopping mechanism. In certain cases, both bipolarons and single polarons can dominate AC conductivity in a-Ch (Shimakawa, 1982; Shimakawa *et al.*, 1994). These will be discussed later.

We should remember that both the QMT and CBH models are based on the pair approximation (PA) in which carriers are confined in a pair of localized states. $\sigma(\omega)$

in this approximation cannot give DC conductivity, $\sigma(0)$, as $\omega \to 0$. These models, therefore, cannot readily account for the experimental data if both DC and AC transports occur by the same mechanism (Long, 1982; Elliott, 1987). A proper approach to hopping AC conductivity seems to be the *continuous-time random-walk* (CTRW) approximation originally developed by Scher and Lax (1973). A simple form of the AC conductivity based on CTRW has been presented as (Dyre, 1985):

$$\sigma(\omega) = \sigma_{\mathrm{DC}} \frac{\mathrm{i}\omega\tau}{\ln(1 + \mathrm{i}\omega\tau)}, \tag{8.38}$$

where ω is the angular frequency of the applied field and τ is the hopping relaxation time. The AC conductivity is strongly correlated to DC conductivity. Note here that σ_{DC} here should be the hopping conductivity from the localized states, but not from the transport in the extended states. The above equation can be applied to a-Ge, a-Si and a-C in which DC and AC conductivity might be attributed to hopping of electrons near E_{F}.

A dispersive AC loss is also found to originate from a macroscopic or mesoscopic scale of inhomogeneities in materials. This is called the Maxwell–Wagner effect. A classical effective medium approximation (EMA) can be useful for such inhomogeneous media (Kirkpatrik, 1973). EMA predicts that the total network conductance σ_{m} in D dimensions follows

$$\left\langle \frac{\sigma - \sigma_{\mathrm{m}}}{\sigma + (D-1)\sigma_{\mathrm{m}}} \right\rangle_{\sigma} = 0, \tag{8.39}$$

where σ is a random value of conductivity. Under the assumption of a random mixture of particles of *two* different conductivities, a simple form of the DC conductivity and Hall mobility can be obtained (Kirkpatrik, 1973; Cohen and Jortner, 1973). EMA has been extended to apply it to the AC conductivity in which σ in Eq. (8.39) is used as a complex conductivity ($\sigma^* = \sigma_1 + \mathrm{i}\sigma_2$) (Springett, 1973). Long *et al.* (1989) have applied this to study the AC loss in a-Si : H, which will be

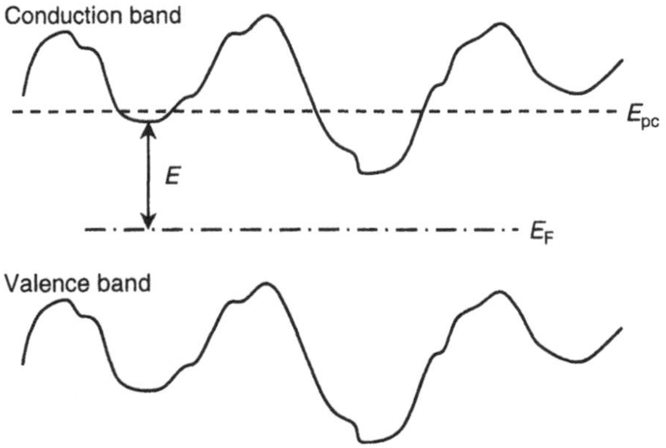

Figure 8.15 Schematic illustration of band-edge fluctuations and the percolation threshold E_{pc}.

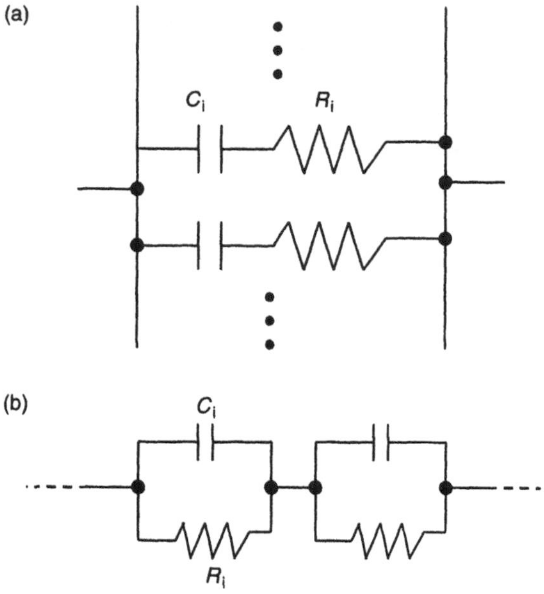

Figure 8.16 Equivalent electrical circuits for: (a) PA and (b) CTRW approximation.

discussed later. Note that the Springett approach can be applied for conductivities only at non-zero frequencies.

A percolation path method (PPM) adapted for zero and non-zero frequencies has been proposed by Dyre (1993). In a macroscopically disordered media, for example, as shown in Fig. 8.15, PPM assumes that both the DC and low-frequency AC currents mainly follow the easiest percolation path between the electrodes. If the local conductivity is thermally activated, the DC conductivity can be given as

$$\sigma_{DC} = \sigma_0 \exp(-E_{pc}/kT), \tag{8.40}$$

where E_{pc} is the percolation energy defined by

$$\int_{-\infty}^{E_{pc}} p(E)\mathrm{d}E = p_c, \tag{8.41}$$

where $p(E)$ is the activation energy probability distribution with p_c being the percolation threshold. Under the low temperature limit, where the local conductivity varies by orders of magnitude for small variations of activation energy, $\sigma(\omega)$ in Eq. (8.38) is obtained as follows (Dyre, 1993, 1997). In one dimension, which is equivalent to the electrical circuit as shown in Fig. 8.16, the conductivity can be presented at the percolation threshold as

$$\frac{1}{\sigma(\omega)} := \int_0^{\infty} \frac{p(\Delta E)}{g(\Delta E) + \mathrm{i}\omega\varepsilon_0\varepsilon_\infty} \mathrm{d}(\Delta E), \tag{8.42}$$

where $g(\Delta E)$ is the local conductivity and $p(\Delta E)$ the distribution function of ΔE. Assuming a uniform distribution, $p(\Delta E) = 1/E_{pc}(0 < \Delta E < E_{pc})$, the distribution of local conductivity is obtained as (Papoulis, 1965):

$$p(g) = \frac{p(\Delta E)}{|dg/d(\Delta E)|} = \frac{kT}{E_{pc}} \frac{1}{g(\Delta E)}. \tag{8.43}$$

Equation (8.38) then becomes

$$\frac{1}{\sigma(\omega)} = \frac{kT}{E_{pc}} \int_{g_{min}}^{g_0} \frac{1}{g(g + i\omega\varepsilon_0\varepsilon_\infty)} dg, \tag{8.44}$$

where $g_{min} = g_0 \exp(-E_{pc}/kT)$. Under the condition of $\omega\varepsilon_\infty \ll g_0$, Eq. (8.44) can be written as

$$\sigma(\omega) = \frac{E_{pc}}{kT} \frac{i\omega\varepsilon_0\varepsilon_\infty}{\ln[1 + i\omega\varepsilon_\infty/g_0 \exp(-E_{pc}/kT]}. \tag{8.45}$$

Letting ω go to zero, one finds $\sigma(0)$ as

$$\sigma(0) = \frac{E_{pc}}{kT} \exp(-E_{pc}/kT). \tag{8.46}$$

Substituting Eq. (8.46) into Eq. (8.45), the latter can be written as

$$\sigma(\omega) = \sigma(0) \frac{i\omega\tau}{\ln(1 + i\omega\tau)}, \tag{8.47}$$

where $\tau = \varepsilon_0\varepsilon_\infty/g_0 \exp(-E_{pc}/kT)$. This is called the dielectric relaxation time. As shown in Eq. (8.47), the AC conductivity is directly related to the DC conductivity. Note, however, that $\sigma_{DC} = \sigma_0$ here is due to the transport in the extended states but not hopping conductivity. Note also that the relaxation time τ here is so called dielectric relaxation time given by $\varepsilon_0\varepsilon_\infty/\sigma_{DC}$ but it is not the hopping relaxation time. PPM is also applied to the AC loss in a-Si : H, which will be discussed later. Although Eq. (8.47) is similar to Eq. (8.38), the underlying physics of deriving them is completely different.

It is interesting to note that $\sigma(\omega)/\sigma_{DC}$ shows a universal scaling function of $\omega\tau$, irrespective of the systems (either the electron hopping or inhomogeneous system). There are many reports on a universal scaling for AC conductivity (Zvyagin, 1980, 1984; Long, 1991; Dyre, 1993, 1997; Shimakawa *et al.*, 1996). A scaling nature is also found in $\sigma(\omega)/\sigma_{DC}$ for ionically conducting glasses (Bunde *et al.*, 1998; Roling *et al.*, 1997; Baranovskii and Cordes, 1999). As the discussion of ionic transport is beyond the scope of this book, we will not discuss it any further in this section.

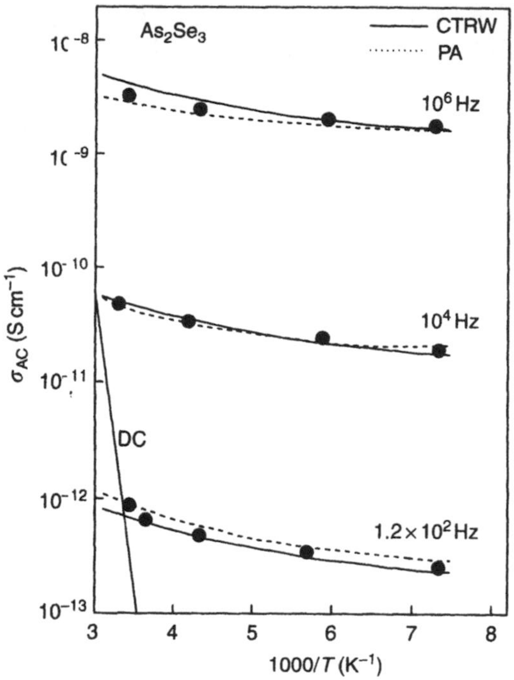

Figure 8.17 Temperature dependence of the AC conductivity at various frequencies in a-As_2Se_3. The solid curves represent the CTRW calculation and dashed curves are the PA calculation (Ganjoo and Shimakawa, 1994).

8.1.1.4.2 EXPERIMENTAL RESULTS

In this section the experimental results obtained in a-Ch, a-Si : H, and other amorphous systems are discussed in terms of the models presented above. First, we will discuss the AC loss in a-Ch. CBH of bipolarons proposed by Elliott (1977) for weak temperature dependences and CBH of single polarons proposed by Shimakawa (1981, 1982) are able to interpret the overall features of AC conductivity. Fitting the experimental data to CBH gives fairly good results and produces the density of charged defects and their energy position within the energy band (Shimakawa, 1981, 1982). Figure 8.17 shows, as an example, $\sigma(\omega, T)$, as a function of temperature in a-As_2Se_3. The experimental data fit very well to $\sigma(\omega, T) \propto \omega^s$ ($s \approx 1.0$). The dashed lines indicate the prediction by PA model. Fitting the experimental data to Eq. (8.34) produces a value for $N_T = [D^+] + [D^-] = 2.4 \times 10^{19}\,\mathrm{cm}^{-3}$, as given in Table 8.2.

The density of defects, estimated from the CBH model, is found in the range of 4×10^{18}–$4 \times 10^{19}\,\mathrm{cm}^{-3}$ in a-Ch (Shimakawa, 1981, 1982). This is, however, always found to be two orders of magnitude larger than that (10^{16}–$5 \times 10^{17}\,\mathrm{cm}^{-3}$) from the other measurements (e.g. drift mobility and light-induced electron spin

Table 8.2 Physical parameters used for estimating the density of defect states and fitting to the experimental data. Other parameters have their usual meanings and are explained in the text

Sample	W_M (eV)	W_1 (eV)	U_{eff} (eV)	$N_T(cm^{-3})$ *CBH based on CTRW*	$N_T(cm^{-3})$ *CBH based on PA*
As_2Se_3	1.80	—	—	2.0×10^{17}	2.4×10^{19}
Se	2.00	0.63	0.32	1.0×10^{18}	2.4×10^{19}
As_2Se_3+0.5 at % Ag	1.85	0.50	0.33	5.0×10^{17}	4.2×10^{19}

resonance (LESR)). If charged defects dominate the drift mobility, LESR, AC conductivity etc., the estimated density of centers should be the same irrespective of the method of measurements. The cause of such a large discrepancy in the estimated density of states is not yet clear. As already mentioned, CBH is based on the PA in which carriers are assumed to be confined on a pair of centers. As usually the carriers are not confined to a pair of centers at lower frequencies, PA may not be valid at lower frequencies. It is therefore suggested that the CTRW approach is moie realistic than PA. It is of great interest to compare the deduced density of defects as estimated from PA and CTRW approaches.

The solid curves in Fig. 8.17 indicate the prediction from CTRW approach (Ganjoo and Shimakawa, 1994). The DC hopping conductivity, calculated for bipolarons, is considerably smaller than the experimental DC conductivity. We should remember that the solid curve of DC conductivity is obtained experimentally and is attributed to the band transport but not hopping transport. From CTRW approach, N_T is estimated to be 2×10^{17} cm^{-3} and seems to be very consistent with that from other measurements as shown in Table 8.2 and very reasonable values of $N_T(2 \times 10^{17}–1 \times 10^{18}$ cm$^{-3})$ are also obtained for other a-Ch (Ganjoo *et al.*, 1996). This clearly supports the suggestion very strongly that the application of CTRW in CBH to estimate the proper density of charged defects in a-Ch is more appropriate. This also leads to the conclusion that the AC conductivity study can be regarded as a proper method for defect spectroscopy in a-Ch.

Next, we discuss AC loss in a-Si : H. Abkowitz *et al.* (1976) first measured the AC loss in a-Si : H and suggested that AC loss can be attributed to the QMT. This analysis, however, was criticized later on, because neither the magnitude of the value of s nor its temperature dependence was found to be consistent with the QMT model (Long, 1982; Elliott, 1987). Subsequently, extensive studies using a sandwich cell, have been done and bulk and interfacial relaxations have been identified by measurements (Shimakawa *et al.*, 1985, 1986). The interfacial relaxation originates from a Schottky barrier which gets easily formed, due to a low density of gap states in a-Si : H. The bulk relaxation was interpreted in terms of the hydrogen related two level systems (atomic relaxation; not electronic hopping). To avoid interfacial relaxation, which does not allow a proper estimation of AC loss, the n^+ undoped n^+ sandwich structure has been used for measurements

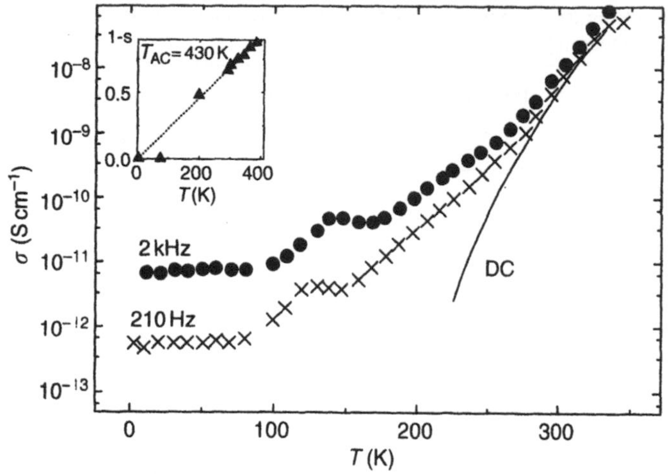

Figure 8.18 Temperature dependence of the AC conductivity for undoped a-Si : H. The inset shows, 1-s as a function of T (Shimakawa *et al.*, 1987).

(Shimakawa *et al.*, 1987). The AC loss measurements down to 2 K have been successfully performed.

Two types of AC losses have been found in a-Si : H as shown in Fig. 8.18. One is a small temperature-independent loss occurring dominantly at low temperatures (2–100 K) and other is a strongly temperature-dependent loss occurring at high temperatures (100–300 K), which merges into the DC conductivity. The nature of these losses is phenomenologically the same as that found in some a-Chs. The low temperature loss is interpreted in terms of CBH process (Shimakawa *et al.*, 1995). However, as will be discussed below, high temperature loss can be interpreted in terms of the so called Maxwell–Wagner type inhomogeneities. EMA adapted for non-zero frequencies by Springett (1973) was first applied to explain the high temperature AC loss in a-Si : H (Long, 1989). Although a qualitative agreement between theory and experiment has been obtained, the magnitude of the AC loss predicted by the theory is found to be two orders of magnitude smaller than the experimental value.

PPM (Dyre, 1993), instead of EMA, adopted for zero and non-zero frequencies has been applied to interpret high temperature loss (Shimakawa *et al.*, 1996) in a-Si : H. The open and closed circles in Fig. 8.19 show the AC loss at 210 Hz and 2 kHz, respectively, which are replotted from the original data in Fig. 8.18. In the high temperature regime, $\sigma(\omega)$ is found to be proportional to ω^s, with $s = 1 - T/T_0$ (Shimakawa *et al.*, 1987). In the PPM, an important parameter is the relaxation time, $\tau = \varepsilon_0\varepsilon_\infty/\sigma_{DC}$, which determines the AC conductivity, provided the DC conductivity σ_{DC} is known (Eq. (8.40)). The calculated results using Eq. (8.39) are shown by the solid curves, by taking $\varepsilon_\infty = 10$, and using the DC conductivity data. The agreement between experiments and calculations is fairly good.

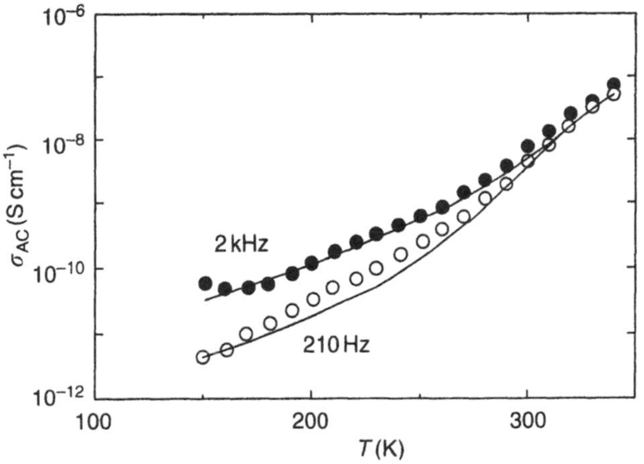

Figure 8.19 Temperature dependence of the AC conductivity replotted from Fig. 8.18. The solid curves show the calculated results using Eq. (8.39).

It is of interest to discuss the parameter T_0 which is deduced from the frequency exponent s. Although the empirical relation, $s = 1 - T/T_0$, cannot be deduced directly from the theoretical prediction of s (given as $1 - 2/\ln(\omega\tau)$), the parameter T_0 is compared with the dispersion parameter, observed in transient photoconductivity measurements, where the photocurrent, $I_p(t)$, is often observed to decay as t^{-s} with $s = 1 - T/T_0$ (Tiedje and Rose, 1980; Orenstein and Kastner, 1981; Murayama and Mori, 1992). However, T_0 observed from this experiment is generally around 300 K, which is smaller than the value (430 K) estimated from the AC loss. Although the transient photocurrent is generally treated in terms of the multiple trapping of electrons in band tail states (MT model) (Tiedje and Rose, 1980; Orenstein and Kastner, 1981), the behavior of $I_p(t)$ can also be interpreted by introducing potential fluctuations (Overhof and Thomas, 1989; Murayama and Mori, 1992) in which the parameter T_0 is used as a measure of the extent of inhomogeneity (Howard and Street, 1991). The difference between the estimates of T_0 from its experimental value may be attributed to the approximate nature of the model used to analyze the AC data. It is concluded therefore that the high temperature AC loss originates from a local conductivity variation, induced by inhomogeneities on a macroscopic and/or mesoscopic scale. This picture is consistent with the model of long range potential fluctuations in a-Si : H (Overhof and Beyer, 1981; Overhof and Thomas, 1989). The reason why long range potential fluctuations dominate in AC loss in a-Si : H may be due to a small density of gap states present in a-Si : H than that in a-Ch.

Finally, we mention the AC losses in other a-semiconductors and insulators. The CBH model has been applied to a-As (Greaves *et al.*, 1979), a-P (Extance, 1985), a-SiO_2 (Shimakawa and Kondo, 1983, 1984), and a-$Si_{1-x}N_x$: H (Shimakawa *et al.*, 1992). The CTRW of electron hopping occuring near E_F has been proposed for a-Ge (Shimakawa, 1989) and a-C (Shimakawa and Miyake, 1988, 1989). These AC losses for a-Ge and a-C will be discussed in Section 8.2, as such electronic transport is expected to occur involving phonons in these materials.

8.1.2 Thermoelectric power

Measurements of thermoelectric power in solids can help to identify whether the carriers involved in the transport are electrons or holes (n- or p-type in other words). Thermoelectric power, S, is related to the Peltier coefficient Π by

$$S = \Pi/T. \tag{8.48}$$

The Peltier coefficient is defined as the energy carried by electrons per unit charge and energy is measured relative to the Fermi energy. Each electron contributes to Π in proportion to its relative contribution to the total conductivity. There is therefore a correlation between S and electrical conductivity σ (Fritzshe, 1974; Mott, 1993) given by

$$\Pi = -\frac{1}{e}\int (E - E_{\mathrm{F}})\frac{\sigma(E)}{\sigma}\frac{\partial f}{\partial E}\,\mathrm{d}E, \tag{8.49}$$

and

$$S = -\frac{k}{e}\int \frac{E - E_{\mathrm{F}}}{kT}\frac{\sigma(E)}{\sigma}\frac{\partial f}{\partial E}\,\mathrm{d}E, \tag{8.50}$$

where σ is given by the energy dependent conductivity $\sigma(E)$ as

$$\sigma = e\int \mu(E)N(E)f(1 - f)\,\mathrm{d}E = -\int \sigma(E)\frac{\partial f}{\partial E}\,\mathrm{d}E. \tag{8.51}$$

Here f is the Fermi distribution function and therefore $f(1 - f) = -kT\,\mathrm{d}f/\mathrm{d}E$ and $\sigma(E) = e\mu(E)N(E)kT$. Note that $S < 0$ for electrons at energies $E > E_{\mathrm{F}}$ and $S > 0$ for holes at energies $E < E_{\mathrm{F}}$.

The thermoelectric power associated with the conduction of electrons in extended states is given as follows. Starting from Eq. (8.50), S is written as

$$\begin{aligned} S &= -\frac{k}{e\sigma}\int \sigma(E)\frac{(E - E_{\mathrm{F}})}{kT}\frac{1}{kT}f(E)[1 - f(E)]\,\mathrm{d}E \\ &= -\frac{k}{e}\frac{1}{kT}\frac{e\int_{E_c}^{\infty} N(E)\mu(E)(E - E_{\mathrm{F}})\exp[-(E - E_{\mathrm{F}})/kT]\,\mathrm{d}E}{e\int_{E_c}^{\infty} N(E)\mu(E)\exp[-(E - E_{\mathrm{F}})/kT]\,\mathrm{d}E}. \end{aligned} \tag{8.52}$$

Assuming $N(E)\mu(E)$ is a constant (independent of energy), we get

$$\begin{aligned} S &= -\frac{k}{e}\frac{1}{kT}\frac{\int_{E_c}^{\infty}(E - E_{\mathrm{F}})\exp[-(E - E_{\mathrm{F}})/kT]\,\mathrm{d}E}{kT\exp[-(E_c - E_{\mathrm{F}})/kT]} \\ &= -\frac{k}{e}\frac{1}{kT}\frac{\int_{E_c}^{\infty} E\exp[-(E - E_{\mathrm{F}})/kT]\,\mathrm{d}E - E_{\mathrm{F}}\int_{E_c}^{\infty}\exp[-(E - E_{\mathrm{F}})/kT]\,\mathrm{d}E}{kT\exp[-(E_c - E_{\mathrm{F}})/kT]}. \end{aligned} \tag{8.53}$$

Let $E - E_c = \varepsilon$, then

$$S = -\frac{k}{e}\frac{1}{kT}\frac{(E_c - E_F)\int_0^\infty \exp(-\varepsilon/kT)\,d\varepsilon + \int_0^\infty \varepsilon \exp(-\varepsilon/kT)\,d\varepsilon}{kT}$$

$$= -\frac{k}{e}\frac{1}{kT}\frac{(E_c - E_F)kT + (kT)^2}{kT}$$

$$= -\frac{k}{e}\left(\frac{E_c - E_F}{kT} + 1\right). \tag{8.54}$$

A comparison of Eqs (8.1) and (8.54) shows that a plot of $\ln \sigma_{DC}$ and S as a function of $1/T$ should have the same slope if conduction takes place within only one band. The value of $E_c - E_F$, estimated from Eq. (8.54), is always smaller than that estimated from Eq. (8.1) in a-semiconductors. This will be discussed later and also in Section 8.2.

Next, let us discuss a degenerate (metallic) case. If E_F lies above a mobility edge, S can be determined considering electrons near E_F. Equation (8.50) is then obtained as

$$S = -\frac{1}{e\sigma(E_F)T}\int \sigma(E)(E - E_F)\frac{\partial f}{\partial E}\,dE. \tag{8.55}$$

Expanding $\sigma(E)$ in a Taylor series at $E = E_F$ as

$$\sigma(E) = \sigma(E_F) + (E - E_F)\sigma'(E_F) + \tfrac{1}{2}(E - E_F)^2\sigma''(E_F) + \cdots,$$

the integral, denoted by I, in Eq. (8.55) is obtained for the first nonvanishing terms

$$I = \int \sigma(E)(E - E_F)\frac{\partial f}{\partial E}dE = \sigma'(E_F)\int_{E_F}^\infty (E - E_F)^2\frac{\partial f}{\partial E}\,dE. \tag{8.56}$$

Changing the variable of integration to ε defined by $(E - E_F)/kT = \varepsilon$, the integral can be evaluated as

$$I = (kT)^2\int_0^\infty \varepsilon^2\frac{\partial f}{\partial E}\,dE = (kT)^2\int_0^\infty \frac{\varepsilon^2}{(e^{\varepsilon/2} + e^{-\varepsilon/2})^2}\,d\varepsilon = \frac{\pi^2}{3}(kT)^2. \tag{8.57}$$

Using Eq. (8.57) in Eq. (8.55), S is obtained as

$$S = -\frac{\pi^2}{3}\frac{k^2T}{e}\frac{\sigma'(E_F)}{\sigma(E_F)} = -\frac{\pi^2}{3}\frac{k^2T}{e}\left\{\frac{d\ln\sigma(E)}{dE}\right\}_{E=E_F}. \tag{8.58}$$

By setting the conductivity as

$$\sigma(E_F) = e\mu(E_F)N(E_F)kT, \tag{8.59}$$

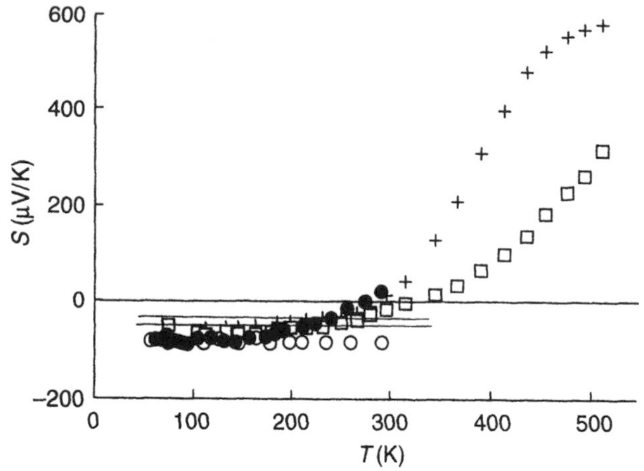

Figure 8.20 Thermopower of sputtered and evaporated a-Ge. Hopping regime may be below 250 K (Theye *et al.*, 1985).

Equation (8.58) can be alternatively represented by

$$S = -\frac{\pi^2}{3}\frac{k^2 T}{e}\frac{N'(E_F)}{N(E_F)} = -\frac{\pi^2}{3}\frac{k^2 T}{e}\left\{\frac{d \ln N(E)}{dE}\right\}_{E=E_F}. \tag{8.60}$$

Here, the mobility $\mu(E)$ is supposed to be independent of energy.

If the states near E_F are localized, then themoelectric power depends dominantly on hopping of electrons at low temperatures. VRH may occur in a narrow energy range $w = \delta E_{opt}$ (see Eq. (8.15)). Several authors have discussed the thermopower under the VRH condition (see, e.g. Mott, 1993). Mott (1993) has derived the following equation

$$S = -\frac{1}{2}\frac{k}{e}\frac{w^2}{kT}\left\{\frac{d \ln N(E)}{dE}\right\}_{E=E_F}. \tag{8.61}$$

When we take $w = \delta E_{opt} \propto T^{3/4}$, we get S proportional to $T^{1/2}$. S should be zero in the Mott VRH regime because $N(E)$ is assumed to be a constant near the Fermi level. An example of experimental data on a-Ge is shown in Fig. 8.20 (Theye *et al.*, 1985). Although the electronic transport at relatively low temperatures is believed to be dominated by VRH, S is not proportional to $T^{1/2}$, suggesting that further refinement of the VRH model is required.

We mentioned before that the value of $E_c - E_F$ estimated from Eq. (8.54) is always smaller than that from Eq. (8.1), for band transport in a-semiconductors. Overhof and Beyer (1983) introduced the relationship between σ (Eq. (8.1)) and

S (Eq. (8.54)) as

$$\ln \sigma + \left| \frac{eS}{k} \right| = \ln \sigma_0 + A \equiv Q(T), \tag{8.62}$$

where $Q(T)$ is called Q function. $Q(T)$ in a-Si : H is found to obey

$$Q(T) = Q_0 - \frac{E_Q}{kT}, \tag{8.63}$$

where $Q_0 = \ln \sigma_0 + \mathrm{A}$ and $E_Q = E_\sigma - E_S$. Figure 8.21 shows the temperature dependence of Q in P-doped a-Si : H. For undoped, P-doped and B-doped a-Si : H, E_Q takes the values between 0.05 and 0.25 eV, depending on doping, and Q_0 is 10, indicating that E_σ and E_S are different in a-Si : H (Overhof and Beyer, 1983). Overhof and Beyer (1983) have suggested that a long-range potential fluctuation of the band edge, as shown in Fig. 8.15, yields the difference between E_σ and E_S: the electrical conduction occurs above a percolation threshold E_th, while thermoelectric power is dominated by carriers in the "valleys." Carriers in valleys surmount "cols" (thermal excitation). This is a type of classical hopping of carriers and hence the energy of ΔW is used to crossover a col as shown in Fig. 8.15. As the total energy to surmount cols is zero $(\Delta W - \Delta W)$, E_S becomes equal to $E_\sigma - \Delta W$. ΔW therefore can be regarded as a measure of the potential fluctuation. This view is phenomenologically similar to the two-channel model of conduction (Nagels *et al.*, 1972; Nagels, 1979). A hopping energy ΔW is required to the transport of charge carriers, when the conduction in the tail states dominates the transport. As the hopping energy ΔW is not involved in the thermoelectric power, as stated above, E_S becomes smaller than E_σ by an amount of ΔW. It is, however,

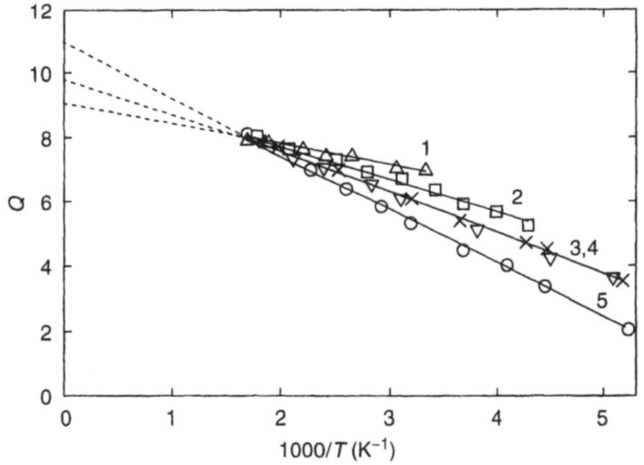

Figure 8.21 Q as a function of the inverse temperature for various n-type a-Si : H (Overhof and Thomas, 1989). (Numbers 1–5 represent different doping concentrations.)

not easy to distinguish these two models, since "valleys" in the potential fluctuation model can be considered phenomenologically to be the same as "band tail states."

The temperature-dependent Q function has also been reported for a-Ge : H (Hauschildt *et al.*, 1982) and a-Ch (Seager *et al.*, 1973; Seager and Quinn, 1975; Emin, 1975). For a-Ge : H $E_Q \approx 0.15\,\text{eV}$ is independent of the doping level but Q_0 depends on doping. These are different from E_Q and Q_0 in a-Si : H, where E_Q depends on doping. For a-Ch $Q_0 \approx 10$, and $E_Q \approx 0.2\,\text{eV}$, which are similar to those for highly doped a-Si : H. Emin (1975) has proposed that Q function in a-Ch can be attributed to small polarons.

8.1.3 Hall effect

Perhaps the most anomalous behavior in the carrier transport in a-semiconductor is that of the Hall effect. This is because, as already stated in Section 8.1.2, the Hall effect has the opposite sign estimated from the thermoelectric power. We call it pn anomaly; holes give a negative sign and electrons positive in the Hall voltage. As shown in Section 8.1.2, there is no sign anomaly in degenerate a-semiconductors in which electronic transport occurs well above the mobility edge. When carrier transport occurs near mobility edge, the mean free path is expected to be very small, which corresponds to the interatomic spacing. As a consequence, the standard transport theory based on the Boltzmann equation cannot be useful.

Applying the concept of hopping polarons developed by Friedman and Holstein (1963), Friedman (1971) has put forward a theory for the Hall effect for carriers moving in a-solids near the mobility edge. It produces interference between two scattering paths involving three atomic sites A, B and C. One path is from A to B direct and the other from A to B via C. The Hall mobility deduced in this way is found to be independent of temperature and in qualitative agreement with the observations, but the Hall coefficient is always found to be negative whether the carriers are electrons or holes. This has been interpreted as Friedman's theory being correct only in predicting the n-type Hall coefficient for p-type material but not the p-type Hall coefficient for n-type material.

For explaining the behavior of n-type a-Si : H, Emin (1977) has suggested a theory by considering that carriers form polarons located on Si–Si bonds. Emin's theory requires that the odd order close loops must be predominant in the structure, because the orbital on each bond is antibonding so that the wavefunction changes sign at each hop. Although polarons are not formed in crystalline silicon, Emin has suggested that they can be formed in amorphous silicon due to its softened structure. The observed activation energy in the mobility is then attributed to the polaron hopping. However, Emin's theory is not widely accepted, because it demands on the electron to move from one bond to an adjacent bond, around an odd numbered ring, which is not always possible in any amorphous structure.

Mott (1991) has suggested that the positive Hall coefficient for n-type silicon can be explained without any assumption of odd-numbered paths, if the centers which scatter electrons are considered to be the stretched Si–Si bonds. Such stretched bonds have electron energies different from the majority of the bonds. Extending then Friedman's theory to such stretched bonds as scattering centers, the interference between two paths, AB and ACB, can lead to a change in the sign of the electronic wavefunction.

Then, applying a perturbative renormalization-group procedure, Okamoto *et al.* (1993) have studied the behavior of weak field Hall conductivity near the mobility edge and found that the anomalous sign in the Hall coefficient can occur if the mean free path of carriers is shorter than a critical value. Accordingly, the microscopic Hall conductivity changes its sign near but above the mobility edge and hence the Hall coefficient also changes its sign. A quantum interference effect of electron transport near the mobility edge has been taken into consideration, which is also taken into account in a metallic conduction regime. The Hall mobility against the carrier mean free path is deduced. This is also consistent with the observation no sign anomaly in degenerate a-semiconductors (Hosono *et al.*, 1996). It should be noted that the quantum interference of free electrons near E_c is important in discussing the electronic transport in the extended states.

8.2 Carrier transport in a deformable lattice (polaronic transport)

An extra electron, or a hole, in crystalline and non-crystalline materials can distort its surroundings (as described in Chapter 6). A carrier accompanied by such distortion is called a polaron. When the spatial extent of the wavefunction of such a carrier is less than or comparable with the inter-atomic or inter-molecular separation, it is called a *small polaron*, otherwise it is a large polaron (see Mott, 1993). It is known that small polarons exist, for example, in alkali halides, molecular crystals, rare-gas solids and some glasses (see Chapter 6). Several comprehensive reviews on small polarons are available (Emin, 1975; Shluger and Stoneham, 1993; Singh, 1994; Firsov, 1995).

While it is believed that small polarons dominate the electronic transport in some transition metal oxide glasses (TMOGs) (Sayer and Mansingh, 1987), the issue still remains a subject of debate (Shimakawa, 1989, 2001). This chapter briefly reviews the current understanding of small polaron transport in a-semiconductors. The two extreme cases for multiphonon assisted hopping of electrons will be discussed below.

8.2.1 Strong carrier–phonon coupling limit

Electrons could couple with both optical and acoustic phonons in a deformable lattice. Let us first discuss the *strong* electron–phonon coupling limit. It is known that the adiabatic approximation cannot explain the overall features of the DC and AC conductivities in glasses (Shimakawa, 1989). Therefore, only the non-adiabatic approximation will be reviewed here. An exact calculation of the non-adiabatic multiphonon transition rate for strong coupling (small polaron) in disordered solids has been carried out by Gorham-Bergeron and Emin (1977), giving

$$\Gamma = \left(\frac{J_{ij}}{\hbar}\right)^2 C(T) \exp\left(-\frac{E_A^{op} + E_A^{ac} + \Delta/2}{kT}\right), \tag{8.64}$$

where J_{ij} and Δ, respectively, are the electron transfer integral and the energy difference between site i and j and $C(T)$ is a weakly temperature-dependent function. E_A^{op} and E_A^{ac} are defined as follows

$$E_A^{op} = \frac{2kT}{\hbar\omega_o} E_b^{op} \tanh\left(\frac{\hbar\omega_o}{4kT}\right), \tag{8.65a}$$

and

$$E_A^{ac} = \frac{1}{N}\sum_g \frac{2kT}{\hbar\omega_{g,ac}} E_b^{ac} \tanh\left(\frac{\hbar\omega_{g,ac}}{4kT}\right), \tag{8.65b}$$

where ω_o is the mean optical frequency (a small dispersion of optical modes is assumed), $\omega_{g,ac}$ is the acoustic phonon frequency at wavevector g, and N is the number of phonon modes. Note that the two energies, E_b^{op} and E_b^{ac}, are the polaronic binding energies related to optical and acoustic phonons, respectively. The non-adiabatic treatment of small polarons requires $J_{ij} < 0.1\,\text{eV}$ (Holstein, 1959).

8.2.2 *Weak carrier–phonon coupling limit*

A simple form of the jump rate of electrons coupled to a mode ω_c in the weak electron–phonon coupling limit (large polaron) can be given approximately as (Shimakawa and Miyake, 1988, 1989):

$$\Gamma = \omega_c \exp(-\gamma p)\left(\frac{kT}{\hbar\omega_c}\right)^p, \tag{8.66}$$

where $p = \Delta/\hbar\omega_c$ (the number of phonons participating) and $\gamma = \ln(\Delta/4E_b) - 1$, where E_b is the polaronic binding energy (a measure of the coupling strength). Note that the condition $kT \gg \hbar\omega_c$ is required in the weak-coupling limit. Equation (8.66) can be obtained as follows (Shimakawa and Miyake, 1989).

The multiphonon jump rate $R(\Delta)$ of localized electrons coupled to one vibrational mode ω_0 has been derived as (Emin, 1975):

$$R(\Delta) := K \sum_{n=-\infty}^{\infty} [I_p(z) \cos(p\phi_n)],$$

$$z = (2E_b/\hbar\omega_0) A_n \operatorname{cosech}(\hbar\omega_0/kT), \tag{8.67}$$

where K is a certain characteristic frequency (probably of the order of ω_0), $I_p(z)$ is the modified Bessel function given by

$$I_p(z) = (z/2)^p \sum_{k=0}^{\infty} \frac{(z^2/4)^k}{k!\Gamma(p+k+1)},$$

where A_n is the lattice relaxation amplitude function and ϕ_n lattice relaxation-phase shift.

In the weak coupling limit ($z \cong 0$), $R(\Delta)$ is given by

$$R(\Delta) \propto [4E_b/\hbar\omega_0)(kT/\hbar\omega_0)]^p/p!, \tag{8.68}$$

where $kT \gg \hbar\omega_0$ is assumed. When p is large, the Stirling formula can be used to give

$$\frac{1}{p!} \cong \frac{1}{\sqrt{2\pi p}} p^{-p} \exp(p). \tag{8.69}$$

Using

$$\left(\frac{4E_b}{\hbar\omega_0}\right)^p = \exp\left[\frac{\Delta}{\hbar\omega_0} \ln\left(\frac{4E_b}{\hbar\omega_0}\right)\right], \tag{8.70}$$

and

$$p^{-p} = (\Delta/\hbar\omega_0)^{-(\Delta/\hbar\omega_0)} = \exp\left[-\frac{\Delta}{\hbar\omega_0} \ln\left(\frac{\Delta}{\hbar\omega_0}\right)\right], \tag{8.71}$$

Equation (8.67) can be rewritten as

$$R(\Delta) \propto \exp(-\gamma p)\left(\frac{kT}{\hbar\omega_0}\right)^p. \tag{8.72}$$

The condition of $G = (4E_b/\hbar\omega_0)(kT/\hbar\omega_0) \ll 1$ needs to be satisfied in the weak coupling limit (Emin, 1975). However, when p is large, the small argument approximation (Eq. (8.67)) can still be valid even for $G \sim 2$ (Shimakawa and Miyake, 1989).

8.2.3 Application of the theories to experimental data

8.2.3.1 DC and AC hopping conductivities

The DC and AC hopping conductivities, in either strong or weak coupling limits, are given by

$$\sigma_{DC} = \frac{N_c(eR)^2\Gamma}{6kT}, \tag{8.73}$$

where N_c is the number of carriers, R hopping distance, and

$$\sigma(\omega) = \sigma_{DC}\frac{i\omega\tau}{\ln(1 + i\omega\tau)}, \tag{8.74}$$

where ω is the angular frequency of the applied field and τ hopping time. Equation (8.74) is a simple form of AC conductivity based on the continuous-time random-walk approximation (Dyre, 1985) and the relaxation time $\tau = 1/\Gamma$ (Shimakawa, 1989a,b).

Glasses containing a high concentration of transition metals (V, W, Fe etc.) called TMOGs, show semiconducting behavior with conductivities in the range 10^{-2} to 10^{-11} S cm^{-1}. It is believed that small polarons dominate the electronic transport in some TMOGs (Sayer and Mansingh, 1987). The Seebeck coefficient between 300 and 500 K generally has the n-type value $\sim$200 μV/K and is almost temperature-independent (Sayer and Mansingh, 1987), suggesting that the number of carriers is almost independent of temperature.

Although TMOGs are not presented in this monograph in detail, we should describe the electronic properties of these glasses to discuss the contribution of small polarons. A typical example of the temperature variation of the DC conductivity σ_{DC} for one of TMOGs, $(V_2O_5)_{80}(P_2O_5)_{20}$, is shown by the solid circles in Fig. 8.22 (Shimakawa, 1989b). One of the characteristic features is that the activation energy decreases gradually with decreasing temperature in most TMOGs. This behavior is, qualitatively, very similar to the prediction by small polarons (Emin, 1974; Gorham-Bergeron and Emin, 1977). It is known that electronic conduction in TMOGs occurs by electron transfer between states associated with transition metal ions: $M^{(n-1)+} - e + M^{n+} \rightarrow M^{n+} + M^{(n-1)+}$, where M^{n+} and $M^{(n-1)+}$ are the normal and reduced valence states of given transition metal ions. In vanadium-oxide systems, hopping of electrons between vanadium ions, V^{4+} to V^{5+}, can dominate the charge transport.

Assuming the formation of small polarons, the temperature dependence of σ_{DC} is calculated using Eq. (8.73). The phonon density of states is approximately $g(\omega) \propto \omega^2$ for acoustic phonons with a cut-off (Debye) frequency ω_D. A mean optical phonon frequency $\omega_o = 3\omega_D$ is assumed (Shimakawa, 1989a,b). One of the important physical parameters is the Debye frequency, which determines the shape of the curve. The details of calculation are given elsewhere (Shimakawa, 1989a). Curves (a)–(c) in Fig. 8.22 show the results of σ_{DC} calculated at three different cut-off frequencies: (a) $\omega_D = 3.1 \times 10^{13}$, (b) 8.2×10^{13} and (c) 1.3×10^{14} s^{-1}. Other physical parameters required for the calculation are taken to be $E_b^{AC} = E_b^{op} = 0.7$ eV, $\Delta = 0.03$ eV, $N_c = 1 \times 10^{21}$ cm^{-3}, $R = 0.4$ nm (average site separation of vanadium ions). The theory fits to the experimental data (curve (b)) excellently well and produces $J_{ij} = 1.2$ eV.

It should be noted, however, that this estimated value of J_{ij} is very much larger than the required value of less than 0.1 eV, predicted from the non-adiabatic treatment of small polarons (Holstein, 1959). Therefore, a large value of J_{ij} obtained above and for many other TMOGs (Shimakawa, 1989a), suggests that

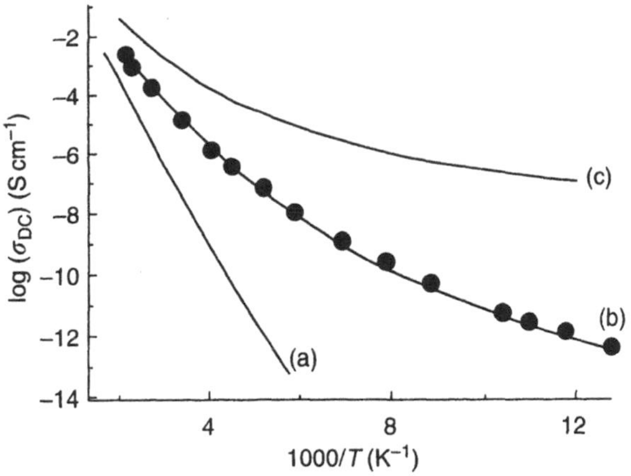

Figure 8.22 Temperature dependence of the DC conductivity in $(V_2O_5)_{80}(P_2O_5)_{20}$ glass calculated at (a) $\omega_D = 3.1 \times 10^{13}$, (b) 8.2×10^{13} and (c) $1.3 \times 10^{14}\,s^{-1}$ (Shimakawa, 1989b).

non-polaronic transport is applicable in this system and the small polaron hopping theory presented is inadequate.

Let us consider briefly the possibility of adiabatic processes. The jump rate of a small polaron in the adiabatic approximation at a high temperature can be expressed as (Austin and Mott, 1969):

$$\Gamma = \Gamma_0 \exp(-E_b/2kT), \tag{8.75}$$

where E_b is the polaron binding enery. Using $N_c = 1 \times 10^{21}\,cm^{-3}$, $R = 0.4\,nm$ for $(V_2O_5)_{80}(P_2O_5)_{20}$, fitting to the experimental data requires $\Gamma_0 = 4.3 \times 10^{13}\,s^{-1}$, corresponding to $T_0 = 2065\,K$. As Eq. (8.75) is valid only for $T > T_0/2$, the applicability of the adiabatic approximation is questioned. Furthermore, the low-temperature behavior of σ_{DC} can be difficult to interpret in terms of this approximation in which σ_{DC} is proportional to $\exp(-\Delta/kT)$ for $T < T_0/4$. It is thus suggested that the adiabatic approximation also cannot easily explain the overall features of the DC conductivity.

Consider a weak coupling multiphonon assisted hopping (WCMH) of conduction electrons. Figure 8.23 shows the temperature variation of DC and AC conductivities in $(V_2O_5)_{80}(P_2O_5)_{20}$ glass. The solid line for the DC conductivity, in the weak-coupling limit is obtained empirically as $\sigma_{DC} = CT^m$ with $m = 13.4$. This is the same temperature dependence for σ_{DC} as predicted by Eq. (8.72). It is known that the relation of $\sigma_{DC} \propto T^m$ is observed in most TMOGs (Shimakawa, 1989b). Before estimating the physical parameters that appear in Eq. (8.64), the AC conductivity will be discussed. The solid lines in Fig. 8.23 (AC conductivity) are the results calculated using Eq. (8.74). The relation $\sigma_{DC} = CT^m$ with $m = 13.4$ has been used for calculating $\sigma(\omega)$. The only fitting parameter here is

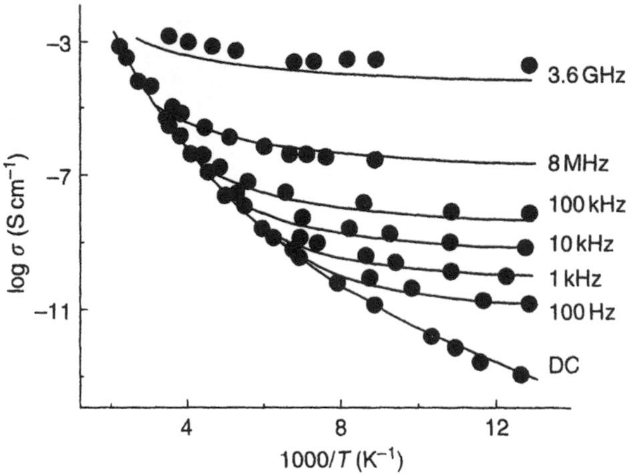

Figure 8.23 Temperature dependence of the DC and AC conductivities in $(V_2O_5)_{80}(P_2O_5)_{20}$ glass (Shimakawa, 1989).

the hopping time τ, which is equal to Γ^{-1}. Fitting the experimental data produces $\Gamma = 2.9 \times 10^{-11}(T/15)^{13.4}\,s^{-1}$ and hence $\omega_c = 2 \times 10^{12}\,s^{-1}$ is deduced from Eq. (8.72), which is very much smaller than $\omega_D = 8.2 \times 10^{13}\,s^{-1}$. The cause of such a small ω_c may originate from some delocalized nature of electrons. When the electron Bohr radius α^{-1} approaches the mean inter-ionic separation of V^{5+}, the wavefunction will be spread out to some extent, producing percolating microclusters (Shimakawa, 1989b). The delocalized electrons across the cluster couple only with long wavelength acoustic phonons, through the relation $\omega_0 = (a_0/\alpha_e^{-1})\omega_D$, where a_0 is the average inter-atomic separation and α_e^{-1} the extent of localization. Taking $a_0 = 0.2$ nm, one gets $\alpha_e^{-1} = 8$ nm, corresponding to the cluster size l_c.

Assuming $R = l_c$ in Eq. (8.73), fitting of the data produces N_c and hence $N(E_F) = N_c/kT$ is estimated to be $3 \times 10^{20}\,cm^{-3}$, which appears to be quite reasonable. The physical parameter γ, appearing in Eq. (8.72), is estimated to be 3.8, which is also a reasonable value in the weak coupling limit. Note that fitting the theory of WCMH to the experimental data is excellent and very reasonable physical parameters are deduced also in other TMOGs (Shimakawa, 1989a).

There are a few reports that suggest that the variable-range hopping (VRH) of small polaron dominates the electronic transport in some of TMOGs (Sayer and Mansingh, 1987). In fact, the experimental data of DC conductivities shown in Fig. 8.22 can be fitted to the relation of ln σ_{DC} vs $T^{-1/4}$ (Shimakawa, 1989b). As already stated, there is a difficulty with the VRH theory, that is, the prefactor predicted by theory becomes considerably smaller than that obtained experimentally. Thus, we should reject the proposal of applying VRH of small polarons in TMOGs.

Although it is difficult to know precisely how the charge carriers interact with the atomic vibrational modes, an unreasonably large value of the transfer energy J_{ij} deduced from the non-adiabatic approach, poses a problem that needs to be overcome. The multiphonon assisted hopping of localized electrons with weak electron-lattice coupling (non-polaron) may dominate electronic transport in TMOGs. The non-polaronic WCMH mechanism can also be applied to a-germanium (Shimakawa, 1989a; Nosaka *et al.*, 1996), a-carbons (Shimakawa and Miyake, 1988, 1989), hydrogenated a-carbons (Helmbold *et al.*, 1995), and some a-Ch (Kumar, 1991). It is still controversial whether or not small polarons exist in some glassy systems. It is predicted, however, that an isolated site itself, for example, V^{4+}, may have polaronic nature and its percolating clusters may interact weakly with phonons. The WCMH mechanism can therefore be considered to be dominant in the present glassy systems.

8.2.3.2 Thermoelectric power

The thermoelectric power for small polarons can be written as (Emin, 1975):

$$S = -\frac{k}{e}\left(\frac{E_0}{kT}\right), \tag{8.76}$$

where E_0 is the energy that characterizes the thermal generation of carriers. The hopping energy ΔW, which is related to the polaron binding energy (see Eq. (8.64)), does not appear in the thermoelectric power, as already discussed. As the contribution to the activation energy arising from the distortion of the initial site is equal to the energy of the distortion at the second site, there is no transfer of vibration energy associated with a small polaron hop.

The electronic conduction, however, requires $\Delta W + E_0$; that is, both the energies for thermal generation of carriers and hopping are required for electronic (not heat) transport. This naturally explains the difference between E_S and E_σ as discussed in Section 8.2. We may remember that the temperature-dependent Q function in Section 8.2 and for some a-Ch is $Q_0 \approx 10$ and $E_Q \approx 0.2\,\text{eV}$ (Seager *et al.*, 1973; Seager and Quinn, 1975; Emin, 1975). Thus, ΔW becomes E_Q here. Emin (1975) has proposed that small polarons dominate electronic transport and hence Q function in a-Ch can be attributed to hopping of small polarons. There is no clear evidence of existence of small polarons in a-Ch, and the band transport based on a standard semiconductor theory is usually used for analyzing the transport data for a-Ch (Mott and Davis, 1979).

Let us discuss the thermoelectric power in TMOGs. As stated earlier, the Seebeck coefficient between 300 and 500 K for TMOGs has a n-type value around 200 μV/K and is almost temperature independent (Sayer and Mansingh, 1987), suggesting that the number of carriers is almost independent of temperature. As mentioned in Section 8.3.1, N_c is given by $N(E_F)kT$ and therefore $E_0 \approx 0$ (no activation type of carrier generation). S under this condition becomes almost temperature independent. This situation is similar to that for a-Ge at low temperatures (see Fig. 8.14). S is also expected to be independent of temperature in the WCMH process, since no particular energy is transported in the hopping process. On the basis of the results of thermoelectric power, therefore, it is difficult to determine which mechanism, strong or weak coupling, dominates the carrier transport in TMOGs.

References

Abdel-Wahab, F. and Shimakawa, K. (1995). *Philos. Mag. Lett.* **71**, 351.

Abdel-Wahab, F., Shimakawa, K. and Hirabayashi, I. (1998). *Philos. Mag. Lett.* **77**, 159.

Abkowitz, M., LeComber, P.G. and Spear, W.E. (1976). *Commun. Physics* **1**, 175.

Ambegaokar, V., Halperin, B.I. and Langer, J.S. (1971). *Phys. Rev. B* **4**, 2612.

Austin, I.G. and Mott, N.F. (1969). *Adv. Phys.* **18**, 41.

Baranovskii, S.D. and Cordes, H. (1999). *J. Chem. Phys.* **111**, 7546.

Bellingham, J.R., Phillips, W.A. and Adkins, C.J. (1990). *J. Phys. C* **2**, 6207.

Beyer, W. and Mell, H. (1977). In: Spear, W.E. (ed.), *Proceedings of the 7th International Conference on Amorphous and Liquid Semiconductors*. CICL, Edinburgh, p. 333.

Böttcher, C.J.F. and Bordewijk, P. (1978). *Theory of Electric Polarization*, Vol. II. Elsevier, Amsterdam.

Brodsky, M.H. and Gambino, R.J. (1972). *J. Non-Cryst. Solids* **8–10**, 739.

Bunde, A., Funke, K. and Ingram, M.D. (1998). *Solid State Ionics* **105**, 1.

Burstein, E. (1954). *Phys. Rev.* **93**, 632.

Carlson, D.E. and Wronski, C.R. (1979). In: Brodsky, M.H. (ed.), *Amorphous Semiconductors*. Springer, Berlin, p. 287.

Cohen, M.H and Jortner, J. (1973). *Phys. Rev. Lett.* **30**, 696.

Davis, E.A., Bayliss, S.C., Asal, R. and Manssor, M. (1989). *J. Non-Cryst. Solids* **114**, 465.

Deutscher, G., Levy, Y. and Souillard, B. (1987). *Europhys. Lett.* **4**, 577.

Dyre, J.C. (1985). *Phys. Lett. A* **108**, 457.

Dyre, J.C. (1988). *J. Appl. Phys.* **64**, 2456.

Dyre, J.C. (1993). *Phys. Rev. B* **48**, 12511.

Dyre, J.C. (1997). *Dynamics of Amorphous Solids and Viscous Liquids*, IMFUFA, Roskilde Universitecenter, text no. 335/97.

Eley, D.D. (1967). *J. Polymer Sci.* **17**, 73.

Elliott, S.R. (1977). *Philos. Mag. B* **36**, 1291.

Elliott, S.R. (1987). *Adv. Phys.* **36**, 135.

Emin, D. (1975). *Adv. Phys.* **24**, 305

Emin, D. (1977). *Phil. Mag.* **35**, 1189.

Extance, P., Elliott, S.R. and Davis, E.A. (1985). *Phys. Rev. B* **32**, 8148.

Firsov, Y.A. (1995). *Semiconductors* **29**, 515.

Fluckiger, R., Meier, J., Goetz, M. and Shah, A. (1995). *J. Appl. Phys.* **77**, 712.

Fortner, J., Karprov, V.G. and Saboungi, M-L. (1995). *Appl. Phys. Lett.* **66**, 997.

Friedman, L. (1971). *J. Non-Cryst. Sol.* **6**, 329.

Friedman, L. and Holstein, T. (1963). *Ann. Phys.* **21**, 494.

Fritzshe, H. (1974). In: Tauc, J. (ed.), *Amorphous and Liquid Semiconductors*. Plenum, London, p. 221.

Ganjoo, A. and Shimakawa, K. (1994). *Philos. Mag. Lett.* **70**, 287.

Ganjoo, A., Yoshida, A. and Shimakawa, K. (1996). *J. Non-Cryst. Solids* **198–200**, 313.

Gorham-Bergeron, E. and Emin, D. (1977). *Phys. Rev. B* **15**, 3667.

Greaves, G.N., Elliott, S.R. and Davis, E.A. (1979). *Adv. Phys.* **28**, 49.

Hamilton, E.M. (1973). *Philos. Mag. B* **26**, 1043.

Hauschildt, D., Stutzmann, M., Stuke, J. and Dersh, H. (1982). *Sol. Energy. Mat.* **8**, 319.

Hayashi, K., Shimakawa, K. and Morigaki, K. (1994). In: Adkins, C.J., Long, A.R. and McInnes (eds), *Hopping and Related Phenomena 5*, World Scientific, Singapore, p. 224.

Helmbold, A., Hammer, P., Thiele, J.U., Rohwer, K. and Meissner, D. (1995). *Philos. Mag. B* **72**. 335.

Holstein, T. (1959). *Ann. Phys.* **8**, 343.
Hosono, H., Yamashita, Y., Ueda, N. and Kawazoe, H. (1996). *Appl. Phys. Lett.* **68**, 661.
Howard, J. and Street, R.A. (1991). *Phys. Rev. B* **44**, 7935.
Kamimura, H. and Aoki, H. (1989). *The Physics of Interacting Electrons in Disordered Systems*, Clarendon Press, Oxford.
Kasuya, T. and Koide, S. (1958). *J. Phys. Soc. Jpn.* **13**, 1287.
Kawabata, A. (1981). *Solid State Commun.* **38**, 823.
Kirkpatrik, S. (1973). *Rev. Mod. Phys.* **45**, 574.
Kishimoto, N. and Morigaki, K. (1979). *J. Phys. Soc. Jpn.* **46**, 846.
Kivelsen, S. and Gellat, C.D., Jr (1979). *Phys. Rev. B* **19**, 5160.
Kondo, M., Chida, Y. and Matsuda, A. (1996). *J. Non-Cryst. Solids* **198–200**, 178.
Levy, Y. and Souillard, B. (1987). *Europhys.* **4**, 233.
Long, A.R. (1982). *Adv. Phys.* **31**, 553.
Long, A.R. (1989). *Philos. Mag. B* **59**, 377.
Long, A.R. (1991). In: Pollak, M. and Shklovskii, B.I. (eds), *Hopping Transport in Solids*. North-Holland, Amsterdam, p. 207.
Lucovsky, G., Wang, C., Williams, M.J., Chen, Y.L. and Mahe, D.M. (1993). *Mater. Res. Soc. Symp. Proc.* **283**, 446.
Meyer, W. and Neldel, H. (1937). *Z. Tech. Phys.* (Leipzig) **12**, 588.
Miller, A. and Abrahams, E. (1960). *Phys. Rev.* **120**, 745.
Morigaki, K. (1999). *Physics of Amorphous Semiconductors*. World Scientific, Imperial College Press, Singapore.
Mott, N.F. (1969). *Philos. Mag.* **19**, 835.
Mott, N.F. (1970). *Philos. Mag.* **22**, 7.
Mott, N.F. (1991). *Phil. Mag. B* **63**, 3.
Mott, N.F. (1993). *Conduction in Non-Crystalline Materials*, 2nd edn, Oxford, Clarendon.
Mott, N.F. and Davis, E.A. (1979). *Electronic Processes in Non-Crystalline Materials*. Clarendon Press, Oxford.
Murayama, K. and Mori, M. (1992). *Philos. Mag. B* **65**, 501.
Nagels, P. (1979). In: Brodsky, M.H. (ed.), *Amorphous Semiconductors*. Springer-Verlag, Berlin, p. 113.
Nagels, P., Callaerts, R. and Denayer, M. (1972). In: Miasek, M. (ed.), *Proceedings of the 11th International Conference of Physics of Semiconductors*. PWN-Polish Scientific Publishers, Warsaw, p. 549.
Nosaka, H., Katayama, Y. and Tsuji, K. (1996). *J. Non-Cryst. Solids* **198–200**, 218.
Okamoto, H., Hattori, K. and Hamakawa, Y. (1993). *J. Non-Cryst. Solids* **164–166**, 445.
Orenstein, J. and Kastner, M.A. (1981). *Phys. Rev. Lett.* **46**, 1421.
Ovadyahu, Z. (1986). *J. Phys. C* **19**, 5187.
Overhof, H. and Beyer, W. (1981). *Philos. Mag. B* **44**, 317.
Overhof, H. and Beyer, W. (1983). *Philos. Mag. B* **47**, 377.
Overhof, H. and Thomas, P. (1989). *Electronic Transport in Hydrogenated Amorphous Semiconductors*. Springer, Berlin, p. 122.
Pankove, J.I. (1975). *Optical Processes in Semiconductors*. Dover, New York, p. 34.
Papoulis, A. (1965). *Probability, Random Variables and Stochastic Processes*. McGraw-Hill Kogakusha Ltd, Tokyo, p. 83.
Pashmakov, B., Clafin, B. and Fritzsche, H. (1993). *J. Non-Cryst. Solids* **164–166**, 441.
Pike, G.E. (1972). *Phys. Rev. B* **6**, 1572.
Pollak, M. and Geballe, T.H. (1961). *Phys. Rev.* **122**, 1742.
Roling, B., Happe, A., Funke, K. and Ingram, M.D. (1997). *Phys. Rev. Lett.* **78**, 2160.

Sayer, M. and Mansingh, A. (1987). In: Pollak, M. (ed.), *Oxide Glasses in Noncrystalline Semiconductors*. CRC Press, Florida, p. 1.
Seager, C.H. and Quinn, R.K. (1975). *J. Non-Cryst. Solids* **17**, 386.
Seager, C.H., Emin, D. and Quinn, R.K. (1973). *Phys. Rev. B* **8**, 4746.
Scher, H. and Lax, M. (1973). *Phys. Rev. B* **7**, 4491.
Shimakawa, K. (1981). *J. Phys.* Paris **42**, C4-167.
Shimakawa, K. (1982). *Philos. Mag. B* **46**, 123.
Shimakawa, K. (1983). *Philos. Mag. B* **48**, 77.
Shimakawa, K. (1989a). *Phys. Rev. B* **39**, 12933.
Shimakawa, K. (1989b). *Philos. Mag. B* **60**, 377.
Shimakawa, K. (1996). *Kotai Butsuri* **31**, 437 (in Japanese).
Shimakawa, K. (2001). *Encyclopedia of Materials: Science and Technology*, Elsevier Science, Amsterdam, p. 3579.
Shimakawa, K., and Abdel-Wahab, F. (1997). *Appl. Phys. Lett.* **70**, 652.
Shimakawa, K. and Kondo, A. (1983). *Phys. Rev. B* **27**, 1136.
Shimakawa, K. and Kondo, A. (1984). *Phys. Rev.* **29**, 7020.
Shimakawa, K. and Miyake, K. (1988). *Phys. Rev. Lett.* **61**, 994.
Shimakawa, K. and Miyake, K. (1989). *Phys. Rev. B* **39**, 7578.
Shimakawa, K., Watanabe, A. and Imagawa, O. (1985). *Solid St. Commun.* **55**, 245.
Shimakawa, K., Watanbe, A., Hattori, K. and Imagawa, O. (1986). *Philos. Mag.* **54**, 391.
Shimakawa, K., Long, A.R. and Imagawa, O. (1987). *Philos. Mag. Lett.* **56**, 79.
Shimakawa, K., Wakamatsu, S., Kojima, M., Kato, H. and Imai, A. (1992). *J. Appl. Phys.* **72**, 2889.
Shimakawa, K., Hayashi, K. and Morigaki, K. (1994). *Kotai Butsuri* **29**, 176 (in Japanese).
Shimakawa, K., Kato, T., Hayashi, K., Masuda, A., Kumeda, M. and Shimizu, T. (1994). *Philos. Mag. B* **70**, 1035.
Shimakawa, K., Kolobov, A.V. and Elliott, S.R. (1995). *Adv. Phys.* **44**, 475.
Shimakawa, K., Kondo, A., Goto, M. and Long, A.R. (1996). *J. Non-Cryst. Solids* **198–200**, 157.
Shimakawa, K., Narushima, S., Hosono, H. and Kawazoe, H. (1999). *Philos. Mag. Lett.* **79**, 755.
Shklovskii, B.I. and Efros, A.L. (1984). *Electronic Properties of Doped Semiconductors*. Springer-Verlag, Berlin.
Shluger, A.L. and Stoneham, A.M. (1993). *J. Phys: Condens. Matter* **5**, 3049.
Singh, J. (1994). *Excitation Energy Transfer Processes in Condensed Matter*. Plenum, New York
Singh, J., Aoki, T. and Shimakawa, K. (2002). *Phil. Mag. B* **82**, 855.
Springett, B.E. (1973). *Phys. Rev. Lett.* **31**, 1463.
Theye, M.L. and Fisson, S. (1983). *Philos. Mag. B* **47**, 31.
Theye, M.L., Gheorghiu, A., Driss-Khodja, K. and Boccara, C. (1985). *J. Non-Cryst. Solids* **77&78**, 1293.
Tiedje, T. and Rose, A. (1980). *Solid St. Commun.* **37**, 49.
Van der Putten, D., Moonen, J.T., Brom, H.B., Brokken-Zijp, J.C.M. and Michels, M.A. (1992). *Phys. Rev. Lett.* **69**, 494.
Yelon, A. and Movaghar, B. (1990). *Phys. Rev. Lett.* **65**, 618.
Yelon, A., Movaghar, B. and Branz, M.H. (1992). *Phys. Rev. B* **46**, 12244.
Zvyagin, I.P. (1980). *Phys. Stat. Sol. (b)* **97**, 143.
Zvyagin, I.P. (1984). *Kinetic Phenomena in Disordered Semiconductors*. Moscow University Press, Moscow (in Russian).

9 Photoconductivity

Two types of photoconductions are defined in solids. One is called the primary photoconduction and the other secondary photoconduction. In the primary photoconduction photocarriers transit from the illuminated electrode to the counter electrode with blocking contacts, and the technique is used for deducing the drift velocity (mobility). In the secondary photoconduction, the whole sample is illuminated and the photocurrent is measured between two electrodes with ohmic contacts. In general, photoconductivity depends on both the mobility of carriers μ and their recombination time τ and hence $\mu\tau$-product becomes a very important factor. The photoconduction depends very strongly on the temperature and excitation intensity. Study of photoconductivity has attracted much attention in a wide range of semiconductors, which is evident from many existing reviews and excellent books (Rose, 1963; Bube, 1992). In the following, we will first describe the secondary photoconduction in hydrogenated amorphous silicon (a-Si : H) and amorphous chalcogenides (a-Ch) and then consider the primary photoconduction.

9.1 Secondary photoconduction

The photoconductivity, denoted by σ_{p}, is defined as

$$\sigma_{\mathrm{p}} = eG\tau\mu, \tag{9.1}$$

where G is the generation rate of photocarriers in the unit of $\mathrm{cm}^{-3}\,\mathrm{s}^{-1}$, τ recombination time measured in seconds (s), and μ the free-carrier mobility ($\mathrm{cm}^{-2}\,\mathrm{V}^{-1}\,\mathrm{s}^{-1}$) when transport occurs in the extended states. If, however, transport occurs in the tail states μ represents the hopping mobility, which is very much smaller than the free-carrier mobility. $G\tau$ is the excess number of photogenerated charge carriers under steady-state illumination. For a thin film of thickness d, neglecting multiple reflection, G can be written as

$$G = \eta N_0(1 - R)\{1 - \exp(-\alpha d)\}/d, \tag{9.2}$$

where η is the quantum efficiency of generation of photocarriers, N_0 number of incident photons per unit area ($\mathrm{cm}^{-2}\,\mathrm{s}^{-1}$), R reflectivity, and α absorption coefficient (cm^{-1}). For $\alpha d \ll 1$, photons are uniformly absorbed throughout the films, then σ_p becomes:

$$\sigma_p \approx e\eta N_0(1 - R)\mu\tau\alpha. \tag{9.3}$$

The photoconductivity can be obtained by both DC (steady-state illumination) and AC (chopping illumination) measurements. In the AC case, photocurrent is usually detected using a Lock-in amplifier, and the amplitude and phase shift of photocurrents thus measured provide information on the recombination and trapping processes (Oheda, 1980; Oheda and Namikawa, 1981; Hattori *et al.*, 1996). In general, the lifetime τ depends on the generation rate as $G^{(\gamma-1)}$, and σ_p varies as G^{γ}. For $\gamma < 1$, the dependence of σ_p on G is "sublinear" and for $\gamma > 1$, it is "superlinear." Therefore, the value of γ provides the background information of recombination. This will be discussed in the following section.

Three basic quantities, (i) photosensitivity, (ii) spectral response and (iii) response time, can be regarded to be important parameters for characterizing a photoconductor. In what follows we will define first these three parameters.

(i) *Photosensitivity*: This is defined by the product of μ and τ, usually denoted by $\mu\tau$-product, and $\mu\tau = \sigma_p/eG$ from Eq. (9.1). Another definiticn of photosensitivity used in the literature is $(\sigma_p - \sigma_d)/\sigma_d$, where σ_d is the conductivity in the dark conditions or also called the dark conductivity (Bube, 1992).

(ii) *Spectral response*: This is defined by the relation between the photoconductivity and wavelength of illumination energy, and is usually presented in the form of spectral response curves. There is a close correlation between optical absorption coefficient α and σ_p [see Eq. (9.3)]. Typical spectral response of photoconductivity curves are shown in Fig. 9.1. In the low-absorption region III, σ_p is proportional

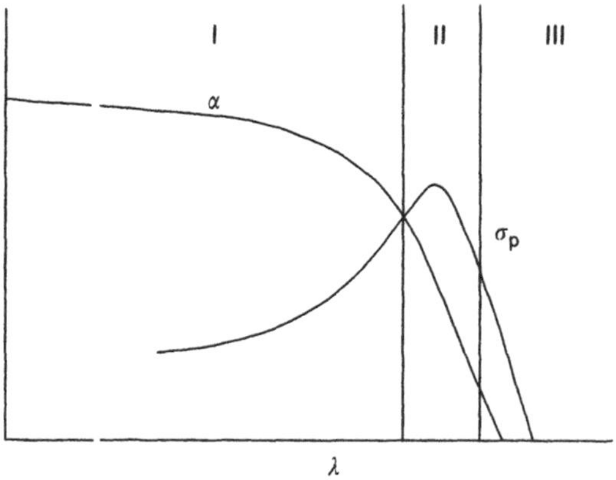

Figure 9.1 Schematic comparison of the shape of the photoconductivity spectral response and absorption coefficient of a photoconductor. Three characteristic regions are illustrated.

to α as given in Eq. (9.3). In the high-absorption region I, as photocarriers are produced near the surface, photoconductivity is dominated by the surface recombination, because the number of recombination centers near the surface becomes larger than in bulk. Therefore, σ_p decreases with photon energy after reaching a maximum. In the intermediate range II, σ_p shows a peak and it can be controlled by the bulk lifetime. Thin films of amorphous semiconductors (a-semiconductors) are usually used as photoconductors and hence high-absorption region I is not of much importance.

In the bulk of chalcogenide glasses, it is easy to observe a peak in the plot of σ_p vs the photon energy $\hbar\omega$. The spectral response has been calculated using the rate of change of the concentration of free holes at a distance z from the illuminated surface, and thus the surface recombination velocity has been estimated to be $s = 1\ \mathrm{m\,s^{-1}}$ for some glasses (Shimakawa *et al.*, 1974).

(iii) *Response time*: When the photoexcitation is turned on or off, photoconductivity reaches its steady-state value or zero, respectively, with the time delay. This time delay is called the response time. In the absence of traps (localized states) the free-carrier lifetime τ (recombination time), determines the response time. The simplest rate equation for the change of the density of photo-excited carriers, n, can be expressed as

$$\frac{\mathrm{d}n}{\mathrm{d}t} = G - \frac{n}{\tau}, \tag{9.4}$$

which gives $n = G\tau$ in the steady state. The rise and decay curves of the density of photo-excited are described by

$$n(t) = G\tau[1 - \exp(-t/\tau)] \tag{9.5}$$

and

$$n(t) = G\tau \exp(-t/\tau), \tag{9.6}$$

respectively.

Such a simple rise or decay curve is usually not observed in a-semiconductors, because of the involvement of electron and hole traps in the recombination processes. In the presence of traps, additional processes of trap filling during the rise and trap emptying during the decay get involved and hence the response time becomes longer than the recombination time τ. Furthermore, the response time is distributed around some peak value, which creates additional complications in the decay and rise curves in a-semiconductors. In this situation, $n(t)$ is given by either a stretched exponential, $\exp[-(t/\tau)^{\beta}]$ or a power law, $t^{-\beta}$. This issue will be addressed later on.

Finally we would like to define the "steady-state Fermi level," also called the "quasi-Fermi level," and "Demarcation level," both of which are very useful in discussing photoconducting behavior of a-semiconductors. The occupancy of all electronic energy levels in a semiconductor is determined by the location of the

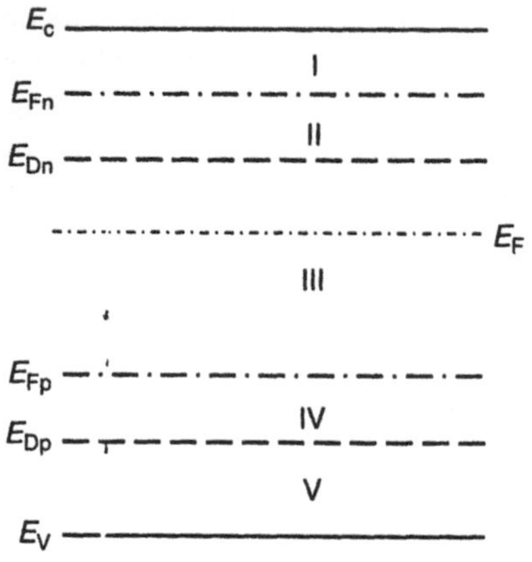

Figure 9.2 Typical energy band diagram showing the quasi-equilibrium Fermi levels, E_{Fn} and E_{Fp}, and the demarcation levels, E_{Dn} and E_{Dp}, for electron and hole, respectively, in an a-semiconductor.

Fermi level, which is the top filled energy level at a temperature of 0 K. The same definition of a Fermi level is used even when a sample is illuminated, and then under the steady state the top filled energy level is called a steady state Fermi level or a quasi-Fermi level.

Figure 9.2 shows a typical energy band diagram showing the thermal equilibrium Fermi level E_F, the quasi-Fermi levels (E_{Fn} for electron and E_{Fp} for hole) and demarcation levels (E_{Dn} for electron and E_{Dp} for hole). E_F is defined as

$$E_c - E_F = kT \ln(N_c/n_0), \tag{9.7}$$

or as

$$E_F - E_v = kT \ln(N_v/p_0), \tag{9.8}$$

where N_c and N_v are the effective density of states in the conduction and valence extended states, and n_0 and p_0 are the number of free electrons and holes in dark conditions, respectively.

Under the illumination also, the quasi-Fermi levels are defined by Eqs (9.7) and (9.8). However, there must be a separate electron quasi-Fermi level E_{Fn} and hole Fermi level E_{Fp}, since $n = n_0 + \Delta n$ and $p = p_0 + \Delta p$, where Δn and Δp are the number of photo-excited electrons and holes, respectively. The quasi-Fermi levels, E_{Fn} and E_{Fp}, are therefore defined as

$$E_c - E_{Fn} = kT \ln(N_c/n) \tag{9.9}$$

and

$$E_{Fp} - E_v = kT \ln(N_v/p). \tag{9.10}$$

Next, we must define the demarcation levels: an electron at the demarcation level E_{Dn} has equal probability of being thermally excited to the conduction extended states and of recombining with a hole in the valence extended states. Thus the demarcation level for electrons is such that states lying above this level act predominantly as traps and those below act as recombination centers. Likewise, a hole at the demarcation level E_{Dp} has equal probability of being thermally excited to the valence extended states and of recombining with an electron in the conduction extended states. Thus for holes all states below E_{Dp} act as hole traps and those above act as recombination centers. Accordingly, as shown in Fig. 9.2, the localized states in regions III and IV act as recombination centers.

Localized states in region I of Fig. 9.2 are in thermal equilibrium with the conduction extended states and their occupancy is determined by the quasi-Fermi level E_{Fn}; and act as electron traps. Likewise, the localized states in region V act as hole traps. States in region II, above E_{Dn}, are mostly occupied by electrons and hence these may act as recombination centers for holes. Both the quasi-Fermi and demarcation levels change their positions with illumination intensity; as a result the recombination kinetics also changes with the illumination intensity. Let us now discuss, in this section, simple kinetics involving trapping and recombination centers, which influence the photoconductivity of a solid. Consider, for example, the case of an insulator, neglecting the thermal generation of carriers, one can write $n = \Delta n$ and $p = \Delta p$ in Eqs (9.9) and (9.10), respectively. This means that there exist no holes in the valence and electrons in the conduction states before the illumination. In this case, the photoexcitation rate G should be balanced by recombination across the bandgap with a rate $R = nSvp$ as

$$G = nSvp = n^2 Sv, \tag{9.11}$$

where S is the recombination cross-section (cm^2) and v is the thermal velocity of carriers (cm s^{-1}). This gives $n \propto G^{1/2}$, and yields $\tau \propto G^{-1/2}$ from Eq. (9.4) in the steady state (see also Section 5.5). The case, $G \propto n^2$, producing $\sigma_p \propto G^\gamma$ with $\gamma = 1/2$, is often called "bimolecular" recombination. Here, direct recombination between electrons and holes, without introducing any recombination centers in the bandgap, is considered. The above kinetics can also be applied to the recombination between electrons and holes in the band tail states.

When we add a sufficient amount of single energy levels N_t of electrons, so that $N_t \gg n$, G should be balanced as

$$G = nSvp + nS_t v(N_t - n_t), \tag{9.12}$$

where S_t is the trapping cross-section (cm^2), N_t is the density of trapping centers (cm^{-3}), and n_t is the density of trapped electrons (cm^{-3}). In this case the condition of $n = p$ cannot be applied. Suppose that the traps are all filled ($n_t \approx N_t$) under illumination, producing $p = n + N_t$. Then we get $n \propto G$ and lifetime becomes a constant as

$$\tau = 1/SvN_t. \tag{9.13}$$

This clearly demonstrates that the number of trapped carriers contributes to the lifetime and G dependence of n. The case of recombination in which, $G \propto n$, producing $\sigma_p \propto G^{\gamma}$ with $\gamma = 1$, is often called the "monomolecular" recombination. It is of interest to note that the introduced traps can induce significant changes in the properties of photoconductivity.

As already stated, γ-value is not 0.5 or 1.0 in most cases. In a-semiconductors, there are recombination and trapping centers which are widely distributed in the bandgap. As a model, we consider an exponential trap distribution (Rose, 1963):

$$N_t(E) = N_0 \exp[-(E_c - E)/kT^*], \tag{9.14}$$

where T^* is a characteristic parameter of distribution. We discuss here only the case of electrons trapped below the Fermi level, E_{Fn}. The total trapped carrier density in this situation is given by

$$\begin{aligned} N_t &\approx \int_{E_c - E_{Fn}}^{\infty} N_0 \exp[-(E_c - E)/kT^*]\,\mathrm{d}E = kT^* N_0 \exp[-(E_c - E_{Fn})/kT^*] \\ &= kT^* N_0 \exp\left[-\frac{E_c - E_{Fn}}{kT}\frac{kT}{kT^*}\right] = kT^* N_0 \left(\frac{n}{N_c}\right)^{T/T^*}, \end{aligned} \tag{9.15}$$

where $n = N_c \exp[-(E_c - E_{Fn})/kT]$ is used. Using Eq. (9.13), then τ can be written as

$$\tau = \left(\frac{n}{N_c}\right)^{-T/T^*} (SvkT^* N_0)^{-1}. \tag{9.16}$$

Substituting $n = G\tau$ into Eq. (9.16) produces

$$\tau \propto G^{-T/(T^*+T)}$$

and

$$n \propto G^{T^*/(T+T^*)}. \tag{9.17}$$

The value of γ, defined by $T^*/(T + T^*)$, therefore depends on the relative magnitudes of T and T^* and hence for a case of $T^* = T$, we get $\gamma = 0.5$ from Eq. (9.17), and for that of $T^* > T$ we get $\gamma = 1.0$.

9.1.1 Photoconductivity in a-Si : H

A typical example of the temperature variation of photoconductivity σ_p as a function of light intensity in a-Si : H is shown in Fig. 9.3. Four distinct regimes appear in the temperature dependence of σ_p (Vanier *et al.*, 1981; Dersh *et al.*, 1983; Vomvas and Fritzshe, 1987; Zhou and Elliott, 1993). Three of these regions are labelled in Fig. 9.3, as I–III. At very low temperatures, corresponding to regime IV, which is not presented in Fig. 9.3, σ_p is almost independent of temperature (Fritzshe, 1989) and as such is universally observed in all a-semiconductors. In the high-temperature region, σ_p is smaller than σ_d and has a maximum value at a specific temperature. However, as σ_p has a small value in this temperature range, it is not of much interest and will not be discussed here.

In the following, we will discuss the behavior of photoconductivity in a-Si : H in each of the other three regimes shown in Fig. 9.3. The transport is dominated by electrons in undoped a-Si : H. The photoconductivity in the high-temperature regime I, above room temperature, has been studied the most. In this regime, σ_p rapidly increases with temperature, which is attributed to an increase in the recombination time and increase in the probability of thermal release of trapped electrons. As described in the previous section, the demarcation level moves towards the deep energy levels, since the probability of thermal release from traps increases with temperature. It is understood in general that a change in lifetime, that is, a change of photoconductivity, is attributed to the temperature dependent variation in the positions of E_{Dn} and E_{Dp}.

In the regime II, a hump and a valley are observed at 120–250 K and 250–350 K, respectively, whose temperatures depend on the generation rate G and the density of dangling bonds (recombination centers). The decrease of photoconductivity

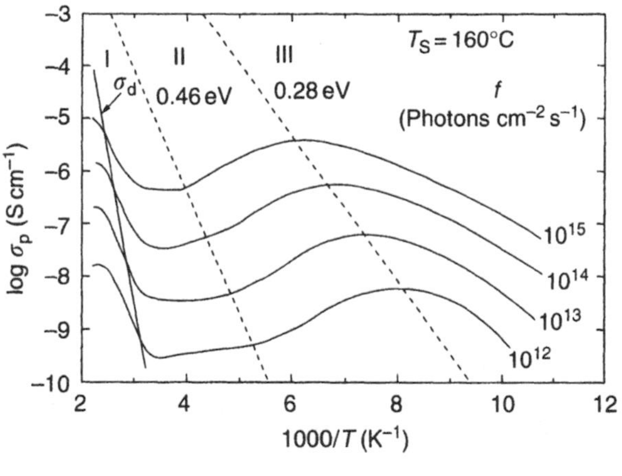

Figure 9.3 Temperature variation of the photoconductivity σ_p at various excitation intensities in a-Si : H (Vanier *et al.*, 1981). The dark conductivity σ_d is also shown.

with increasing temperature is called the "thermal quenching" (Pearsans and Fritzshe, 1981; Carius *et al.*, 1987), which is not observed in doped a-Si : H. This can be interpreted as follows: in regime II, there are two relevant recombination steps: (i) the tunneling of band tail electrons to neutral dangling bonds D^0, producing D^- states and then (ii) the diffusion of holes in the tail states to recombine with D^- states formed in step (i). With increasing temperature holes can hop more easily in the localized tail states, resulting in an increase in the recombination rate and hence decrease in photoconductivity (Carius *et al.*, 1987). Alternatively, if holes are self-trapped at lower temperatures, then with increasing temperature holes can be released from self-trapped states to the tail states enhancing the recombination (Morigaki, 1999).

In regime III, σ_p decreases rapidly with decreasing temperature with an activation energy of 0.04–0.11 eV. This activation energy appears to be independent of the excitation intensity G. The recombination is bimolecular in nature, that is, γ in $\sigma_p \propto G^\gamma$ is 0.5. Unlike the photoconductivity at high temperatures (regimes I and II), it is little affected by light soaking, which is known to create recombination centers in region III (see Chapter 10). The interpretation of the photoconductivity data in this regime is still a matter of debate. Zhou and Elliott (1993) have suggested that the recombination is dominated by direct tunneling between trapped electrons and trapped holes in the respective band-tail states and free electrons which are assumed to be at quasi-thermal equilibrium dominate the photoconductivity. Since this model does not involve deep recombination centers, no light soaking in this regime can be easily accounted for. The temperature dependence of photoconductivity is predicted from the number of free electrons calculated using some physical parameters in which an exponential distribution of tail states is assumed (Zhou and Elliott, 1993). Qualitatively, the temperature dependence of photoconductivity in this regime is well explained by the Zhou–Elliott model (1993). Quantitatively, however, one obtains the absolute magnitude of the calculated photoconductivity significantly larger than that of the measured one. An alternative model for the photoconductivity in the regime III has been proposed by taking into account the self-trapping of holes in the tail states (Morigaki, 1999). The recombination occurs after the self-trapped holes are released to the tail states. The activation energy for the photoconductivity in this model is predicted to be 0.03–0.08 eV, which is close to that observed experimentally.

Baranovskii *et al.* (1994) have criticized the band-transport model and suggested an alternative model. Hopping transport of electrons in the tail states, at a particular energy level, which is called the "transport energy," is suggested to be dominant at lower temperatures (Shklovskii *et al.*, 1989, 1990). In their alternative model, the temperature-dependent hopping mobility mostly dominates photoconductivity. The low-temperature photoconductivity has a universal nature in a-semiconductors as already stated and will be discussed in a separate section together with the photoconductivity in a-Ch.

Next, we discuss the dynamical response of the photoconductivity after stopping steady illumination. Through the study of photoconducting decays we can understand how the quasi equilibrium states under illumination return to the

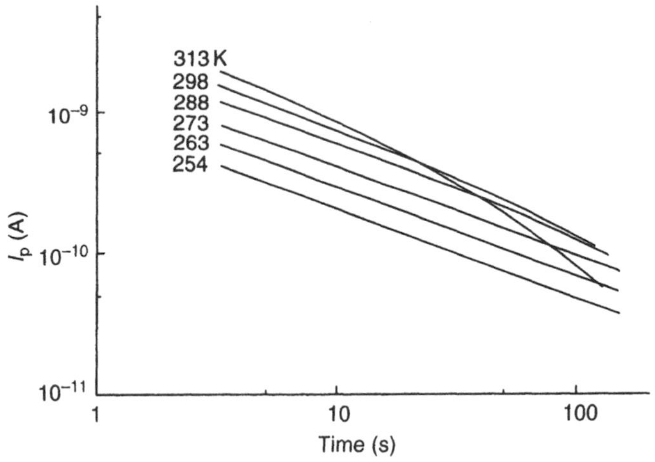

Figure 9.4 Decays of photocurrent $I_p(t)$ in a-Si : H after stopping the steady illumination, as a function of temperature (Shimakawa and Yano, 1986).

equilibrium states. Deep states dominate the decay kinetics. The non-exponential long-term photocurrent decay in $I_p(t)$ for $t > 1$ s observed in a-Si : H (Andreev *et al.*, 1984; Shimakawa and Yano, 1984) has been quantitatively analyzed (Shimakawa and Yano, 1984; Shimakawa *et al.*, 1986; Shimakawa, 1998), and found to follow a power law. It is known that the transient photocurrent (TP) with pulsed photo-excitation also decays as $t^{-\beta}$, where $0 < \beta < 1$ (Hvam and Brodsky, 1981; Orenstein and Kastner, 1981; Tiedje and Rose, 1981). The discussion of TP, together with transient photocurrent time of flight (TOF) measurement, will be described later and hence we only discuss here the long-term decays.

The decays of photocurrent $I_p(t)$ for a-Si : H after cessation of steady illumination at several different temperatures are shown in Fig. 9.4. I_p is nearly proportional to $t^{-\beta}$ with $\beta \approx 0.6$, except at the highest temperature of 313 K. At other temperatures <313 K, β is almost temperature-independent. In this time range, $I_p(t)$ is independent of the illumination intensity used in the range of 4×10^{13}–2×10^{15} photons cm^{-2} s^{-1}, while the steady photocurrent $I_p(0)$ at 273 K is proportional to $G^{0.8}$. The long-time decay can be interpreted as follows (Shimakawa *et al.*, 1986): under the steady-state illumination, as discussed in Eq. (9.9), E_{Fn} for electrons is closer to the mobility edge of conduction states, resulting in an increase of the number of trapped electrons in the tail localized states. E_{Fp} for holes, likewise, is closer to the mobility edge of valence states. The number of trapped holes in the tail states also increases. After stopping illumination, the trapped electrons and holes recombine and their quasi-Fermi levels return to the original equilibrium value E_F. This situation is schematically shown in Fig. 9.5. The tail states are described for simplicity as single levels, that is, E_{tc} and E_{tv}.

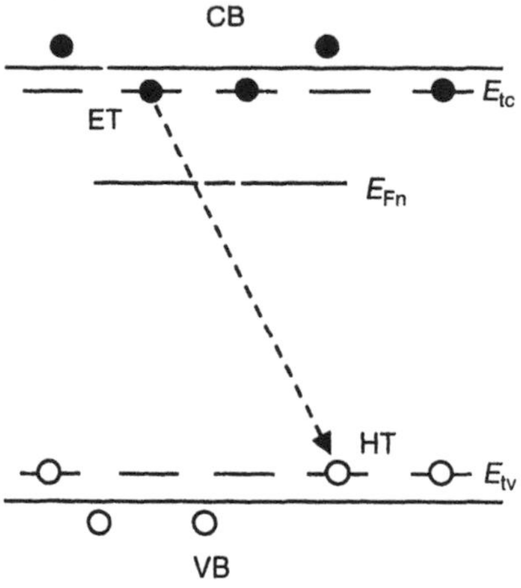

Figure 9.5 Schematic illustration of the recombination process between trapped electrons denoted by ET and holes denoted by HT in the tail states.

The density of trapped electrons n'_t is given by

$$n'_t = \Delta n_t + n_t = N_{tc} \exp(-\varepsilon/kT), \tag{9.18}$$

where n_t is the equilibrium density before illumination and Δn_t the increase in the excess density of trapped electrons due to illumination, $\varepsilon = E_{tc} - E_{Fn}$, and N_{tc} effective density of the conduction tail states. The density of free electrons n' is given by

$$n' = \Delta n + n_0 = N_c \exp[-(E_c - E_{Fn})/kT], \tag{9.19}$$

where Δn is the excess density of free electrons, n_0 the equilibrium density of free electrons. Δn is then obtained from Eqs (9.18) and (9.19) as

$$\Delta n = \Delta n_t N_c \exp[-(E_c - E_{tc})/kT]/N_{tc}. \tag{9.20}$$

Equation (9.20) predicts that the photocurrent ($\propto \Delta n$) is proportional to the excess density of trapped electrons.

Assuming a dispersive diffusion-controlled bimolecular recombination, the rate equation for excess Δn_t can be given by

$$\frac{d\Delta n_t}{dt} = -A(T)t^{-(1-\beta)}\Delta n_t^2, \tag{9.21}$$

where $A(T)$ is a temperature-dependent parameter, β is a dispersion parameter ($0 < \beta < 1$), and Δn_t is assumed to be equal to the number of trapped holes Δp_t.

The excess density of free electrons is then obtained from Eqs (9.20) and (9.21)

$$\Delta n = \frac{\Delta n_t(0)\beta N_c \exp[-(E_c - E_{tc})/kT]}{N_{tc}[\Delta n_t(0)A(T)t^\beta + \beta]}, \tag{9.22}$$

where $\Delta n_t(0)$ is the initial density of trapped electrons, which depends on illumination intensity. $I_p(t)$ at long times $(\Delta n_t(0)A(T)t^\beta \gg \beta)$ becomes

$$I_p(t) \propto \frac{\beta N_c}{N_{tc}A(T)} t^{-\beta} \exp[-(E_c - E_{tc})/kT]. \tag{9.23}$$

Equation (9.23) shows that $I_p(t)$ can be described by a power-law dependence of time and $I_p(t)$ is independent of illumination intensity at long time, which explains the experimental results quite well. Activation energies of $I_p(t)$ at $t = 5$ and 50 s are estimated to be 0.19 eV, which is consistent with the energy of well-defined states near band edges obtained from the TOF (Spear and LeComber, 1976; Marshall *et al.*, 1983) and the transient photocurrent (Oheda, 1985) measurements. It is not clear, however, whether the well-defined activation energy (0.19 eV) can be attributed to the tail states. Negative-U defects, T_3^+ and T_3^-, originally predicted by Adler (1978) and Elliott (1978), may be considered to be a candidate for trapping centers of electrons and holes, instead of tail states, as the energy levels of paired T_3^+–T_3^- centers are anticipated to lie near the band edges from the optically-detected magnetic resonance (ODMR) (Morigaki and Yoshida, 1985).

The above arguments imply the assumption of homogeniety in the material on a mesoscopic or macroscopic scale. If inhomogeneities on such scales are found in a material, then we should take account of such effects into the photoconductivity. In fact, there are a number of supporting evidences, and the influence of inhomogeniety in thermodynamic properties (Branz and Silver, 1990), the DC transport (Overhof and Beyer, 1981; Overhof and Thomas, 1989), TOF (Murayama and Mori, 1992; Liu *et al.*, 1994), AC transport (Long, 1989, Shimakawa *et al.*, 1996), etc. have been discussed in this context. The long-term decay (Shimakawa *et al.*, 1996), DC photoconductivity (Shimakawa, 2002) and AC photoconductivity (Shimakawa and Ganjoo, 2002) in a-Si : H have been discussed in modulated band edges by an electrostatic potential fluctuation, as shematically shown in Fig. 9.6.

Illumination creates free carriers in the conduction and valence extended states and electrons become concentrated in deep wells of the potential (holes in humps). The excess electrons and holes created in this way become spatially separated. An electron and a hole may not directly recombine if they are not located at the same point in space and hence the recombination may take place through the extended states E_{pcn} or E_{pcp} (critical percolation path (a)) (Tkach, 1975). If a certain number of deep localized states exists, the recombination can occur via such centers (process (b) or (c)). When it is assumed for simplicity that a discrete level exists below E_{pcn} or above E_{pcp}, the situation becomes the same, at least phenomenologically, as the recombination between localized electrons and holes, which has been discussed above; that is, the same equations (Eqs (9.18)–(9.23)) can be used. Again by assuming dispersive reaction rate, non-exponential long-term

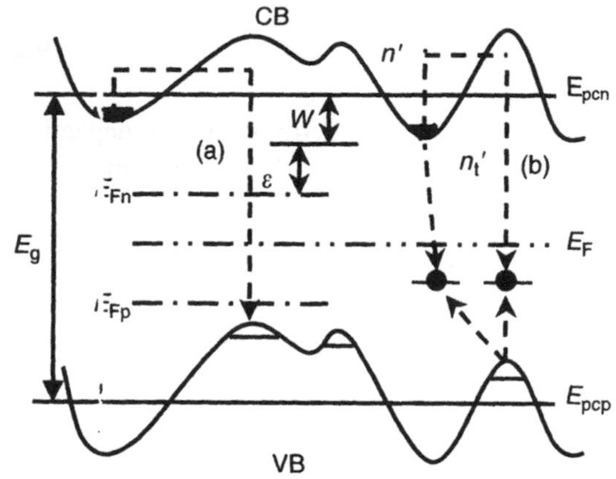

Figure 9.6 Schematic illustration of symmetric potential fluctuations: CB and VB extended states, E_{pcn} and E_{pcp} are the percolation threshold for electrons and holes, respectively. E_{Fn} and E_{Fp} are the quasi-Fermi levels for electrons and holes, respectively. E_F is the equilibrium Fermi level.

photocurrent decay near room temperature is explained in terms of the dispersive diffusion-controlled bimolecular recombination of electrons and holes in the fluctuated band states. These carriers recombine through recombination centers in the bandgap. The activation energy, $W = 0.19\,\text{eV}$, for an inhomogeneous system is a measure of the magnitude of potential fluctuations in undoped a-Si : H.

An alternative explanation for the origin of the thermal quenching already discussed for regime II in Fig. 9.3 may also be possible in terms of potential fluctuation model, since the localized electrons cannot be distinguished phenomenologically from the electrons in potential wells.

Through the study of photocurrent decay after stopping steady illumination, the role of deep trapping centers has been identified. Finally, we discuss the transient response of photocurrent using short light pulses. This technique is called the "transient photocurrent (TP)" method, which is different from the so-called the TOF method classified into the primary photoconduction category. TP will be discussed in what follows.

Transient photocurrent measurements are the most direct ways of probing the charge transport in the vicinity of the mobility edge. In the first few picoseconds or even in fractions of a picosecond, the photo-excited carriers generated in the band states thermalize down to the mobility edge and further down to the tail states. However, a dielectric relaxation time of about 10^{-10} s may not allow to obtaining precise information on the picosecond-scale photoconductivity (Tiedje, 1983). Hvam and Brodsky (1981) measured the TP in the time range between 10^{-8} and 10^{-2} s in a-Si : H, as shown in Fig. 9.7. The photocurrent exhibits a power law decay, $t^{(\beta-1)}$, with $\beta = T/T_c$ in some temperature range, where T is the temperature and

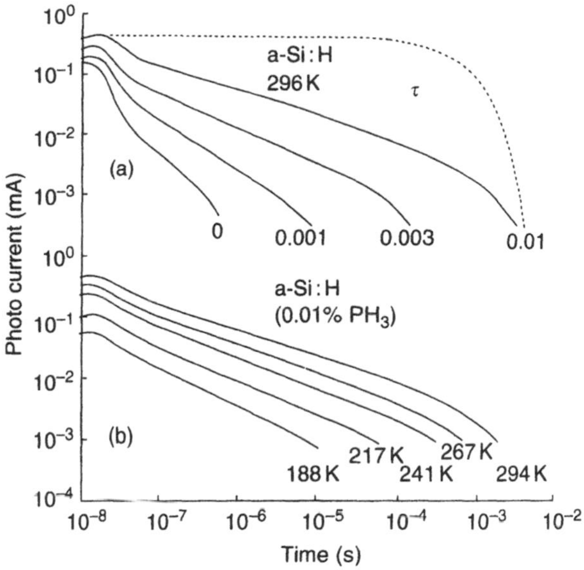

Figure 9.7 Photocurrent decays in undoped and P-doped a-Si : H after short-pulse excitation. (a) at room temperature and different doping ratios; (b) at doping ratio of 0.01 % PH_3 and different temperatures (Hvam and Brodsky, 1981).

T_c (≈460 K) is a characteristic temperature, which may correspond to the tailing parameter, 40 meV, of localized tails. The power law behavior can be explained by the multiple-trapping (MT) model which will be discussed below (Orenstein and Kastner, 1981; Tiedje and Rose, 1981).

The photo-excited electrons drop down through the extended states to the localized tail states in a time of the order of 10^{-12} s. For an exponential distribution of tail states, the DOS is given as

$$g(\varepsilon) = \frac{N_0}{kT_c} \exp(-\varepsilon/kT_c), \tag{9.24}$$

where T_c is the characteristic tailing parameter, considering the zero energy ($\varepsilon = 0$) at the mobility edge so that the positive energy points are toward the midgap. It is assumed that each localized state has the same probability of capturing an electron. Therefore, the initial energy distribution of electrons trapped in the tail states is proportional to the DOS, that is, $fg(\varepsilon)$ with $f < 1$. Then the electrons trapped at energy ε are thermally excited into the extended states above the mobility edge after a time t given by

$$t = \nu_0^{-1} \exp(\varepsilon/kT), \tag{9.25}$$

where ν_0 is the order of a phonon frequency. Each time an electron is thermally excited to a state above the mobility edge, it has a chance to fall into a deep state from which reemission takes a longer time. At some time t, all the deep states with energy $\varepsilon > kT \ln(\nu_0 t)$ will be accumulating electrons without reemitting them, since their mean time for thermal emission is much longer than t (see Eq. (9.25)). On the other hand, states of energy $\varepsilon < kT \ln(\nu_0 t)$ will have many thermal emission and recapture events before the time t, and hence the occupation of these shallow states will follow the Boltzmann statistics, that is, the occupation can be described by $g(\varepsilon)\exp(\varepsilon/kT)$. Thus the density of electrons occupying the tail states has a peak at ε_t^*, which is the deepest level from where an electron can be thermally emitted in time t, and then it decreases at energies further down towards the midgap with time. Thus ε_t^* can be regarded as a demarcation energy. This is shown schematically in Fig. 9.8 and is called the MT model.

Since the electron density is peaked at ε_t^*, the trap-limited drift mobility can be discussed effectively with a single trapping level, as given by (Spear, 1969):

$$\mu_d = \mu_0 \frac{n_f}{n_f + n_t}, \tag{9.26}$$

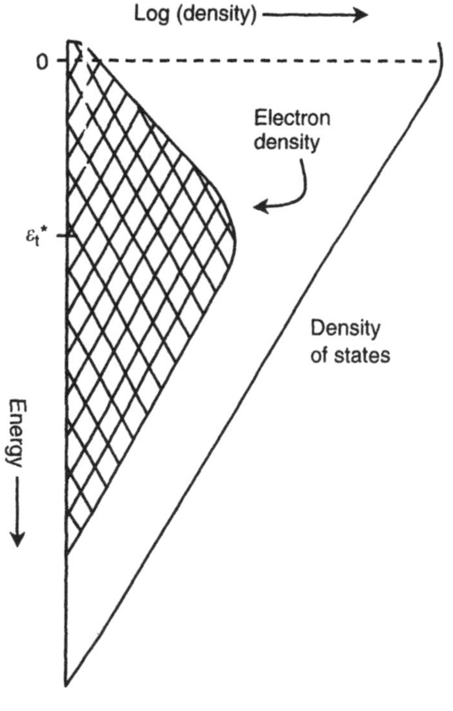

Figure 9.8 Trapped electron distribution at a time $t \gg \nu_0^{-1}$ after the initial excitation pulse.

where μ_0 is the microscopic mobility and n_f and n_t are the number of free and trapped electrons, respectively. Under the condition of $n_f \ll n_t$, μ_d is

$$\begin{aligned}
\mu_d &\approx \mu_0 \frac{n_f}{n_t} = \mu_0 \frac{N_c \exp(-\varepsilon_t^*/kT)}{\int_{\varepsilon_t^*}^{\infty} g(\varepsilon)\,d\varepsilon} = \mu_0 \frac{N_c \exp(-\varepsilon_t^*/kT)}{N_0 \exp(-\varepsilon_t^*/kT_c)} \\
&= \frac{\mu_0 N_c}{N_0} \exp\left[-\varepsilon_t^* \left(\frac{1}{kT} - \frac{1}{kT_c}\right)\right] \\
&= \frac{\mu_0 N_c}{N_0} \exp\left[-kT \ln(\nu_0 t) \left(\frac{1}{kT} - \frac{1}{kT_c}\right)\right] \\
&= \frac{\mu_0 N_c}{N_0} (\nu_0 t)^{\beta - 1}, \qquad (9.27)
\end{aligned}$$

where $\beta = T/T_c$ and N_c is the effective DOS of the conduction extended states. Thus the TP observed can be interpreted in terms of the MT model. It is found that the time-dependent photocurrent is dominated by drift mobility and recombination is not important in these time scale measurements.

Finally, in this section, we should also discuss some controversial TP data reported for undoped and phosphorous-doped a-Si : H (Oheda, 1985). Using a weak illumination of intensity $\sim 1 \times 10^{13}$ photons cm^{-2} to avoid saturation of the gap states, different shapes of the transient current as predicted from the MT model mentioned above have been observed. The decay rates are found to depend on temperature only weakly. From these results Oheda (1985) has concluded that TP cannot be explained by the simple conventional MT model. Instead of assuming an exponentially distributed tail states, he has suggested that there exist distinct trap levels at around 0.2 eV below the conduction mobility edge.

There are some attempts made at calculating DOS directly from the decay curve (e.g. see Marshall, 1999; Naito, 1999; Reynolds *et al.*, 1999). Reynolds *et al.* (1999) have also observed incomplete power-law decay in the TP for weak illumination intensity, which is similar to what Oheda (1985) has obtained. A careful adoption of illumination intensity may be required to get the proper results. A general method for determining the localized DOS in highly resistive semiconductors, using TP and Laplace transformation, is shown by Naito (1996).

9.1.2 *Photoconductivity in a-Ch*

A typical example of the temperature variation of photoconductivity σ_p as a function of light intensity in a-Sb_2Te_3 is shown in Fig. 9.9. Unlike in a-Si : H, there is no hump in the temperature dependence of σ_p (Bube, 1992). The photoconductivity has a maximum at a specific temperature T_m with its magnitude generally larger than the dark conductivity for $T < T_m$, and smaller than the dark conductivity for $T > T_m$. In a medium temperature range, σ_p decreases with decreasing temperature (see, e.g. Enck and Pfister, 1976). At very low temperatures, σ_p becomes almost independent of temperature, as mentioned in the previous section, which will be discussed later on.

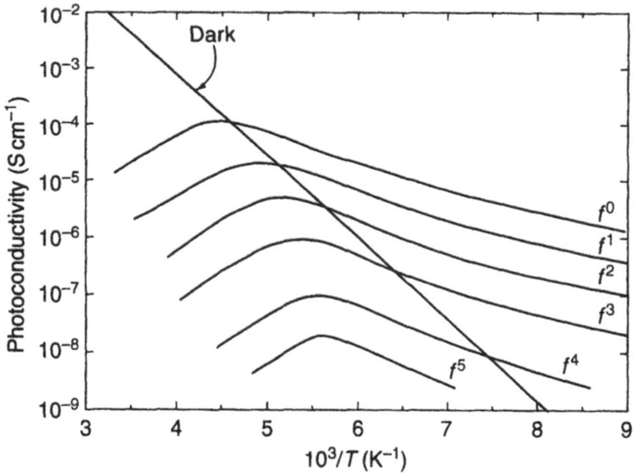

Figure 9.9 Temperature variation of DC-photoconductivities in sputtered Sb_2Te_3 films. Highest light intensity f^0 corresponds to 2.5×10^{17} photons $cm^{-2}\,s^{-1}$. Other f superscripts indicate the optical density of the corresponding attenuation factor, for example, $f^1 = 2.5 \times 10^{16}$ photons $cm^{-2}\,s^{-1}$ (Bube, 1992).

In the following, we will discuss the photoconductivity in each of the temperature regimes. Here also we will not discuss the photoconductivity for $T > T_m$, because then $\sigma_p < \sigma_d$, which is not of much interest for "photoconductors." Interested readers in the photoconductivity in this temperature range, may like to refer to excellent reviews and books (see, e.g. Spear and LeComber, 1976; Bube, 1992).

For $T < T_m$, the photoconductivity is observed to possess the following features: $\sigma_p \propto G^{1/2}$ at high G and $\sigma_p \propto G$ at low G. σ_p decreases with decreasing temperature as $\sigma_p \propto \exp(-W/kT)$ with $W \approx 0.1$ eV at lower temperatures, for example, in a-Sb_2Te_3. The activation energy appears to be independent of the excitation intensity. However, the temperature dependence of σ_p is not always dominated by the thermally activated process, that is, no straight lines are found in the $\ln \sigma_p$ vs $1/T$ plots (Bube, 1992).

Traditionally, the photoconductivity in this temperature range can be explained as follows: for high G, the quasi-Fermi levels move towards the respective band tail states. Then tail-to-tail recombination, which is bimolecular in nature, becomes dominant. Let us denote Δp_t and Δn_t as localized holes and electrons in their respective band tails, which gives $G \propto \Delta p_t \Delta n_t \approx \Delta p_t^2$ and hence $\Delta p_t \propto G^{1/2}$, because holes are the dominant carriers in a-Ch, which is due to higher mobility of holes than that of electrons. It should be remembered here that as Eq. (9.20) represents Δn for free electrons, Δp for free holes is given by

$$\Delta p = \Delta p_t N_v \exp[-(E_{tp} - E_v)/kT]/N_t. \tag{9.28}$$

This gives the required photoconductivity dependence on G as

$$\sigma_p \propto G^{1/2} \exp[-(E_{tp} - E_v)/kT], \tag{9.29}$$

or

$$\sigma_p \propto G^{1/2}.$$

Now it remains left to explain the linear dependence of photoconductivity on G in the lower G range. At lower G, the quasi-Fermi levels are close to the equilibrium Fermi level, E_F. Thus, the trapped excess holes, Δp_t, can recombine with electrons near E_F, because the population of excess electrons, Δn_t, in the tail states is relatively very small as compared to that of holes near E_F. As will be discussed in Chapter 10, a positively charged center becomes a neutral center, denoted by D^0, by capturing a photogenerated electron according to $D^+ + e \rightarrow D^0$. Note that D^0 states are located near E_F. Let us denote by ΔD^0 to the increase in D^0 due to this reaction and then one gets $G \propto \Delta p_t(\Delta n_t + \Delta D^0)$. Assuming $\Delta n_t \ll \Delta D^0$, produces $G \propto \Delta p_t \Delta D^0$ and hence $\Delta p_t \propto G$, which is the monomolecular dependence on G as observed experimentally. The photoconductivity is then given by

$$\sigma_p \propto G \exp[-(E_{tp} - E_v)/kT]. \tag{9.30}$$

Assuming that the photoconductivity in this regime is dominated by free holes in a-Ch, the above argument is primarily similar to the discussion given by Zhou and Elliott (1993) (see also Section 9.2). Alternatively, hopping of holes in the valence tails are suggested to contribute to σ_p in this regime. This will be discussed later, together with the photoconductivity at low temperatures.

One important remark should be made on the temperature dependence of photoconductivity in a-Ch. We have discussed above the case of the small activation energy, for example, $W \approx 0.1$ eV in a-Sb_2Te_3. Some materials, for example, a-As_2Se_3 and a-GeSe, have $W \approx 0.52$ and 0.48 eV, respectively (Shimakawa, 1986). That means deep centers, rather than shallow states (like tail states), are expected to be involved in these materials. This leads to the conclusion that the tail-to-tail recombination cannot in general dominate the recombination processes in a-Ch.

Next, we discuss the dynamical response of the photoconductivity after stopping illumination as discussed in a-Si : H. Non-exponential long-term photocurrent decays ($t > 1$ s) after stopping the steady illumination have been observed in a-Ch as well as a-Si : H (Mamontova and Nadolny, 1973; Fuhs and Meyer, 1974; Oheda and Namikawa, 1976; Chamberlain and Moseley, 1982; Kumeda *et al.*, 1985). It has been found that the photocurrent for most a-Ch decays as $\exp(-Ct^{\beta})$, where $0 < \beta < 1$, while in a-Si : H it decays as $t^{-\beta}$ (Shimakawa, 1985, 1986).

An example of a-As_2Se_3 is shown in Fig. 9.10. To explain the long-term photocurrent decay in a-Ch, Shimakawa (1985, 1986) has made the following assumptions: (i) The concentration of charged defects D^+ and D^- is greater than that of other centers in the bandgap. (ii) The majority of excess electrons and

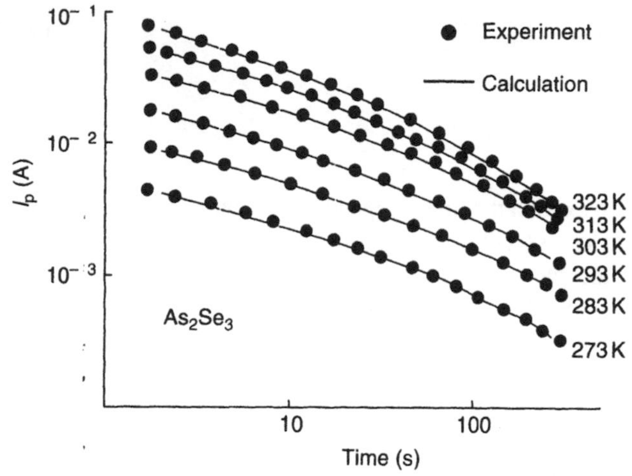

Figure 9.10 Decays of photocurrent $I_p(t)$ after stopping the steady illumination in a-As_2Se_3 (Shimakawa, 1986).

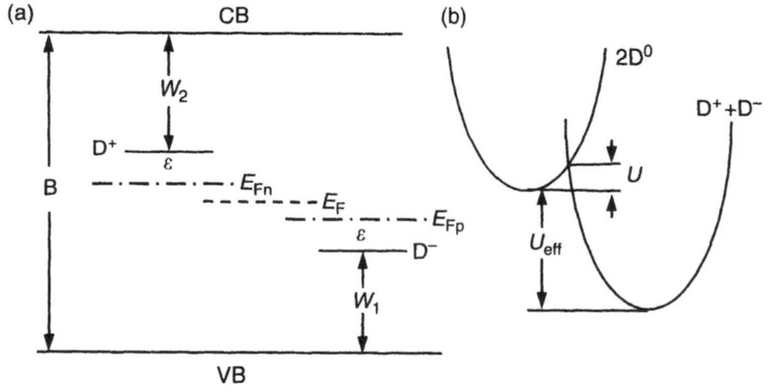

Figure 9.11 (a) Thermal energy levels and (b) configurational coordinate diagram for tunneling assocciated with D^+, D^0 and D^-. E_{Fn} and E_{Fp} are the quasi-Fermi levels for electrons and holes, respectively. U_{eff} is the effective negative correlation energy ($B - W_1 - W_2$).

holes are trapped by D^+ and D^- according to the reactions $D^+ + e \rightarrow D^0$ and $D^- + h \rightarrow D^0$, respectively. As shown in Fig. 9.11(a), quasi-Fermi levels, E_{Fn} and E_{Fp}, move towards their respective bands due to these captures. W_1 is the energy needed to capture an electron from the valence band by a D^0 to become D^-, and W_2 is the energy needed to remove an electron from D^0 and transfer it to the conduction band.

We discuss the recombination processes in a-Ch in a manner similar to that in a-Si : H, discussed earlier. The density of trapped electrons n'_t is given by

$$n'_t = \Delta n_t + n_t = N_t \exp(-\varepsilon/kT), \tag{9.31}$$

where Δn_t is the excess density of D^0, n_t the equilibrium density of D^0, $\varepsilon = E_{tc} - E_{Fn}$, and N_t the total density of charged defects ($=[D^+]+[D^-]$). Holes can dominate photocurrents in a-Ch, as E_{Fp} of holes is closer to the valence band than E_{Fn} of electrons to conduction band. Here $W_2 > W_1$ is implicitly assumed. The density of free holes p' is given by

$$p' = \Delta p + p_0 = N_v \exp[-(E_{Fp} - E_v)/kT], \tag{9.32}$$

where p_0 is the equilibrium density and Δp the excess density of free holes. Δp is then obtained from Eqs. (9.31) and (9.32) as

$$\Delta p = \Delta n_t N_v \exp[-W_1/kT]/N_t, \tag{9.33}$$

which predicts that the photocurrent ($\propto \Delta p$) is proportional to the excess density of trapped electrons Δn_t.

After cessation of illumination, Δn_t decreases due to the recombination following: $2D^0 \rightarrow D^+ + D^-$. Before the recombination, D^0 can hop due to tunneling of a hole from D^0 to D^- and tunneling of an electron from D^0 to D^+. The location of D^0 is thus converted into that of D^- and D^+. This type of recombination can be presented by the bimolecular reaction. However, if the interpair separation (random pair separation) is larger than the intrapair separation, a geminate-like monomolecular recombination can predominate (Shimakawa, 1986). It is hence assumed that both the bimolecular and monomolecular reactions of excess D^0 are dominant in this case. Assuming the dispersive recombination rate, the rate equation can be written as

$$\frac{d\Delta n_t}{dt} = -A(T)t^{-(1-\beta)}\Delta n_t^2 - B(T)t^{-(1-\beta)}\Delta n_t, \tag{9.34}$$

where $A(T)$ and $B(T)$ are temperature-dependent parameters, and β is a dispersion parameter ($0 < \beta < 1$). Here the first and second terms represent the bimolecular and monomolecular processes, respectively. The solution of Eq. (9.34) is obtained as

$$\Delta n_t = \frac{C(B(T)/A(T))\exp\left[-(B(T)/\beta)t^{\beta}\right]}{1 - C\exp\left[-(B(T)/\beta)t^{\beta}\right]}, \tag{9.35}$$

where C is given by

$$C = \frac{\Delta n_t(0)}{(B(T)/A(T)) + \Delta n_t(0)}, \tag{9.36}$$

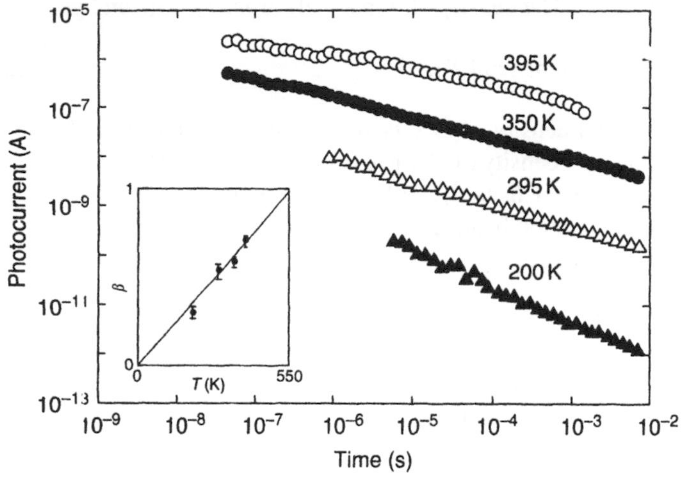

Figure 9.12 Decays of photocurrent after pulsed excitation with band-gap light measured at several temperatures. The temperature-dependent β is shown in the inset (Orenstein and Kastner, 1981).

with $\Delta n_t(0)$ being a constant that depends on the illumination intensity. Combining Eqs (9.33) and (9.35), the photocurrent for long time delay is obtained as

$$I_p(t) \propto \Delta p = C' \exp(-W_1/kT) \exp[-B(T)t^{\beta}/\beta], \tag{9.37}$$

solid curves in Fig. 9.10 are the calculated results using Eq. (9.37). Fitting of Eq. (9.37) to the experimental data is excellent, producing some physical parameters W_1 and β for several systems of a-Ch (Shimakawa, 1986). The potential barrier U for the reaction, $2D^0 \rightarrow D^+ + D^-$, is also deduced to be around 40 meV from fitting the experimental data. The different decay kinetics between a-Si : H and a-Ch can be attributed to the natures in defects, i.e. the principal defects are negative-U charged defects in a-Ch and positive-U neutral defects in a-Si : H, which are discussed in Chapter 7.

Figure 9.12 shows the transient photocurrent (TP) as a function of temperature in a-As_2Se_3 (Orenstein and Kastner, 1981). The power law decay following $I_p(t) \propto t^{\beta-1}$, predicted from Eq. (9.27), is observed as found in a-Si : H as well. The temperature dependence of the dispersion parameter $\beta = T/T_c$ is shown in the inset and the slope of a linear fit produces $kT_c \approx 50$ meV, which corresponds to the tailing parameter of the valence tail states.

9.1.3 Low temperature photoconductivity

A particular energy level in the band tail, called the "transport energy" level, plays a significant role in the hopping transport of carriers in both equilibrium and

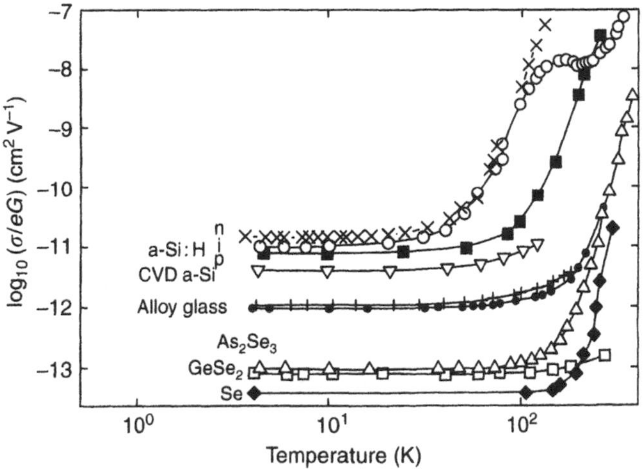

Figure 9.13 Temperature dependence of normalized photoconductivity for typical a-semiconductors (Fritzshe, 1989).

non-equilibrium conditions for both steady-state and transient phenomena. In this section, through the study of steady-state photoconductivity at low temperatures for both a-Si : H and a-Ch, the role of the transport energy level in the description of the carrier transport and recombination phenomena is briefly reviewed.

Figure 9.13 shows the temperature dependence of the normalized photoconductivity σ_p/eG for some a-semiconductors. A similar behavior is also observed in a-Ge : H (Shimakawa and Aoki, 1992). All curves show the same general features. The quantity σ_p/eG is constant at low temperatures, and its value depends on the material. In this temperature regime, the exponent γ in the relation, $\sigma \propto G^\gamma$, is about 1.0 and it decreases to $\gamma \approx 0.5$ when σ_p/eG begins to rise with temperature. The nearly universal behavior of the low temperature photoconductivity suggests that in this region the transport process is related to a common general feature of a-semiconductors (Fritzsche, 1989).

A theory developed by Shklovskii *et al.* (1989, 1990) suggests that the low-temperature photoconductivity is dominated by the energy-loss (downward) hopping of photocarriers (electrons in a-Si : H and holes in a-Ch) in the band-tail states. An importance of a particular energy level in the localized tail states for a-semiconductors was first pointed out by Grünewald *et al.* (1979), in which a numerical analysis of the exponentially decreasing density of states was presented. Shapiro and Adler (1985) reached the same conclusion through an analytical formula, that is, the vicinity of a particular level in the band tail dominates the hopping transport and its position in energy is independent of the Fermi level.

It should be noted that so called the "transport energy (TE)" has also been proposed by Monroe (1985) through the study of non-equilibrium energy relaxation of electrons by hopping. It turned out that the relaxation process changes at the transport energy. Now, it is understood that the transport energy proposed by Monroe in non-equilibrium process coincides with the energy level discussed by Grünewald *et al.* (1979) and Shapiro and Adler (1985). It is shown that TE determines both recombination and transport of electrons in the non-equilibrium steady-state under illumination (Shklovskii *et al.*, 1989, 1990). It is therefore understood that TE dominates both equilibrium and non-equilibrium and both transient and steady-state transport phenomena in a-semiconductors (Baranovskii *et al.*, 1994). TE is derived as follows (Shkolvskii *et al.*, 1989, 1990). At very low temperatures, energy-loss hopping transport is only possible with emission of phonons, that is, the carriers are hopping down the tail states, whereas at intermediate temperatures, transport can occur by hopping involving both up and down in tail states. TE is defined as the energy at which the probability of upward hops equals to that of downward hops.

As considered in the transient photoconductivity (Section 9.1.1), here also an exponential density of states distribution is assumed for both electrons and holes, given by

$$g(E) = \frac{N_0}{\varepsilon_0} \exp\left(-\frac{\varepsilon}{\varepsilon_0}\right), \tag{9.38}$$

where the energy ε is measured positive from the mobility edge ($\varepsilon = 0$) towards the midgap, N_0 the total density of tail states, and ε_0 the tailing parameter. As already described in Chapter 8, the hopping rate is given as

$$\nu = \nu_0 \exp\left(-\frac{2r}{a} - \frac{\delta E}{kT}\right), \tag{9.39}$$

where r is the separation between an occupied site and an empty site with energy difference δE, a is the decay length of the localized wavefunction and ν_0 the characteristic phonon frequency. On the basis of Eqs (9.38) and (9.39), the concept of the TE can be derived by introducing the differential conductivity as

$$\mathrm{d}\sigma(\varepsilon) = \frac{e^2}{kT} \mathrm{d}n(\varepsilon) D(\varepsilon), \tag{9.40}$$

for $0 < \varepsilon < \varepsilon_F$, where ε_F is the Fermi level. Here $\mathrm{d}n(\varepsilon)$ is the carrier concentration in the energy interval ε and $\varepsilon + \mathrm{d}\varepsilon$ given by

$$\mathrm{d}n(\varepsilon) = g(\varepsilon) \exp\left(\frac{\varepsilon - \varepsilon_F}{kT}\right) \mathrm{d}\varepsilon, \tag{9.41}$$

and $D(\varepsilon)$ is the diffusion constant at energy ε written as

$$D(\varepsilon) = \tfrac{1}{3} r^2(\varepsilon) \nu(\varepsilon). \tag{9.42}$$

The diffusion constant $D(\varepsilon)$ increases as $n(\varepsilon)$ decreases when ε moves towards the mobility edge. The average hopping length r between states lying below the energy ε is given by

$$r(\varepsilon) \cong \left[\frac{4\pi}{3}\int_{\varepsilon}^{\infty}\frac{N_0}{\varepsilon_0}\exp\left(-\frac{\varepsilon}{\varepsilon_0}\right)\mathrm{d}\varepsilon\right]^{-1/3} = \left(\frac{4\pi N_0}{3}\right)^{-1/3}\exp\left(\frac{\varepsilon}{3\varepsilon_0}\right). \tag{9.43}$$

Using Eq. (9.43) in Eq. (9.42), and then substituting Eqs (9.41) and (9.42) into Eq. (9.40) we find that the differential conductivity $\mathrm{d}\sigma(\varepsilon)$ has a sharp maximum at

$$\varepsilon_t = 3\varepsilon_0 \ln\left[\frac{3\varepsilon_0}{kT}\frac{a}{2}\left(\frac{4\pi N_0}{3}\right)^{1/3}\right], \tag{9.44}$$

which is obtained by setting $\mathrm{d}\sigma(\varepsilon)/\mathrm{d}\varepsilon = 0$. This energy, ε_t, is called TE and the total conductivity is given by

$$\sigma = \int_0^{\infty}\mathrm{d}\sigma(\varepsilon) = \frac{e^2}{kT}D(\varepsilon_t)\,n(\varepsilon_t), \tag{9.45}$$

where

$$n(\varepsilon_t) = Wg(\varepsilon_t)\exp\left(\frac{\varepsilon_t - \varepsilon_F}{kT}\right), \tag{9.46}$$

is the concentration of electrons contributing to hopping transport in the range of energies of width $W = (6\varepsilon_0 kT)^{1/2}$ and centered at ε_t (Shklovskii *et al.*, 1990).

Equation (9.44) predicts that, as temperature increases, ε_t moves toward the mobility edge. At some particular temperature T^* the transport energy approaches the mobility edge. Thus the transport occurs in the extended states above T^*, and in tail states at $T < T^*$. In fact, the temperature-dependent photoconductivity in a medium temperature range (e.g. regime III in Fig. 9.3) can be well replicated by Eq. (9.45) (Shklovskii *et al.*, 1990). As discussed in the previous section, the hopping model, using the concept of the transport energy, can equally explain the temperature-dependent photoconductivity in a-semiconductors including a-Ch.

Next, we discuss zero temperature photoconductivity. At low temperatures, the energy-loss (downward) hopping to a nearest localized state of the tail at distance r with the rate

$$\nu_d(r) = \nu_0 \exp(-2r/a). \tag{9.47}$$

The competing process of recombination also takes place with a similar expression of rate

$$\nu_r(R) = \nu_{rad}\exp(-2R/a), \tag{9.48}$$

where ν_{rad} is the dipole radiation rate, R is the electron–hole separation, and a is taken as the decay length of the localized electron. Here it is assumed that the localization radius of the hole is smaller than that of electron. The values of $\nu_0 \approx 10^{12}$ and $\nu_{rad} \approx 10^8\ s^{-1}$ are usually taken as the prefactors, which define a characteristic length R_c as

$$R_c = \frac{a}{2} \ln\left(\frac{\nu_0}{\nu_{rad}}\right). \tag{9.49}$$

At $R = R_c$, the rates of recombination and hopping are equal. When R becomes greater than R_c, a photo-excited carrier does not recombine geminately, since electrons can diffuse before recombining with holes. Shklovskii *et al.* (1990) defined a "survival" probability $\eta(R)$ that a photo-excited pair does not recombine geminately, and they found $\eta(R) \approx R_c/R$ for $R \gg R_c$. Remember that the geminate recombination cannot contribute to the photoconductivity. Non-geminate recombination may occur when the electron–hole pair separation becomes about half the geminate pair distance, which can approximated by the half of the average electron (or hole) separation, that is, $R' = ((4\pi n_0/3)^{-1/3})/2$, where n_0 is the steady-state electron concentration at $T = 0$. The value of n_0 is obtained by equating as

$$\frac{G\eta(R')}{\nu_r \exp(-2R'/a)} = n_0, \tag{9.50}$$

and the current density j is given by

$$j = G\eta(R')p_{av}(R'), \tag{9.51}$$

where $p_{av}(R')$ is the average dipole moment given by

$$p_{av}(R') = (eR')\frac{1}{3}\frac{eER'}{\varepsilon_0} = \frac{1}{3}\frac{(eR')^2}{\varepsilon_0}E, \tag{9.52}$$

which is obtained as follows: the dipole moment, with the angle θ that the applied field E makes with the line joining the hopping sites, $p'(R')$, is written by $p'(R') = eR'\cos\theta$. By taking into account an exponential distribution of sites (see Eq. (9.38)), the difference of occupation probability due to energy difference between two sites is given as

$$\exp\left[\frac{eE(R'\cos\theta/2)}{\varepsilon_0}\right] - \exp\left[-\frac{eE(R'\cos\theta/2)}{\varepsilon_0}\right] \approx \frac{eER'\cos\theta}{\varepsilon_0}, \tag{9.53}$$

for $eER'\cos\theta \ll \varepsilon_0$. Then the average dipole moment is

$$p_{av} = p'\frac{eER'\cos\theta}{\varepsilon_0} = \frac{(eR')^2}{\varepsilon_0}E\cos^2\theta = \frac{(eR')^2}{3\varepsilon_0}E. \tag{9.54}$$

Note that averaging $\cos^2\theta$ over all directions gives 1/3.

Using Eq. (9.54) in Eq. (9.51), the photoconductivity is then obtained as

$$\sigma_p = j/E = G\eta(R')\frac{e^2 R'^2}{3\varepsilon_0}. \tag{9.55}$$

Using Eq. (9.49) in Eq. (9.55), we get

$$\sigma_p = GR_c\frac{e^2 R'}{3\varepsilon_0} = G\frac{e^2 a}{12\varepsilon_0}\left(\frac{4\pi}{3}\right)^{-1/3} n_0^{-1/3}\ln\left(\frac{\nu_0}{\nu_{rad}}\right). \tag{9.56}$$

Comparing Eq. (9.56) with $\sigma_p \propto G^\gamma$, we get

$$\gamma \approx 1 - an_0^{1/3} \approx 1, \tag{9.57}$$

which is consistent with the experimental observation (Fritzshe, 1989). Using reasonable values, for example, for a-Si : H, as $G = 10^{20}\,\mathrm{cm^{-3}\,s^{-1}}$, $a = 1\,\mathrm{nm}$, $\varepsilon_0 = 0.025\,\mathrm{eV}$, and $\nu_0/\nu_{rad} = 10^4$, and $n_0^{-1/3}/a = 13$, we obtain $\sigma_p/eG = 2.5 \times 10^{-12}\,\mathrm{cm^2\,V^{-1}}$ and $\gamma = 0.93$, which is in close agreement with the experimental result of $\sigma_p/eG \approx 10^{-11}\,\mathrm{cm^2\,V^{-1}}$ and $\gamma = 0.95$ (Fritzshe, 1989).

The σ_p/eG curves of different a-semiconductors, for example, a-Si : H, a-Ge : H, and a-Ch, have very similar shapes at relatively low temperatures. The temperature-independent photoconductivity is explained in terms of the hopping down motions (before recombination) in tail states. An important role of the transport energy in intermediate temperature range has already been suggested in this section. It is of interest to note that the down-hop motion of electrons in the tail states is argued to occur under the influence of potential fluctuations and the corresponding alternative expression for the low-temperature photoconductivity in a-semiconductors is also presented (Shimakawa and Ganjoo, 2002).

Experimentally in a-As_2Se_3, the hopping relaxation time of holes in the valence tails is obtained through the AC photoconductivity measurements (Ohno and Shimakawa, 2000). Frequency dependent photoconductivity is observed, suggesting that the hopping of carriers dominates the transport in a-As_2Se_3. The relaxation time is almost constant at 5×10^{-2} s between 20 and 150 K and then decreases with increasing temperature to 5×10^{-5} s at 300 K. The temperature-dependent steady-photoconductivity has also been calculated using the hopping relaxation time estimated from the AC photoconductivity, which agrees with the experimental data below room temperature. This may support that the model of hopping of photocarriers using the concept of the transport energy dominates the photoconduction, while the band transport is suggested to be dominant in the intermediate-temperature region, as discussed earlier (Zhou and Elliott, 1993). Note that the conductivity is found to be frequency-independent in the band transport.

9.2 Primary photoconduction

The TOF experiment is completely different from the secondary photoconduction experiment described in Section 9.1. In the TOF technique, photocarriers transit

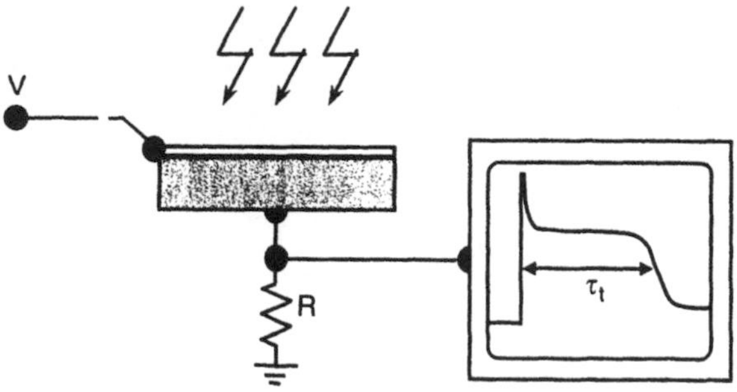

Figure 9.14 Measurement system for the transient photoresponse of insulating thin films by the so called time-of-flight technique.

from the illuminated electrode to the counter electrode with blocking contacts and carriers are far from thermal equilibrium. So, this type of transport is called the primary photoconduction and a sample of sandwich configuration is used in the experiment as shown in Fig. 9.14. Both electrons and holes are created as a thin sheet below the illuminated transport electrode using a flash of highly absorbable illumination. Therefore, light of energy much larger than bandgap is required. One of the carriers, electron or holes, drifts across the sample under the external applied voltage V. The response is shown schematically in Fig. 9.14 as expected from the conventional recombination-free Gaussian dispersion of the charge packet during transit in an ideal case. While these carrier packets are moving through the sample, a current I_D is induced in the external circuit. The drift mobility is then estimated by

$$\mu_d = \frac{v_d}{V/d} = \frac{d/\tau}{V/d} = \frac{d^2}{V\tau_t}, \tag{9.58}$$

where τ_t is called the transit time. The Gaussian dispersion means that the carrier packet is broadened by a Gaussian profile with root-mean-square (RMS) deviation from the mean position $\Delta l = (2Dt)^{1/2}$, where D is the diffusion coefficient (Marshall *et al.*, 1983).

Transit pulses of the above form are observed in many crystalline and some a-semiconductors, for example, room temperature hole transport in a-Se (Marshall and Owen, 1972) and room temperature electron transport in device quality a-Si : H (Marshall *et al.*, 1986). However, in the majority of circumstances, including the low-temperature measurements on the above materials, TOF studies reveal markedly different characteristics, as shown in Fig. 9.15 (Marshall *et al.*, 1983). The observed transients show no plateau region, indicating a spread in the arrival times (anomalous dispersion). However, a plot of log (current) vs log (time), reveals a break in the curve and the current decreases with time as

$$I(t) \approx t^{\beta_i - 1} (t < \tau_t), \tag{9.59}$$

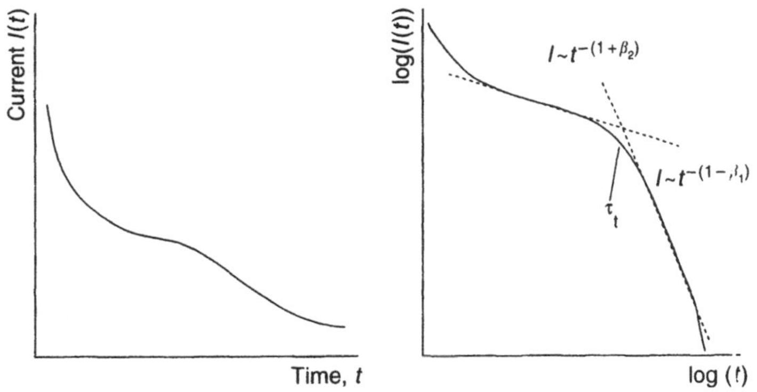

Figure 9.15 Anomalously dispersive time-of-flight current, plotted on linear and logarithmic scales.

and

$$I(t) \approx t^{-(1+\beta_2)}(t > \tau_t), \tag{9.60}$$

where $\beta_1(<1)$ and $\beta_2(<1)$ are called the dispersion parameters, and τ_t is the transit time. The above transient form of current is called dispersive or anomalously dispersive, which is attributed to non-Gaussian profile of carrier packets.

The first theoretical description of this phenomenon was given by Scher and Montroll (1975) and anomalous dispersion was suggested to be caused by a very broad distribution of transit time of each photo-excited carrier. Pollak (1977) then suggested that only hopping does not induce an anomolous dispersion. The most popular one to explain the anomolous dispersion, now, is the MT model (Orenstein and Kastner, 1981; Tiedje and Rose, 1981), which has already been discussed (see Eq. (9.27)).

Following MT, the power-law decay can be regarded as a decay of free-carrier density or as a decay of the drift mobility. The usual definition of the transit time τ_t leads to

$$\int_0^{\tau_t} \mu_d F \, dt = d, \tag{9.61}$$

where F is the applied electric field ($= V/d$). Substituting Eq. (9.27) into Eq. (9.61) produces τ_t as

$$\tau_t = \nu_0^{((1/\beta)-1)} \left(\frac{N_0 d}{\mu_0 N_c F} \right)^{1/\beta}. \tag{9.62}$$

Note that the transit time has a nonlinear dependence on d and F in the dispersive transport.

Let us discuss, finally, the current decay for $t > \tau_t$. After the transit time, an electron thermally emitted from a trap near ε_t^* will be extracted at the back electrode without being retrapped below ε_t^*. As a result, the photocurrent is dominated by the rate of thermal emission of trapped electrons below ε_t^*. $I(t)$ is therefore given by

$$\begin{aligned} I(t) = \frac{\Delta Q}{\Delta t} &= \frac{efkT(N_0/kT_c)\exp\left(-\varepsilon_t^*/kT_c\right)}{\nu_0^{-1}\exp\left(\varepsilon_t^*/kT\right)} \\ &= ef\beta N_0\nu_0 \exp\left[-\varepsilon_t^*\left(\frac{1}{kT_c}+\frac{1}{kT}\right)\right] \\ &= ef\beta N_0\nu_0 \exp\left[-kT\left(\frac{1}{kT_c}+\frac{1}{kT}\right)\ln(\nu_0 t)\right] \\ &= ef\beta N_0\nu_0 (\nu_0 t)^{-(\beta+1)}. \end{aligned} \tag{9.63}$$

Note here that $\Delta Q = efg(\varepsilon_t^*)kT(f < 1)$ is the density of trapped charges at the energy ε_t^* which become free at a rate of $\nu(\varepsilon_t^*) = \nu_0\exp(-\varepsilon_t^*/kT)$. Comparing Eq. (9.63) with Eqs (9.59) and (9.60), we get $\beta = \beta_1 = \beta_2$. Figure 9.16 shows an example of the TOF results for electrons measured at 160 K in a-Si : H (Tiedje, 1984), giving $\beta_1 \approx \beta_2 = 0.50$. The transit time, which is clearly evident as a knee, was measured as a function of applied voltage. The resulting drift mobility defined by Eq. (9.58) should have a power-law field dependence as

$$\mu_d \propto F^{(1/\beta_1)-1}, \tag{9.64}$$

where Eqs (9.58) and (9.62) are used. The experimental data show $F^{0.8}$ at 160 K, which produces $\beta_1 = 0.56$, and may be regarded consistent with $\beta_1 = 0.50$

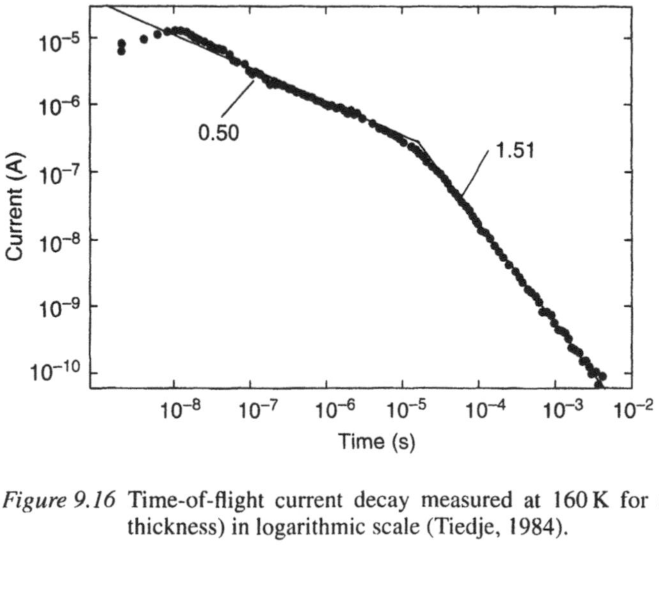

Figure 9.16 Time-of-flight current decay measured at 160 K for a-Si : H (3.8-μm thickness) in logarithmic scale (Tiedje, 1984).

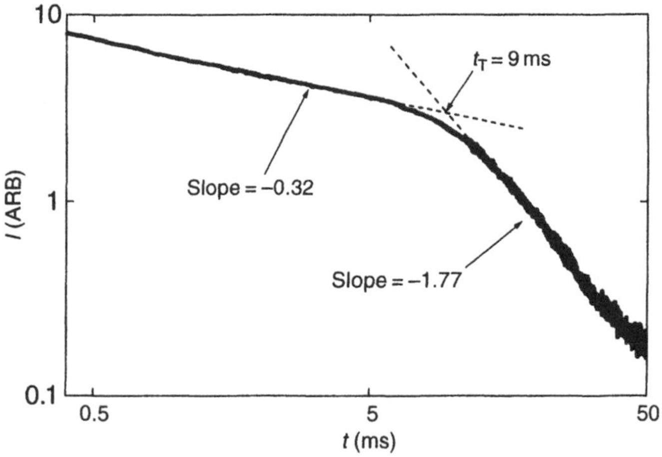

Figure 9.17 Time-of-flight current decay measured at room temperature for a-As_2Se_3 in logarithmic scale (Pfister and Sher, 1975).

obtained from Fig. 9.16. The temperature-dependent dispersion parameter obtained experimentally is approximately $\beta = T/T_c$, which produces $T_c \approx$ 310 K($\approx$27 meV). T_c obtained from TOF here is smaller than T_c (460 K) obtained from TP method (see Section 9.1.1). The simple model based on MT is in good agreement with the experimental data. As the electron drift mobility is reported to be of the order of 1 $cm^2\,V^{-1}\,s^{-1}$ at room temperature, then $\mu_0 \approx 10\,cm^2\,V^{-1}\,s^{-1}$ is deduced (Tiedje, 1984). The hole transport has been discussed in a similar way. The drift mobility for holes, $\mu_d \approx 10^{-3}\,cm^2\,V^{-1}\,s^{-1}$ near room temperature is obtained and $\mu_0 \approx 0.5\,cm^2\,V^{-1}\,s^{-1}$ is estimated. T_c for valence band tails is estimated to be 490 K (42 meV) which is larger than that for the conduction band tails (Tiedje, 1984).

Let us now move to TOF in a-Ch. Figure 9.17 shows an example of the TOF results for holes in a-As_2Se_3 (Pfister and Scher, 1975). There is a comprehensive review on the photoconductivity in a-Ch (Enck and Pfister, 1978). The power-law dependence of $I(t)$ is found with $\beta_1 = 0.68$ and $\beta_2 = 0.77$. Unlike a-Si : H, the values of β_1 and β_2 are not generally the same in a-Ch. It is found experimentally that the transit time depends on the electric field as $F^{-1.85}$, producing $\mu_d \propto F^{0.85}$in which $\beta_1 = 0.54$ should be taken in the MT model. Unlike a-Si : H, $\beta_1 = 0.68$ estimated from the $I(t)$ curve does not seems to be consistent with $\beta_1 = 0.54$. Furthermore, β_1 and β_2 are mostly temperature-independent. It is therefore concluded that MT model cannot be applied to a-Ch.

The drift mobility at room temperature estimated from Eq. (9.58) for holes is typically of the order of $10^{-5}\,cm^2\,V^{-1}\,s^{-1}$ at field of around $10^5\,V\,cm^{-1}$ in a-Ch. Note that unlike in a-Si : H the drift mobility for electrons is much smaller than that for holes in a-Ch. The low value of the drift mobility in a-Ch can be attributed to a deep-trap limited transport mechanisms. In fact, thermally activated hole drift mobility is reported with an activation energy of ~0.55 eV in a-As_2Se_3 and ~0.25 eV in a-Se. The contribution of the band-tail states to the drift mobility in a-Ch is small compared with deep localized states. Why MT model is applicable to a-Si : H but not to a-Ch, may be due to a small density of deep localized states in device quality a-Si : H. In fact, the number of deep states (D^+ and D^-) is expected to be 10^{17}–10^{18} cm^{-3}, as discussed in Chapter 7.

The following points are also to be noted: although TOF data seem to agree with MT model as stated above (Oheda, 1985), a question on the applicability of MT for interpreting TOF results arises. In MT model, a demarcation-like energy ε_t^* is defined over the whole specimen. In TOF experiments, on the other hand, excess carriers are initially created near the surface under quasi-equilibrium. Thus the time evolution of excess carriers in energy cannot be characterized by an identical value of ε_t^*. Therefore, the adaptability of MT model to TOF results is not yet clearly understood and it is controversial.

MT model has been applied to a-Si : H, but it has also been criticized (Murayama *et al.*, 1992, 1998). To explain the electric field dependence of the transit time in a-Ch and a-Si : H, Murayama *et al.* (1992) claimed the hopping distance of 2–5 nm, which is clearly inconsistent with MT model. Alternatively, the hopping of electrons or self-trapped electrons at the band edge in the presence of long-range potential fluctuations has been treated on a fractal structure by Murayama and Mori (1992) and Murayama *et al.* (1998). Using Monte Carlo simulation, they have shown that the temperature and field dependent $I(t)$ can be well explained by the random walk of electrons or holes on a fractal structure for a-Si : H and a-Ch. Hence, it may be concluded that TOF study is still incomplete in a-solids.

References

Adler, D. (1978). *Phys. Rev. Lett.* **41**, 1755.

Adler, D. (1980). *Solar Energy Mater.* **2**, 199.

Andreev, A.A., Zherzev, A.V., Kosarev, A.I., Koughia, K.V. and Shlikak, I.S. (1984). *Solid State Commun.* **52**, 589.

Baranovskii, S.D., Thomas, P. and Adriaenssens, G.J. (1994). In: Marshall, J.M., Kirov, N. and Vavrek, A. (eds), *Electronic, Optoelectronic and Magnetic Thin Films*. John Wiley & Sons Inc., New York, p. 35.

Branz, M.H. and Silver, M. (1990). *Phys. Rev. B* **42**, 7420.

Bube, R.H. (1992). *Photoelectronic Properties of Semiconductors*. Cambridge University Press, Cambridge.

Carius, R., Fuhs, W. and Weber, K. (1987). In: Kastner, M.A., Thomas, G.A. and Ovshinsky, S.R. (eds), *Disordered Semiconductors*. Plenum Press, New York, p. 369.

Chamberlain, R.V. and Moseley, A.J. (1982). *Jpn. J. Appl. Phys.* **21**, 13.

Dersh, H., Shwietzer, L. and Stuke, J. (1983). *Phys. Rev. B* **28**, 4678.

Elliott, S.R. (1978). *Philos. Mag. B* **38**, 325.

Enck, R.G. and Pfister, G. (1976). In: Mort, J. and Pai, D.M. (eds), *Photoconductivity and Related Phenomena*. Elsevier, Amsterdam, p. 215.

Fritzshe, H. (1989). *J. Non-Cryst. Solids* **114**, 1.

Fuhs, W. and Meyer, D. (1974). *Phys. Stat. Sol. A* **24**, 275.

Grünewald, M., Thomas, P. and Würtz, D. (1979). *Phys. Stat. Sol. (b)* **94**, k1.

Hattori, K., Okamoto, H. and Hamakawa, Y. (1996). *J. Non-Cryst. Solids* **198–200**, 288.

Hvam, M. and Brodsky, M.H. (1981). *Phys. Rev. Lett.* **46**, 371.

Kumeda, M., Kawachi, G. and Shimizu, T. (1985). *Philos. Mag. B* **51**, 591.

Liu, E.Z., Wickboldt, A.E., Pang, D., Chen, J.H. and Paul, W. (1994). *Philos. Mag. B* **70**, 109.

Long, A.R. (1989). *Philos. Mag. B* **59**, 377.

Mamontova, T.N. and Nadolny, A.J. (1973). *Phys. Stat. Sol. A* **18**, K103.

Marshall, J.M. (1983). *Rep. Prog. Phys.* **46**, 1235.

Marshall, J.M. (1999). In: Marshall, J.M., Kirov, N., Vavrek, A. and Maud, J.M. (eds), *Thin Film Materials and Device-Development in Science and Technology.* World Scientific, Singapore, p. 175.

Marshall, J.M. and Owen, A.E. (1972). *Phys. Stat. Sol.* **12**, 181.

Marshall, J.M., Michil, H. and Adriaenssens, G.J. (1983). *Philos. Mag. B* **47**, 211.

Marshall, J.M., Street, R.A. and Thompson, M.J. (1986). *Philos. Mag. B* **54**, 51.

Morigaki, K. (1999). *Physics of Amorphous Semiconductors.* World Scientific & Imperial College Press, London.

Morigaki, K. and Yoshida, M. (1985). *Philos. Mag. B* **52**, 289.

Murayama, K. and Mori, M. (1992). *Philos. Mag. B* **65**, 501.

Murayama, K.,Oheda, H., Yamasaki, S. and Matsuda, A. (1992). *Solid State Commun.* **81**, 887.

Murayama, K., Ikeuchi, R., Kuwabara, J. and Hiramoto, H. (1998). *Phys. Stat. Sol. (b)* **205**, 129.

Naito, H. (1997). In: Marshall, J.M., Kirov, N., Vavrek, A. and Maud, J.M. (eds), *Future Directions in Thin Film Science and Technology.* World Scientific, Singapore, p. 96.

Oheda, H. (1980). *Solid State Commun.* **33**, 203.

Oheda, H. (1981). *J. Appl. Phys.* **52**, 6693.

Oheda, H. (1985). *Philos. Mag. B* **52**, 857.

Oheda, H. and Namikawa, H. (1976). *Jpn. J. Appl. Phys.* **8**, 1465.

Ohno, T. and Shimakawa, K. (2000). *J. Non-Cryst. Solids* **266–269**, 894.

Orenstein, J. and Kastner, M. (1981). *Phys. Rev. Lett.* **46**, 1421.

Overhof, H. and Beyer, W. (1981). *Philos. Mag. B* **44**, 317.

Overhof, H. and Thomas, P. (1989). *Electronic Transport in Hydrogenated Amorphous Semiconductors.* Springer, Berlin.

Pearsans, P.D. and Fritzshe, H. (1981). *J. Physique* **42**, C4-597.

Pfister, G. and Scher, H. (1975). *Bu. Am. Phys.* **20**, 322.

Pollak, M. (1977). *Philos. Mag. B* **36**, 1157.

Raynolds, S., Main, C., Webb, D.P. and Rose, M.J. (1999). In: Marshall, J.M., Kirov, N., Vavrek, A. and Maud, J.M. (eds), *Thin Film Materials and Device-Development in Science and Technology.* World Scientific, Singapore, p. 183.

Rose, A. (1963). *Concepts in Photoconductivity and Allied Problems.* Pergamon, Oxford.

Scher, H. and Montroll, E.W. (1975). *Phys. Rev. B* **12**, 2455.

Shapiro, F.R. and Adler, D. (1985). *J. Non-Cryst. Solids* **74**, 189.

Shimakawa, K. (1985). *J. Non-Cryst. Solids* **77–78**, 1253.

Shimakawa, K. (1986). *Phys. Rev. B* **34**, 8703.

Shimakawa, K. (2002). *Phil. Mag. Lett.* **82**, 635.

Shimakawa, K. and Aoki, T. (1992). *Phys. Rev. B* **46**, 12750.

Shimakawa, K. and Ganjoo, A. (2002). *Phys. Rev. B* **65**, 165213.

Shimakawa, K. and Yano, Y. (1984). *Appl. Phys. Lett.* **45**, 862.
Shimakawa, K., Yoshida, A. and Arizumi, T. (1974). *J. Non-Cryst. Solids* **16**, 258.
Shimakawa, K., Yano, Y. and Katsuma, Y. (1986). *Philos. Mag. B* **54**, 285.
Shimakawa, K., Kondo, A., Goto, M., Long, A.R. (1996). *J. Non-Cryst. Solids* **198–200**, 157.
Shklovskii, B I., Fritzsche, H. and Baranovskii, S.D. (1989). *Phys. Rev. Lett.* **62**, 2989.
Shklovskii, B.I., Levin, E.I., Fritzshe, H. and Baranovskii, S.D. (1990). In: Fritzshe, H. (ed.), *Advances in Disordered Semiconductors*. World Scientific, Singapore, p. 161.
Spear, W.E. (1969). *J. Non-Cryst. Solids* **1**, 197.
Spear, W.E. and LeComber, P.G. (1976). In: Mort, J. and Pai, D.M. (eds), *Photoconductivity and Related Phenomena*. Elsevier, Amsterdam, p. 185.
Tiedje, T. (1983). In: Joannopoulosm, J.D. and Lucovsky, G. (eds), *The Physics of Hydrogenated Amorphous Silicon II*. Springer, Berlin, p. 261.
Tiedje, T. (1984). In: Pankove, J.I. (ed.), *Semiconductors and Semimetals*, Vol. 21, Part C. Academic Press, Orland, p. 207.
Tiedje, T. and Rose, A. (1981). *Solid State Commun.* **37**, 49.
Tkach, Yu.Ya. (1975). *Soviet Phys. Semicond.* **9**, 704.
Vanier, P.E., Delahoy, A.E. and Griffith, R.W. (1981). *J. Appl. Phys.* **52**, 5235.
Vomvas, A. and Fritzshe, H. (1987). *J. Non-Cryst. Solids* **97–98**, 823.
Zhou, J.H. and Elliott, S.R. (1992). *Philos. Mag. B* **66**, 801.
Zhou, J.H. and Elliott, S.R. (1993). *Phys. Rev. B* **48**, 1505.

10 Reversible photoinduced effects

Illumination of amorphous semiconductors (a-semiconductors) with light, having a photon energy comparable with or smaller than that of the bandgap, induces various changes on structural and electronic properties of a material. Such photoinduced phenomena can be associated with the presence of a high degree of structural flexibility in amorphous solids (a-solids). Those photoinduced changes, which get reversed and the material returns back to the original state by thermal annealing, are called the reversible effects. The current understanding of reversible photoinduced changes in the hydrogenated amorphous silicon (a-Si : H) and amorphous chalcogenides (a-Ch), as case examples, is reviewed in this chapter. Two types of photoinduced changes, defect creations and structural changes, will be discussed here.

10.1 Defect creation

10.1.1 a-Si : H

A decrease in the photoconductivity after photoillumination was first discovered in a-Si : H by Staebler and Wronski (1977, 1980). This type of photoinduced change, referred to as photodegradation, is usually accompanied by the defect creation and is known as the Staebler–Wronski effect (SW effect). Although a great deal of interest and effort have been devoted to understanding SW effect, the origin for the defect creation process seems still to be unclear (Shimakawa and Elliott, 1995). The SW effect has been reviewed in several excellent books (Street, 1991; Redfield and Bube, 1996; Morigaki, 1999), which are valuable sources for surveying the overall features of photodegradation. In this section, first we plan to briefly discuss previous models of defect creation by illumination in a-Si : H, and then to consider the case of pairing of like charge carriers, particularly holes, due to strong interaction with lattices first predicted by Anderson (1975). The pairing of excited holes on a covalent bond, presents an alternative mechanism of bond breaking due to illumination in a-Si : H and a-Chs (Singh and Shimakawa, 2000), which will be discussed later in this section.

A generally accepted model for breaking weak Si–Si bonds is that the breaking is induced by non-radiative recombination of electrons and holes, producing

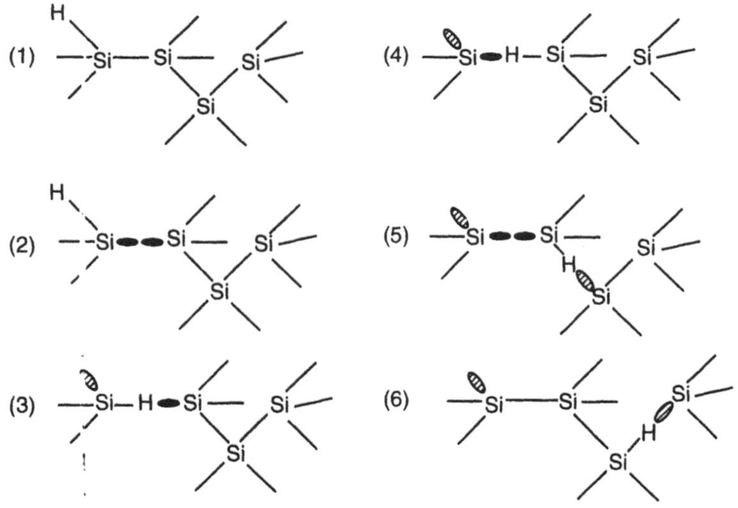

Figure 10.1 Schematic illustration of a model for the light-induced well-separated T_3^0 centers, which is formed in successive steps of Si–H bond switching (Morigaki, 1988).

a pair of neighboring dangling bonds (DB) (Stutzmann *et al.*, 1985). This model is, however, not consistent with the electron spin resonance (ESR) results, which evidently suggests the presence of well separated DB's (Dersh *et al.*, 1981; Yamasaki *et al.*, 1987). This bond-breaking model (BBM) was then modified to require a Si–H bond switching to stabilize the created DB's and it was extended by taking into account long range diffusion of hydrogen (Street, 1991; Morigaki, 1988).

Morigaki (1999) has proposed that the following processes occur successively during illumination: (1) Self-trapping of a hole stretches a specific weak Si–Si bond adjacent to a Si–H bond. (2) Switching of a Si–H bond toward a weak Si–Si bond occurs by non-radiative recombination of an electron and the self-trapped hole. (3) Hopping and/or tunneling of hydrogen occurs to stabilize the a-Si : H network. These processes are repeated for further movement of the created DB sites (shown in Fig. 10.1). It is interesting to note that the evidence of hole self-trapping on weak bonds has been presented through the optically detected magnetic resonance (ODMR) studies (Morigaki, 1999). For the involvement of hydrogen, other long-range hydrogen diffusion models have been proposed, for example, hydrogen is emitted from a Si–H bond and traps at a strained Si–Si bond to form Si–H and a DB (Jackson and Zhang, 1990). The light enhanced H diffusion, in fact, has also been reported (Street, 1991). This requires the photoinduced DB to be close to hydrogen (DB and H complex). There is no signature of such complex from ESR measurements (Yamasaki and Isoya, 1993), while the ENDOR measurements indicate evidence for DB–H complex (Morigaki, 1999). This scheme is still a controversial matter.

The hydrogen collision model (Branz, 1999) may overcome the shortcomings of the above models by proposing a spin-inactive final state for the two mobile hydrogen atoms that were removed from Si–H bonds. As shown in Fig. 10.2(a), recombination induced emission of H from Si–H bonds creates both mobile H and

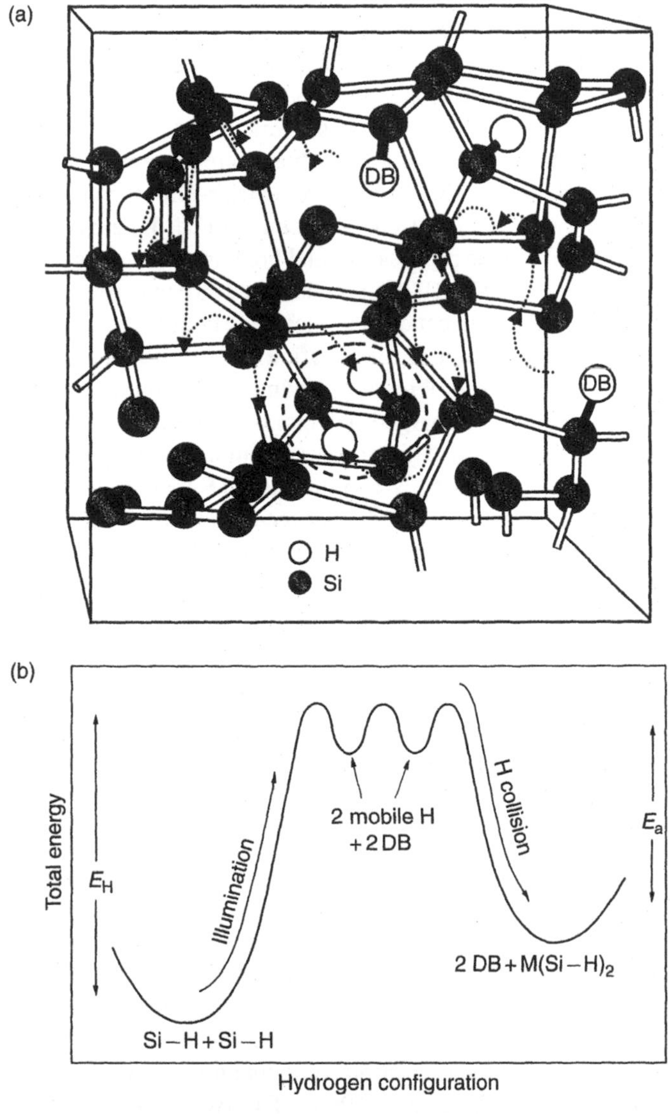

Figure 10.2 (a) Schematic illustration of the reaction of defect creation. H atoms diffuse along the dotted lines and then reach a metastable complex $(Si–H)_2$ inside the dashed oval (Branz, 1999). (b) Configurational coordinate diagram of the reaction of defect creation. E_H and E_a represent the energies of hydrogen diffusion and metastability annealing, respectively.

DB's behind them. When two mobile H associate in a metastable two H complex, newly created two DB's become stable. Fig. 10.2(b) shows the configurational coordinate diagram for such a metastable Si–H complex denoted by $M(Si\text{–}H)_2$. The energies for hydrogen diffusion E_H and annealing of metastability E_a are also indicated in Fig. 10.2(b). Note that in this model DB's are created by emission of H from Si–H bonds, while in BBM DB's are created by cessation of Si–Si bonds. Apparently, there are two problems with the hydrogen collision model: (1) it breaks only a Si–H bond, that means the presence of hydrogen is essential for the applicability of the model, and (2) a single photon is considered to be adequate to break a Si–H bond by exciting an electron from bonding to anti-bonding orbitals. This means that while the model is regarded as successful in Si–H, it cannot be applied to a-Chs, where neither H atoms nor Si–H bonds are present.

The mechanisms stated above should involve carrier–lattice interactions. Although the involvement of vibrational energy is mentioned, its exact role or any quantitative consideration in the process of bond breaking has not been taken into account. Shimakawa *et al.* (1993, 1995) have assumed strong carrier–phonon interactions and proposed an alternative model applicable to both a-Ch and a-Si : H. The primary event of photoexcitation is the creation of a self-trapped exciton (STE) and then its non-radiative recombination, leading either to the creation of metastable STE (or intimate pair of charged defects) or its return to the ground state through radiative recombination. This process is similar to that proposed in $a\text{-}SiO_2$ (Shulger and Stefanovich, 1990). According to STE model, the primary event is an optical excitation, not a non-radiative recombination. Weakening a bond with STE creation eventually leads to bond breaking (Song and Williams, 1996). It may, however, be noted that the creation of metastable STE is the result, not cause, of bond breaking (Shulger and Stefanovich, 1990).

If a pair of holes localizes on a weak bond due to a strong carrier–phonon interaction (Anderson's negative-U) (Singh and Shimakawa, 2000), a weak bond will break, because energetically such an excited state is shown to be more favorable, and has a larger bond length. For calculating the energy eigenvalue of a pair of holes localized on a bond, a linear chain of amorphous material is considered in a simplified way (see Chapter 6). The main results can be described as: (1) The gained energy ΔE with a pair of holes localized on the same bond in a-Si : H and $a\text{-}As_2Se_3$ is between 0.16 and 0.50 eV depending on the kind of phonon modes involved. (2) The bond on which the two holes are localized gets enlarged twice as large as the corresponding bond with a single localized hole. As soon as that occurs the bond gets broken, which agrees with the simulation work (Fedders *et al.*, 1992). As already argued, the metastable DB's should be separately located, DBs created by this process should diffuse by bond switching mechanism, which may accompany hydrogen motion until stabilization (Stutzmann *et al.*, 1985; Jackson and Zhang, 1990; Branz, 1999; Morigaki, 1999). The two neutral DB's then become stable and have positive-U in nature, while the creation processes involve negative-U nature.

One of the excited holes is localized in the bonding orbital σ_{wb} of a weak bond in the valence band tails. Then, this orbital will move upward (self-trapping)

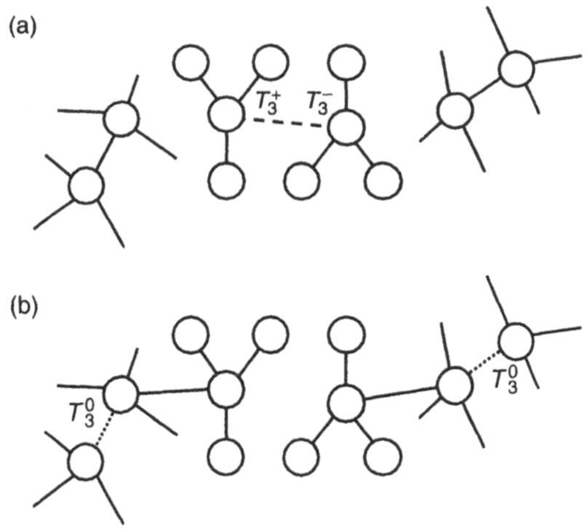

Figure 10.3 Schematic illustration of (a) an IP of charged T_3^+(sp) − T_3^- centers from a weak bond and (b) isolated and neutral T_3^0 defects which are created by bond-switching reactions (Shimakawa *et al.*, 1995; Singh and Shimakawa, 2000).

due to strong hole and lattice interaction. As the pairing of holes is energetically favorable, a second hole gets localized on it and the bond is ultimately broken due to removal of both covalent electrons. The criterion of strong carrier and lattice interaction required for creating negative-U is met in a-Si : H through hydrogenation, which makes the a-Si network a more flexible structure. One of the two DBs thus created can be positively charged and the other negatively charged by capturing the excited electrons. These configurations, known as intimate pairs (IP) of $T_3^+ - T_3^-$, are energetically favorable but may be unstable (Shimakawa *et al.*, 1995). A schematic illustration of this mechanism is presented in Fig. 10.3. The released excess energy due to pairing of holes (negative-U) can be used for bond switching and hydrogen diffusion to stabilize the network (metastability), producing $2T_3^0$ DB [Fig. 10.3(b)]. There is only circumstantial evidence of the above model. Using the Schottky type cell, when the electric field is applied to increase the number of holes, relative to that of electrons, the cell shows more degradation in its I–V characteristics, that is, yielding a less diode nature due to an increase in inner resistance of the cell (Imagawa *et al.*, 2000). This suggests that holes play more dominant role in the photodegradation than electrons. BBM by non-radiative recombination energy, on the other hand, requires in principle that the holes and electrons play equal roles.

Holes are known to play a similar role in carrier-induced degradation, in which a gradual ON-current degradation during continuous operation under accumulation of holes in thin-film transistors (TFTs) has been observed (Schropp *et al.*, 1987). Carrier-induced metastabilities in the p-layer of a-Si : H p–n junction devices

(Crandall *et al.*, 1991) and p–i–n devices (Pfleiderer *et al.*, 1984) have also been reported. Metastability induced by excess carriers due to rapid thermal quenching in doped a-Si : H may also fall into the category of carrier induced metastability (Kakalios *et al.*, 1987; Street *et al.*, 1987). It may be noted that there is a significant difference between photoinduced and charge induced carriers: absorption of a photon creates a pair of electron and hole, while charge injection introduces only one type of carrier into the system. This may imply that the defect creation occurs when sufficient excess holes are introduced into a-Si : H. Electrons may have a secondary effect on degradation, while holes have the primary role.

Now let us look at the kinetics of creation of light induced metastable defects (LIMD). In BBM model, DBs are directly created by non-radiative recombination of photocarriers, n and p, and the rate equation for inducing the defect is given by (Stutzmann *et al.*, 1985)

$$\frac{dN}{dt} \propto np \propto \left(\frac{G}{N}\right)^2, \tag{10.1}$$

where N is the number of DBs and G is the light intensity, leading to the well-known relation

$$N(t) \propto G^{2/3} t^{1/3}. \tag{10.2}$$

In the hydrogen collision model, the same relation holds, since DBs are created through bimolecular recombination of mobile H that has a density proportional to G/N (Branz, 1999). The pairing of holes model (Singh and Shimakawa, 2000) also predicts the same relation, because the relation involves p^2, p being the concentration of excited holes. Therefore, Eq. (10.1) can be applied to any pair of two particles: pair of electron and hole (BBM), pair of two holes (pairing hole model) or pair of two H atoms (hydrogen collision model). Fitting Eq. (10.2) to the experimental data is good in the medium time range.

Redfield and Bube (1996), on the other hand, have proposed a rate equation in which time dispersive generation and annihilation rates of defects are taken into account as

$$\frac{dN}{dt} \propto t^{\alpha-1}(G - DN), \tag{10.3}$$

where D is a constant and $\alpha (0 < \alpha < 1)$ is a dispersion parameter, which will be discussed later. Equation (10.3) leads to a stretched exponential form as

$$N(t) = N_s - [N_s - N(0)] \exp\left[-\left(\frac{t}{\tau}\right)^{\alpha}\right], \tag{10.4}$$

where $N_s (= N(\infty))$ is the saturated density of defects and τ is an effective time constant. Experimentally, the number of induced defects saturated at a certain value depends on the temperature and light intensity. Fitting Eq. (10.4) to the experimental data is excellent in the full time range (Redfield and Bube, 1996).

Similar kinetics, taking into account the time dispersive reaction rate, have been proposed in the STE model (Shimakawa *et al.*, 1995).

Having discussed the power law and the stretched exponential time dependence of $N(t)$ in Eqs (10.2) and (10.4), respectively, an alternative kinetic equation may be considered of interest. The increase in the number of defects with time observed in a-Si : H and a-Ch (Section 10.1.2) is, phenomenologically, very similar to the increase of population with time in biological systems (e.g. time evolution of bacteria). In biological systems, this is called the logistic growth of population (Murray, 1989). A self-limiting process should operate when the population becomes too large and it can be expressed by the following non-linear rate equation

$$\frac{dN}{dt} = (k - \lambda N)N, \tag{10.5}$$

where k and λ are constants. In this model, the effective growth rate is described by $k - \lambda N$. We apply the logistic equation to the defect creation kinetics in a-Si : H (Senda *et al.*, 1999). Now, the reaction rate is assumed to be a time-dependent (dispersive reaction) as $k = At^{\alpha-1}$, although its origin is not clear. Introducing $k = At^{\alpha-1}$ into Eq. (10.5), $N(t)$ is given as

$$N(t) = \frac{N(0)N_s \exp(t/\alpha)^\alpha}{N_s + N(0)[\exp(t/\alpha)^\alpha - 1]}, \tag{10.6}$$

where $N(0)$ and N_s are the initial and saturated values of defects, respectively, and $\tau \equiv (\alpha/A)1/\alpha$ is the effective creation time. We call this the stretched logistic function (SLF), since the exponential term is stretched by the factor of α.

One of the examples for fitting of SLF to the experimental data (ESR spin density) is shown in Fig. 10.4 and such a fitting produces the parameters $N(0)$, N_s, τ, and α. The value of N_s thus obtained approaches $10^{17}\ \mathrm{cm}^{-3}$. The most interesting parameter is $N(0)$, which can be regarded as the "seed" for defect creation and is obtained as $\approx 3 \times 10^{15}\ \mathrm{cm}^{-3}$ close to the number of pre-existing neutral DBs. The parameter A is found to be proportional to $G^{0.5}$, suggesting that a bimolecular process dominates the defect creation. The remarkable point that we can deduce here is that the pre-existing thermal-equilibrium defects facilitate LIMD creation. This may be supported by the molecular-dynamics simulation, which suggests that LIMD are nucleated by a localized defect state (Fedders *et al.*, 1992).

Let us also discuss the kinetics of the rate equation [Eq. (10.5)]. The term kN describes the creation process and λN^2 describes annihilation. These are completely different from the physical situations characterized by Eqs (10.1) and (10.3). The striking factor is the term $+kN$, which implies that LIMD creation is proportional to the number of existing LIMD at a time. This means that a defect induced by illumination, itself, produces other defects, which is phenomenologically similar to a biological system, for example, the growth of bacteria. The self-limiting process is described by the term $-\lambda N^2$. This term describes the "bimolecular recombination" of two defects, which means that a return to a normal bond may occur during bond switching (defect diffusion), leading to the stabilization of the DBs. This annihilation term is therefore easy to understand.

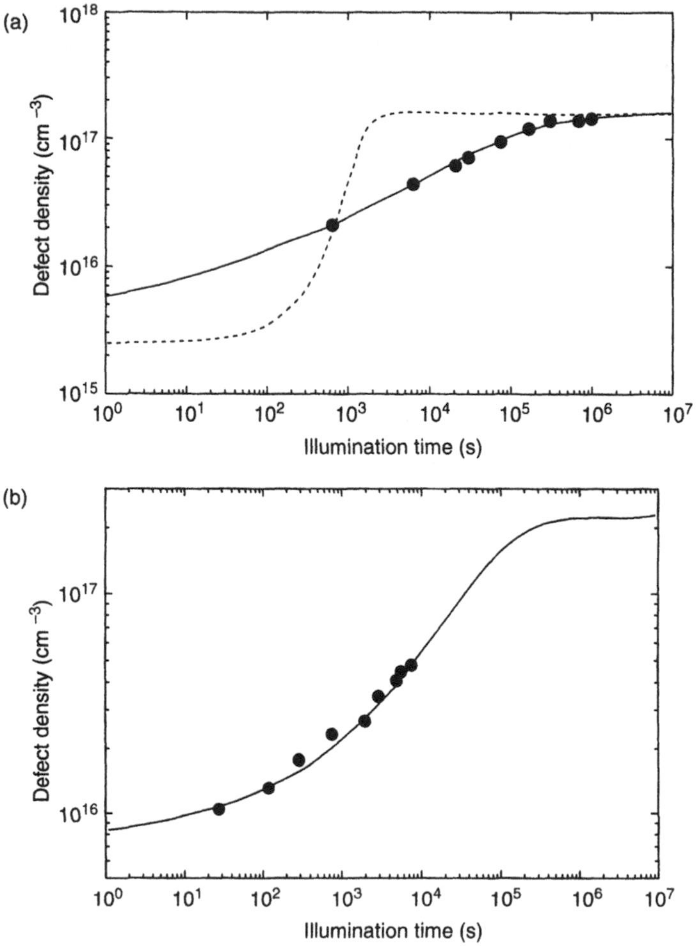

Figure 10.4 Fitting of the time-dependent spin density data in a-Si : H (Senda *et al.*, 1999). (a) Solid circles are the experimental data (Shimizu, 1993), the dashed and solid curves represent the results of logistic and stretched logistic equations, respectively. $N(0) = 2.5 \times 10^{15}\,\text{cm}^{-3}$, $N_s = 1.6 \times 10^{17}\,\text{cm}^{-3}$, $\alpha = 0.15$, and $\tau = 2.95$ s are taken as the fitting parameters. (b) Solid circles are the experimental data (Stutzmann, 1985), and solid curves represent the results of stretched logistic equations. $N(0) = 7.2 \times 10^{15}\,\text{cm}^{-3}$, $N_s = 3.0 \times 10^{17}\,\text{cm}^{-3}$, $\alpha = 0.27$, and $\tau = 400$ s are used as the fitting parameters.

However, the question about the physical origin of kN arises here. Do defects themselves produce new defects during illumination? This will be discussed later together with the creation of LIMD in a-Ch (Section 10.3). Thus, the main conclusions of the logistic model are: (1) LIMD creation is initiated around pre-existing thermal-equilibrium defects. (2) A created defect, itself, induces "seeds," which facilitate further defect creation. (3) The self- limiting process for LIMD creation is the recombination of two DBs and hence the number of induced defects saturates at a certain value, depending on the illumination conditions.

Finally, we discuss the quantum efficiency η for producing defects, which is defined as $\eta = N/n_p$, where n_p is the total number of illuminating photons ($\propto Gt$). Figure 10.5 shows the relation between η and n_p estimated for undoped a-Si : H. $N(t)$ (or $N(n_p)$) here is estimated conventionally from the photocurrent as $I_p(t) \propto 1/(N(0) + N(t)) \cdot \eta$ decreases with the total photon number, n_p, and becomes 10^{-8} after prolonged illumination. This means that one defect is created by 10^8 photons, which is consistent with the estimates obtained from ESR measurements (Morigaki, 1999). Thus, LIMD creation can be regarded as a very low efficiency process. As η is directly related to Eqs (10.2) or (10.4) or (10.6), it can also be described by some functional forms, depending on the dynamics that dominates the defect creation. Of particular interest here is η, which follows a single universal curve, irrespective of excitation energy (Shimakawa *et al.*, 2003). For example, we may expect smaller η for smaller excitation energy. Surprisingly, η seems to be independent of the energy of exciting photons. If BBM by non-radiative process is dominant in the LIMD creation, η should increase with increasing energy, since with higher excitation energy the released energy by non-radiative recombination for defect creation is larger than that with smaller excitation energy. This may also be regarded as indirectly supporting evidence for the pairing of the hole model in which only exciting free holes in a band state or localized holes in the tail states are required to create defects (Singh and Shimakawa, 2000).

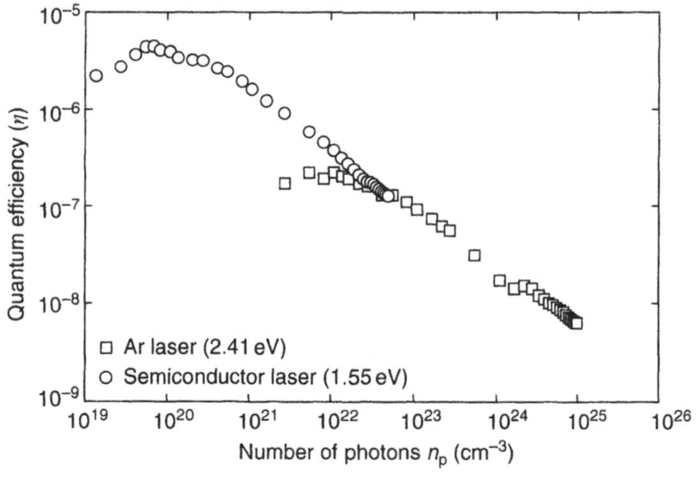

Figure 10.5 Relation between the quantum efficiency η of defect creation and total number of incident photons, n_p, in a-Si : H. The definition of η is N/n_p where N is the number of induced defects (Shimakawa *et al.*, 2003).

As the electronic transport and optical properties are greatly affected by the LIMD creation, these will be discussed further in Section 10.3.

10.1.2 a-Chalcogenides

A decrease in the photoconductivity after illumination has also been found in a-Ch (Shimakawa *et al.*, 1990, 1992, 1995). This might be linked with an increase in the light-induced ESR (LESR) spins by prolonged illumination (Guillaume *et al.*, 1977; Biegelsen and Street, 1980; Hautala *et al.*, 1988). First, a microscopic model for the mechanism of creating LIMD, for example, for a-$As_2S(Se)_3$ (Shimakawa *et al.*, 1990, 1992, 1995), is shown in Fig. 10.6. As already described in the previous chapter on defects, it is widely accepted that defects in a-Ch have negative-U in nature, contrary to those in a-Si : H. Hence, only charged defects exist in this class of material (Mott and Davis, 1979). The creation of widely separated random pairs (RPs) of positively and negatively charged centers is responsible for the decrease in the photoconductivity. Such RPs result from defect conserved bond switching reactions at optically induced IPs (conjugate pairs of charged defects, for example, $P_2^+ - C_1^-$ and $P_4^+ - C_1^-$, where P and C refer to pnictogen and chalcogen centers, respectively). It may be noted that the photoinduced IP can be regarded as the optically created STE (Shimakawa *et al.*, 1995; Song and Williams, 1996), and

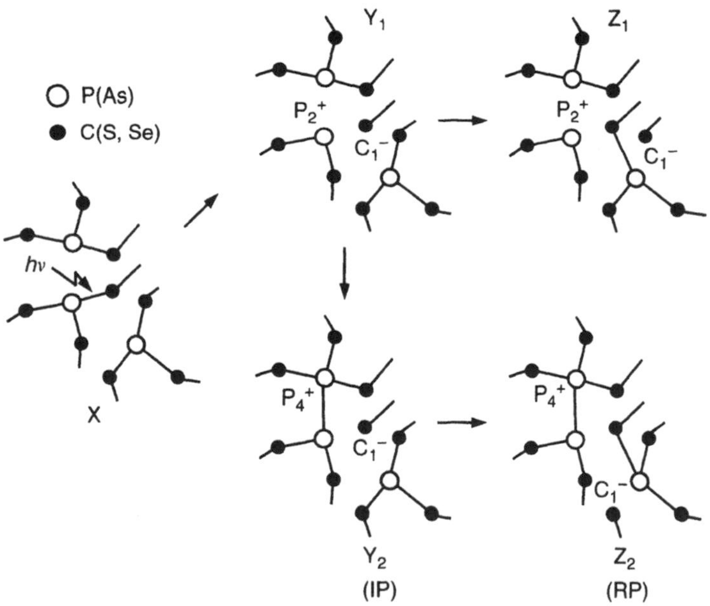

Figure 10.6 Schematic illustration of the optical generation of IP's of charged defect states (Y_1 and Y_2) from the ground state structure of a-$As_2S(Se)_3$. Subsequent bond-switching reactions can lead to a large separation between the charged defects, that is, RP ($Z_1 \cdot Z_2$) (Shimakawa *et al.*, 1990, 1995).

IP and RP can coexist. It is also known that both IPs and RPs dominate electronic and optical properties in a-Ch (Shimakawa *et al.*, 1995).

LESR measurements (see Chapter 7) provide information on the microscopic nature of charged defects. There are two types of LESR centers, which are termed LESR-I and LESR-II (Hautala *et al.*, 1988). The type-I center is unstable and can only be annealed at lower temperatures, while the type-II center is stable and can be annealed at room temperature. The microscopic origin of these two centres has been discussed by Elliott and Shimakawa (1990) and will be described later in Section 10.3.

Next, we should discuss how STEs, as a result of bond breaking, are induced in a-Ch by illumination. Note that as pointed out in Section 10.1.1, the creation of metastable STE is the result of bond breaking (Shulger and Stefanovich, 1990). We must, once again, remember the model for a pair of holes localized on a weak bond due to a strong carrier–phonon interaction. This model can be more favorable in a-Ch than in a-Si : H, because the flexibility in a-Ch network is expected to be larger than that in a-Si : H. This leads to a stronger carrier–phonon interaction in the former, as also described in Chapter 6. In a-Ch, as discussed in Chapter 7, the chalcogen derived lone pair (LP) forms the top of the valence states, while the p-bonding forms the lower occupied valence states and these two overlap to some extent (Elliott, 1991).

A schematic illustration of the mechanism of bond breaking, for example, for a-$As_2S(Se)_3$ (Singh and Shimakawa, 2000), is presented in Fig. 10.7. Let us assume that one of the photoexcited holes is self-trapped on a As–S bonding orbital. The excited electrons can either be in the conduction extended or tail states. A second excited hole will get paired with the first one, because such a paired hole state is energetically more favorable. Subsequently the bond will get further enlarged (Fig. 10.7(b)) and broken ultimately. Two DBs are thus created in this way. Due to the effect of negative-U, one of the DBs (As site) can be positively charged and the other (S site) negatively charged by capturing the excited electrons. This process is just the $X - Y_1$ transformation described in Fig. 10.6, in which P_2^+–C_1^- configuration as a negative-U system is easy to realize, that is, two electrons go to C_1 after breaking of a P–C covalent bond and P_2 center is positively charged. Bandgap illumination or subgap illumination as well can therefore excite bonding electrons as well as LP electrons. A pair of holes is therefore expected to break a bond as proposed for a-Si : H (see Section 10.1.1).

It may also be possible that two electrons are self-trapped (instead of holes) on As–S anti-bonding states because the electrons are more easy to localize than holes in a-Ch, resulting in the ultimate bond breaking. It is to be noted that both pairing of holes in the bonding orbital and that of electrons in the anti-bonding orbital result in the breaking of the bond.

Let us discuss the dynamics of LIMD creation in a-Ch. Time evolution of LIMD is not followed by the power-law dependence such as $t^{1/3}$ and is found to be well replicated by Eq. (10.4) (Shimakawa *et al.*, 1995). Note also that SLF (Eq. (10.6)) fits well to the experimental data (Senda *et al.*, 1999). It is therefore hard to distinguish which kinetics dominate LIMD creation. As SLF seems to have a

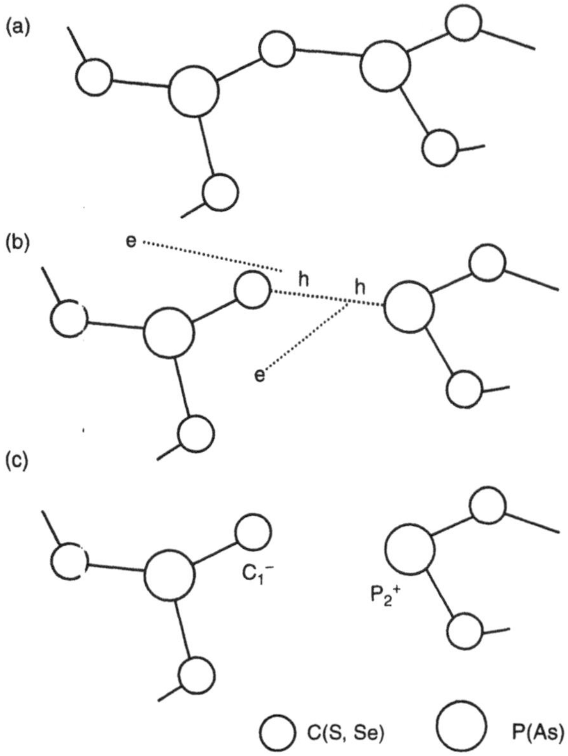

Figure 10.7 Schematic illustration of the mechanism of optical generation of IPs due to pairing of excited holes from the ground state structure of $As_2S(Se)_3$. P and C refer to pnictogen (As) and chalcogen (S or Se), respectively. The superscripts and subscripts refer to the charge state and coordination number, respectively.

universal nature, it is of interest to discuss it in more detail. It is reported that the fitting of Eq. (10.6) produces $N(0) = 7 \times 10^{17}\ \mathrm{cm}^{-3}$ and $N_s = 1 \times 10^{20}\ \mathrm{cm}^{-3}$ in a-As_2S_3. This value of $N(0)$ is close to the pre-existing defects (Mott and Davis, 1979), and is similar to that for a-Si : H. The ratio, $N_s/N(0)$, for As_2S_3 is almost the same as that for a-Si : H. As the electronic transports and optical properties are greatly affected by the LIMD creation, these will be discussed in Section 10.3.

10.2 Structural changes

10.2.1 a-Si : H

It is of interest to detect the photoinduced structural changes in a-Si : H, since the structural changes are predicted to be related to the LIMD creation

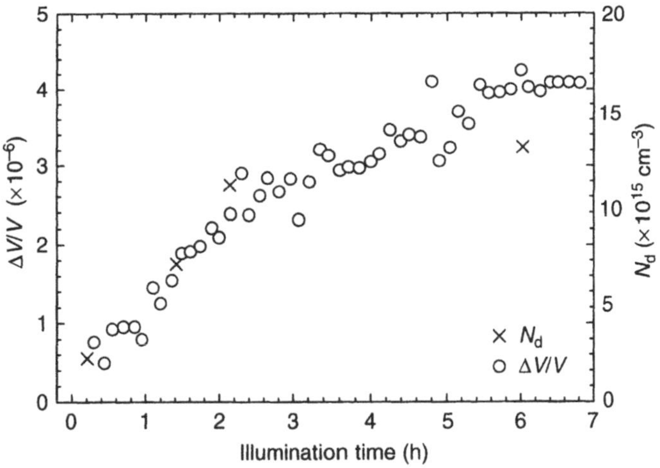

Figure 10.8 The time evolution of volume change $\Delta V/V$ and number of induced defects N_d in a-Si : H (Gotoh *et al.*, 1998).

(Fritzshe, 1995a). Although the change in the infrared absorption spectra (Zhao *et al.*, 1995) and changes in polarized electroabsorption (Shimizu *et al.*, 1997) have been reported, these can only offer indirect evidence for structural changes. The direct experimental proof of the photoinduced volume expansion (PVE) has been reported by employing a bending detected optical lever technique, with a detection limit of $\Delta V/V \approx 2 \times 10^{-7}$ (Gotoh *et al.*, 1998). The experiments have been carefully done to avoid any thermal expansion. The time evolution of PVE together with LIMD creation by an Ar-ion laser (300 mW cm^{-2}) is shown in Fig. 10.8. The value of PVE, $\Delta V/V$, in a-Si : H approaches 4×10^{-6}, which is very much smaller than 4×10^{-3} found in a-As_2S_3. From the volume change per unit defect, it is suggested that PVE occurs on a medium range scale. An intimate correlation of dynamics between PVE and LIMD is observed (Nonomura *et al.*, 2000). It is not clear, however, whether PVE is an origin of LIMD or LIMD is an origin of PVE, or these are independent of each other. It is also shown that the PVE becomes maximum in device quality materials, but it is not yet understood why?

10.2.2 a-Chalcogenides

Unlike a-Si : H, a-Chs exhibit a variety of photostructural changes in density, hardness, elastic constants and optical properties (Shimakawa *et al.*, 1995). This can be attributed to the more flexible nature of an a-Ch network than that of a-Si : H. Among these changes, the most prominent characteristic is the decrease in the Tauc bandgap [photodarkening (PD)] of a-Ch, and hence PD and related effects have recently been a topic of comprehensive reviews (Tanaka, 1980, 1990, 2001; Pfeiffer *et al.*, 1991). As the structural changes, for example, macroscopic volume expansion and the first sharp diffraction peak (FSDP), in X-ray diffraction pattern are accompanied by PD (Tanaka, 1980), PD and structural changes cannot be separately identified.

Recently, however, Tanaka (1998) has discovered that the time evolution of PD and PVE are different, suggesting that PD is not directly related to PVE. In fact, PVE does not occur when a sample is illuminated under hydrostatic compression, despite the appearance of PD (Pfeiffer *et al.*, 1991; Tanaka, 1998). This means that PD is not a direct consequence of PVE. We will discuss PD and PVE separately.

10.2.2.1 Photodarkening

A decrease in the optical bandgap observed, for example, $\Delta E_o/E_o \approx -2\%$ in a-As_2S_3, is commonly known as photodarkening. It was first reported nearly 25 years ago that the bandgap illumination causes reversible PD in a-As_2Se_3 and a-As_2S_3 films (De Neufville *et al.*, 1973). An example for PD, together with PVE recently observed in a-As_2S_3 is shown in Fig. 10.9 (Tanaka, 1998), which illustrates the time evolution of PD and PVE by bandgap illumination. Also, by illumination with light of energy below the bandgap, a-As_2S_3 shows the occurrence of PD (Hisakuni and Tanaka, 1994). PD saturates after a certain time, depending on the illumination conditions like temperature and light intensity. One of the important features shown in Fig. 10.10 is the temperature dependence of PD given by Tanaka (1983), where temperature is normalized to the glass transition temperature T_g. PD is promoted at lower temperatures and hence it is understood to be a purely photon-related effect. It is believed that a change in the interaction of chalcogen

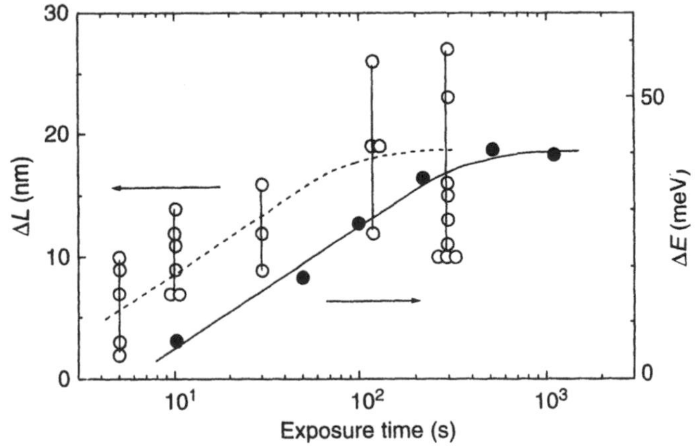

Figure 10.9 The time evolution of volume change ΔL and bandgap change ΔE in a-As_2S_3 (Tanaka, 1998).

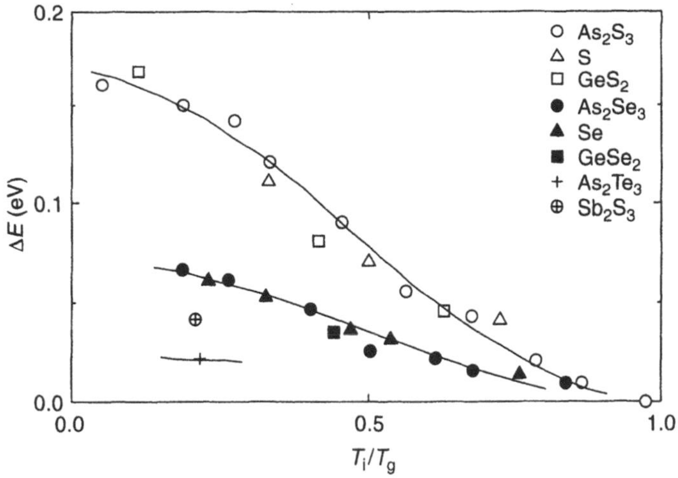

Figure 10.10 Temperature dependence of PD observed in various chalcogenides normalized to the glass transition temperature T_g (Tanaka, 1983).

LP electrons is needed for PD to occur. An increase in LP–LP interaction yields a broadening in the valence band (VB), causing a reduction in the bandgap.

There are two models, which have been put forward in support of the above mechanism. A prolonged illumination causes a change in the atomic (chalcogen) positions (Tanaka, 1980, 1990, 2001) or bond breakings and/or bond alternations between chalcogenides (Elliott, 1986; Kolobov *et al.*, 1997). A basic problem with the above model is that it is difficult to understand how particular atoms can be excited in a solid. The top of the VB is formed by LP bands, and hence there is no reason why particular atoms should be excited.

Shimakawa *et al.* (1998) then proposed an alternative model for PD and PVE. Accordingly, the macroscopic or mesoscopic interaction in layered structures is dominant for PD and PVE to occur in a-Ch, because all LP and also bonding electrons have equal probabilities of being excited. Therefore, electrons or holes in the extended states (or band tail states) can be regarded as being responsible for PD or PVE, but not the individual atoms. It is well known that $As_2Se(S)_3$ has basically layered structures as shown in Fig. 10.11. As the mobility of electrons is much lower than that of holes, the photocreated electrons reside mostly in the conduction-band (CB) tail states, while holes diffuse away to the unilluminated region through valence extended and tail states, which are regarded to be the origin of the Dember type photovoltage.

During the illumination, therefore, layers that absorb photons become negatively charged, giving rise to a repulsive inter-layer Coulombic interaction, which produces a weakening in the van der Waals forces. Hence, the interlayer distance increases (PVE). This process is indicated by the arrow E. A slip (S) motion

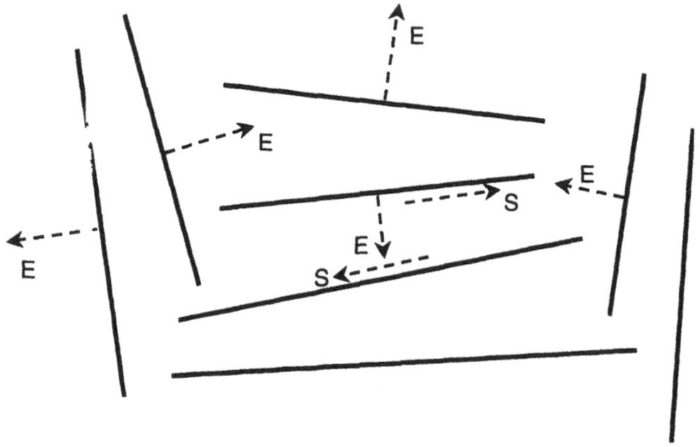

Figure 10.11 Schematic illustration of layered clusters expected in a-$As_2S(Se)_3$. The expansion and slip motions are indicated by the arrows E and S, respectively.

along the layers, which is shown by the arrow S, should also take place with the occurrence of the E process between neighboring clusters. As the energy required for a slip motion along layers is expected to be larger than that for the expansion normal to layers, the rate of S motion may be lower than that of process E. The slip motion from the equilibrium position may produce an increase in the energy of the highest occupied states owing to an increase in the total LP–LP interactions, which is supported by a tight binding calculation in crystalline As_2S_3 (Watanabe *et al.*, 1988). This leads to a widening of the valence states, but conduction states remain unchanged, resulting in PD.

The PD is known to decrease or disappear when a certain amout of a group I metal or other metallic additive is introduced into a-As_2Se_3 (Liu and Taylor, 1987; Iovu *et al.*, 2000). A group I metal, such as copper, may act as bridging atoms between layers and hence reduce the flexibility of the layer network, reducing the ability of both E and S motions. On the contrary, as shown in Fig. 10.12, giant changes in the PD (dashed line) and PVE (solid line) have been reported in obliquely deposited well-annealed a-As_2S_3 films (Kuzukawa *et al.*, 1999). This can be due to a large flexibility of the structure with a large free volume and then $\Delta E_o/E_o$ approaches -8%.

In order to understand the dynamics of PD in more details, *in situ* measurement of the time evolution of PD are required and a few preliminary works have already been done (Tanaka, 1976; Naito *et al.*, 1987; Iovu *et al.*, 2000). Successively, detailed *in situ* measurements of the time evolution of PD have also been done in a-As_2S_3 films (Ganjoo *et al.*, 2000, 2002). Figure 10.13 shows the change in the absorption coefficient with incident number of photons at different temperatures in a-As_2S_3. The symbols of the experimental data are explained in the corresponding figure caption. The magnitude of changes decreases with increasing temperature.

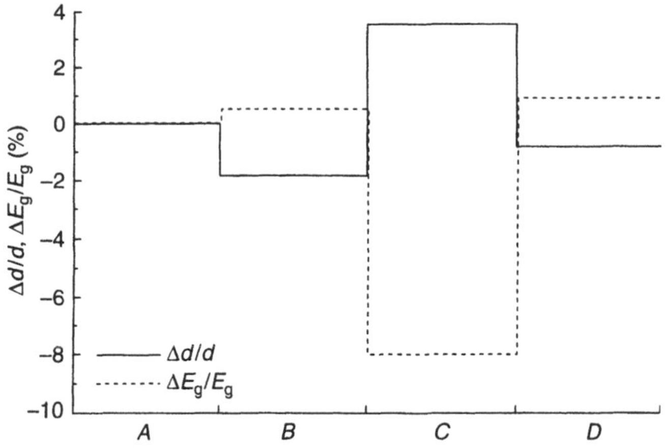

Figure 10.12 Variation in relative changes in thickness d and optical bandgap E_g of a-As_2S_3 (Kuzukawa *et al.*, 1999). A: as-deposited, B: thermally annealed, C: illuminated and D: thermally annealed again.

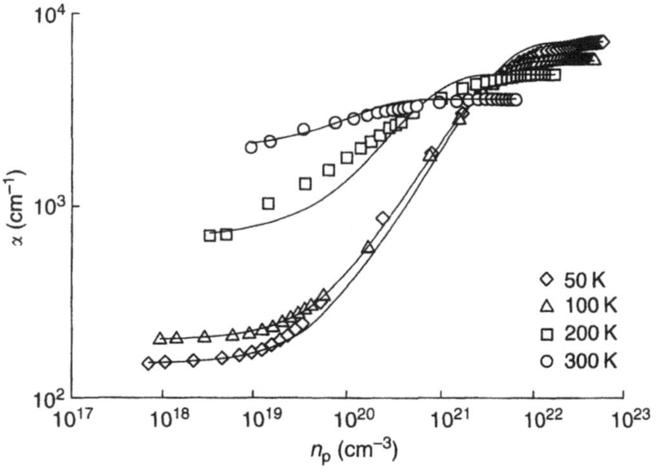

Figure 10.13 Variation of the absorption coefficient ($\alpha = \alpha_0 + \Delta\alpha$) as a function of the number of absorbed photons n_p at various temperatures. Symbols represent the experimental data and solid curves are the calculated results using the effective medium percolation model (Ganjoo *et al.*, 2000).

The solid curves in Fig. 10.13 show the theoretical predictions based on the effective medium percolation theory (EMPT). Here, the optical absorption coefficient is calculated through the AC conductivity; $\alpha(\omega) = \sigma(\omega)/nc\varepsilon_0$, where n is the refractive index, c the light velocity, and ε_0 the permittivity of vacuum

(Shimakawa, 2000). Assuming a random mixture of particles of two different optical conductivities, a volume fraction C having a conductivity $\sigma_1(\omega)$ and the rest of the material $\sigma_2(\omega)$ can be obtained. It is assumed that the system has two distinct sites, that is, one is the cluster-like prospective site for PD and other is PD site. It was assumed that the potential sites N_T can be transformed into the PD sites according to the following phenomenological equation

$$\frac{\mathrm{d}N}{\mathrm{d}n_p} = k_p(N_T - N) - k_r N, \tag{10.7}$$

where n_p is the total photon number (cm^{-3}), k_p and k_r are assumed to be time dispersive rates of promotion and recovery, respectively, given as $k_p = An_p^{\beta-1}$ and $k_r = Bn_p^{\beta-1}$, where $\beta < 1$ is the dispersion parameter. N is then obtained as

$$N = N_s[1 - \exp\{-(n_p/N_p)^{\beta}\}], \tag{10.8}$$

where $N_s = AN_T/(A + B)$ and $N_p = [\beta/(A + B)]^{1/\beta}$ are the saturated number of PD sites and effective number of photons, respectively. Both N_s and N_p depend on temperature. Note that the volume fraction of PD occurrence is N/N_s.

A good fit of the experimental data to theoretical predictions suggests that the mechanism of percolative growth of PD potential sites dominates the dynamics of PD in a-Ch

10.2.2.2 Photoinduced volume expansion

As stated in the previous section, usually PVE always occurs with PD. As shown in Fig. 10.11, $\Delta V/V$ reaches +6% after annealing (see the states B and C) obliquely deposited As_2S_3 films. Note that the value of $\Delta V/V$ for a normal (flat) As_2S_2 is 0.5%. A change in the FSDP can be related to the PVE. Based on the alternative "repulsive and slip motion" model (Shimakawa *et al.*, 1998), the energy stored in a layer is estimated to be 6–60 meV. This repulsive energy can induce PVE by reducing the attractive van der Waals energy, which is estimated to be about 750 meV for each layer. Therefore, the energy introduced into the layers is about 1–10% of the van der Waals energy, which seems to be quite reasonable for inducing PVE by about 0.5% for a flat sample. The experimentally observed widening of the valence angle subtended at sulphur atoms within a layer, causing a subsequent increase in the distance between two arsenic atoms bridged by a chalcogen atom, on illumination can also be explained by the repulsive force involved in the process E. The inter layer repulsive force acts as a compressive force on each layer.

In situ measurements of PVE have been performed to understand the dynamics of PVE, as we have observed so far only the metastable changes after illumination. Figure 10.14 shows the relative changes of thickness, $\Delta d/d$, with time for obliquely deposited a-As_2S_3 film (Ganjoo *et al.*, 1999, 2000). As soon as the light is switched on, the thickness increases rather rapidly, reaches a maximum after approximately 30 s, and then decreases slowly with increasing time. This decrease continues during the rest of the illumination. This behavior of decrease

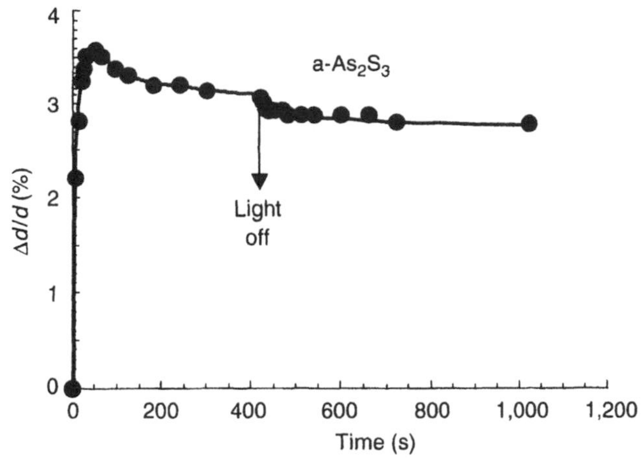

Figure 10.14 Relative changes in thickness $\Delta d/d$ with time in obliquely deposited a-As_2S_3 films (Ganjoo *et al.* 1999, 2000).

in the thickness change during illumination is similar to that of the degradation of photocurrent (Shimakawa *et al.*, 1990) suggesting that the number of photo-carriers is responsible for PVE to occur as already suggested. By turning off the illumination, the thickness decreases slightly. As time passes, PVE decays slowly, due to relaxation of the structure and reaches a metastable state, which can be observed as metastable PVE. It is of interest to point out that two distinct behaviors in PVE, transient and metastable, have been observed in a-As_2S_3 films.

10.2.2.3 Photoinduced anisotropy

If some a-Chs are illuminated by *linearly-polarized light*, like from Ar or He–Ne lasers, then light-induced anisotropic effects, such as dichroism and birefringence, are also observed along with PD (Zhdanov *et al.*, 1979). Figure 10.15 shows the optical transmittance anisotropy at the spectral region near the fundamental absorption in a-As_2S_3 (Kimura *et al.*, 1985). The dashed and dotted curves show the characteristics probed with light having parallel and perpendicular polarization to that of excitation illumination. The dash–dotted curve is obtained using unpolarized light. The absorption coefficient $\alpha(\perp)$ is larger than $\alpha(\|)$, where $\perp$ and $\|$ denote the components perpendicular and parallel to the polarization direction of light, respectively. The refractive index $n(\perp)$ is also larger than $n(\|)$. The photoinduced dichroism defined by $\Delta\alpha = [\alpha(\|) - \alpha(\perp)]$ and birefringence defined by $\Delta n = n(\|) - n(\perp)$, for example, in a-$As_2S_3$, are estimated to be $10^2\,\text{cm}^{-1}$ at $\alpha \sim 10^4\,\text{cm}^{-1}$ and 0.002 at $n \sim 2.6$, respectively.

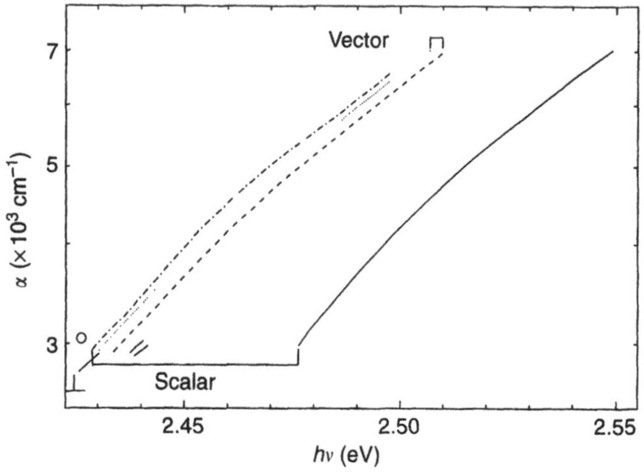

Figure 10.15 Optical absorption coefficient vs photon energy in annealed (solid curve) and illuminated (dash, dotted, and dash–dotted curves) a-As_2S_3 (Kimura *et al.*, 1985). The scalar and the vectoral effects are also indicated. See the text for details.

The first systematic study of these phenomena has been performed by Lyubin and Tikhomirov (1989, 1990, 1991) and subsequently numerous reports on this topic (see recent reviews by Lyubin and Klebanov, 1998; Tanaka, 2001). This type of phenomenon is now called the photoinduced vectoral effect and PD is called the scalar effect. PD and photodichroism seem to appear with different kinetics. The principal experimental results in this area are summarized below.

(i) The photodichroism reaches saturation much earlier than PD. (ii) PD (scalar effect) occurs most effectively at low temperatures, while the vectoral effect occurs at higher temperatures, for example, dichroism in a-As_2S_3 becomes maximum when illuminated at room temperature. (iii) Excitation energy dependence of the vector effect, for example, in a-As_2S_3, shows a maximum at around 2.3 eV, while the scalar effect increases with excitation energy at 2.0–2.5 eV. These are mostly constant at 2.5–3.5 eV. (iv) The vector effect occurs even in materials that do not exhibit PD. (v) The magnitude of vector effect is approximately 1/10 of the scalar effect. (vi) The thermal annealing of the vectoral effect is believed to occur at a temperature lower than that of the scalar effect: the vector effect, for example, in a-As_2S_3, disappears with thermal annealing at 370 K, while PD disappears at 450 K (Lyubin and Tikhomirov, 1989). However, this result is in contrast with that obtained by Tanaka *et al.* (1996), who demonstrated that the birefringence introduced by 2.3 eV excitation disappears with annealing near the glass transition temperature (450 K). As far as annealing properties are concerned, the data are controversial.

From the above observations, it can be concluded that the origin of the vector effect can be different from that of the scalar effect. There are a number of models to explain the vector effect. Fritzsche (1995), phenomenologically, has offered an explanation for the negative optical anisotropy, that is, $\Delta n < 0$. The refractive index of illuminated glasses along the electric field becomes smaller, since structural elements having higher refractive indices are preferentially excited by electric field and the elements are thermally relaxed to other orientations. Accordingly, the dielectric constant along the electric field vector decreases. This is a very clear explanation. However, microscopic structures giving rise to the vector effect remain controversial. They are, however, classified into two categories: (i) photoinduced orientation of atomic bonds or charged defects, which occur on a microscopic scale or atomic scale (Janossy *et al.*, 1987; Tikhomirov and Elliott, 1995; Kolobov *et al.*, 1997); (ii) photoinduced orientation of layer like clusters, which occur on a mesoscopic scale (Zhdanov *et al.*, 1979; Tanaka, 1996, 2001; Tanaka and Ishida, 1998).

A change in the direction of covalent bonds and/or LP electrons can cause optical anisotropy. A model by Kolobov *et al.* (1997) is based on this concept, that is, photoinduced conversion between bonding and non-bonding electrons is assumed. Charged DB may also participate in the vector effects, since these defects have dipole moments (Lyubin, 1990, Tikhomirov and Elliott, 1995). Lyubin and Klebanov (1998) have also suggested that many kinds of structural changes are responsible for the anisotropy.

Murayama (1987) considered that the orientational change of large molecular units such as $AsS_{3/2}$ in As_2S_3 is responsible for the vector effect. The scale length in Murayama's suggestion is mesoscopic. Following the idea of Zhdanov *et al.* (1979), Hajto *et al.* (1982) and Tanaka *et al.* (1996) have suggested that an orientational change in layer clusters ~2 nm size is responsible for the anisotropy. The layer may have the raft-like structure (Phillips, 1981). The layer cluster model may require thermal energy to rotate or re-orient the layer clusters on a mesoscopic scale. In fact, the vector effect occurs at higher temperatures, in contrast with the scalar effect. The layer cluster model is also consistent with the X-ray diffraction measurements (Tanaka, 2001).

Although, it is not easy to conclude which model is valid to explain the over all features of the vector effect, the layer cluster model seems to be easy to access the experimental data. The layer clusters have also been assumed to explain PD as discussed in a previous section (charged layer model). Thus, a model proposed for PD (Shimakawa *et al.*, 1998) may be extended to explain also the vector effect, in which layer clusters can be preferably charged owing to a direction of electric field of polarized light.

Finally, we should discuss the generation of photoinduced stress. Linearly polarized light also induces dilation and contraction in thickness, suggesting a correlation between the optical and mechanical anisotropy, as observed in PD and volume changes (Krecmer *et al.*, 1997). This was originally predicted by Fritzsche (1995). AsSe film was evaporated onto a cantilever (0.6 μ m in thickness) which monitored the bending. A linearly polarized He–Ne laser beam, used to illuminate

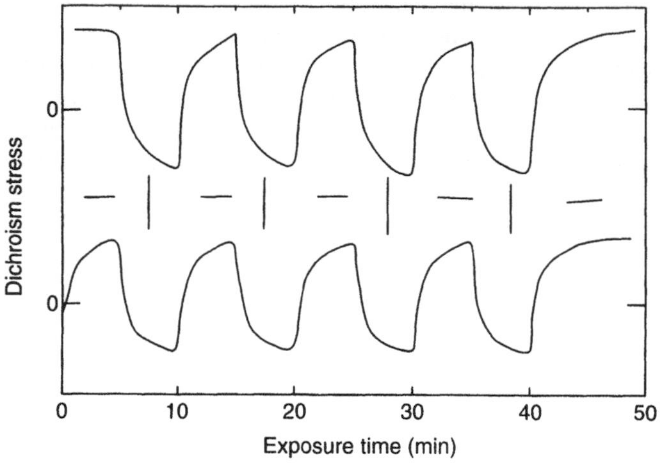

Figure 10.16 Photoinduced stress and dichroism in an AsSe film caused by He–Ne laser illumination with repeated orthogonal polarizations (Krecmer *et al.*, 1997).

AsSe, induces bending. Figure 10.16 shows the photoinduced stress and dichroism in an AsSe film caused by He–Ne laser illumination with repeated orthogonal polarizations (Krecmer *et al.*, 1997). Depending upon the direction of the electric field vector of light, the bimetallic cantilever bends upward and downward. Contraction occurs along the direction of the electric field vector, while a dilation occurs perpendicular to that direction. The mechanical motion and dichroism are induced in such a way repeatedly with similar time scales. This indicates that a linearly polarized light can induce controllable nano-scale contraction and dilation. Therefore, a potential application of a new type of positioning devices for nanotechnology is suggested, in which only polarized light is used without using any electrical signals.

10.3 A universal feature of photoinduced changes

There are certain universal common features in the photoinduced changes observed in a-semiconductors (a-Si : H and a-Chs). These are evident in properties such as photoluminescence, ESR, photoconductivity, AC conductivity, and midgap optical absorption, all of which are directly related to LIMD creation (Shimakawa *et al.*, 1995). These are summarized in Table 10.1. The notations, +, − and 0 denote increase, decrease and no change, respectively in a photoinduced effect after illumination. PL2 is usually not observed before the illumination, but it appears only after the illumination. However, in a poor quality a-Si : H, with higher density of pre-existing DBs, PL2 is observed without the illumination. The DC conductivity in a-Ch seems to remain unchanged after illumination. Since the Fermi level is

Table 10.1 Photoinduced changes observed in a-AsS_3 and a-Si : H

	a-As_2S_3	a-Si : H
PL1 (main band)	–	–
PL2 (subband)	+	+
Optical absorption		
Urbach slope	–	–
Subgap absorption	+	+
ESR	Undetected	+
LESR	+	+
Photoconductivity	–	–
AC conductivity	+	+
DC conductivity	0	–

Note
The notations, +, – and 0, denote increase, decrease and no change, respectively in the photoinduced effect after bandgap illumination (Shimakawa *et al.*, 1995).

considered to be always pinned midway between D^+ and D^- centers. However, recent measurements in a-Sb_2Se_3 show a decrease in DC conductivity, similar to that observed in a-Si : H (Aoki *et al.*, 1999), suggesting that the distribution of photoinduced charged defects may be broadened in energy. This may push the Fermi level toward the conduction band.

The annealing temperatures for these changes are shown in Fig. 10.17. Its behavior is almost the same for both, chalcogenides and a-Si : H, while there exist slight differences in ESR and DC conductivity, which can be attributed to a difference in sign of the effective correlation energy of the principal defects, negative-U in chalcogenides and positive-U in a-Si : H. Note that the annealing behavior of fatigued PL1 with short time irradiation is different from that of prolonged irradiation for both a-As_2S_3 and a-Si : H, suggesting that there are two distinct inducing mechanisms depending on the irradiation time.

Note also that LIMD creation, known as E′ center and oxygen hole center (OHC), and PVC occur by illumination in vitreous SiO_2(v-SiO_2) (Song and Williams, 1996). The occurrence of photoinduced changes therefore seems to be a universal feature in a-semiconductors and insulators.

As PL, a radiative recombination process, always competes with non-radiative recombination processes, PL decreases with increasing the number of defects, if defects act as non-radiative centers. Photoconductivity, on the other hand, is dominated by both radiative and non-radiative recombinations, since the density of steady-state free carriers, G/τ, is determined by the recombination time, where G is the carrier generation rate and τ the recombination time (see Chapter 9). Thus the decrease in PL, actually PL1, and that in photoconductivity are due to increase in the number of defects, which also enhance AC conductivity and subgap optical absorption. Thus at least phenomonologically, defect-related properties should have similar behaviors independent of materials.

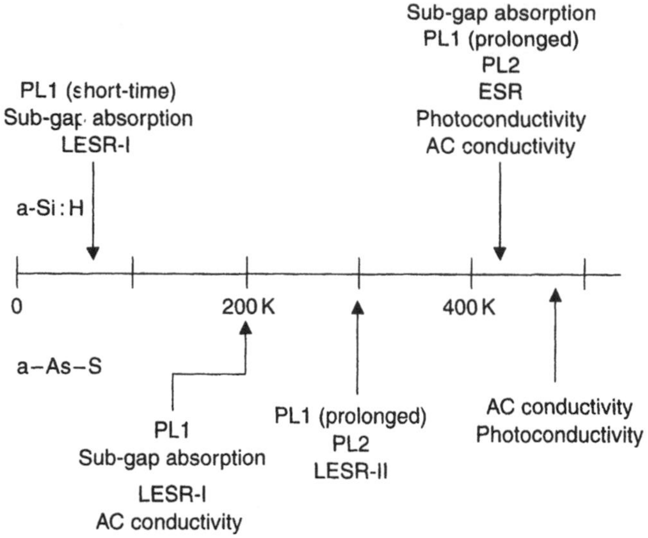

Figure 10.17 Annealing temperatures for various photoinduced changes in a-As_2S_3 and a-Si : H (Shimakawa and Elliott, 1995).

The overall feature of photoinduced effects observed in a-Ch can be explained in terms of the charged and neutral defects. Figure 10.18 shows, for example, for a-As–S system, photoinduced defects and related phenomena (Shimakawa and Elliott, 1995; Shimakawa *et al.*, 1995). The primary initiation event is believed to be the photoinduced creation of an IP of oppositely charged defects. Such defects are relatively unstable and can be annealed at low temperatures. Defect conserved bond switching (DCBS) processes can lead to the spatial separation of such defects, because defects of lower stability will anneal out at lower temperatures. Those of higher stability will anneal out at much higher temperatures. Hence, the following represent the current status of understanding of defect related properties in a-As–S systems.

1 The PL1 center is a STE localized on an As–S bond.
2 Fatigue of PL1 is due to breaking of As–S bonds (broken line); the fatigue, recovered by annealing at 200 K, is attributed to broken As–S bonds (Fig. 10.18(a)) and those recovered by annealing at 300 K are broken As–S bonds (Fig. 10.18(b)).
3 PL2 is related to the As_2^0 center or As–As_2^0 bonds (Fig. 10.18(b)).
4 Electron occupying As_2^+ and holes S_1^-, contribute to unstable LESR-I signals (Fig. 10.18(a)).
5 As_2^0 and S_1^0 are LESR-II centers (Fig. 10.18(b)).

6 Midgap absorption is attributed to optical excitation of electrons and holes from LESR-I centers.
7 Two electron hopping between IPs of As_2^+–S_1^- centers causes the AC loss induced by the low-temperature illumination. These centers are annealed at lower temperatures.
8 Two electron hopping between randomly distributed As_2^+ and S_1^- centers causes the AC loss induced by high-temperature illumination. These centers are annealed at higher temperatures.
9 Principal recombination centers contributing to the decrease in photoconductivity are randomly distributed S_1^- centers, since holes are the dominant photocarriers in a-Ch.

This microscopic view can also be applied to other a-Ch systems. Owing to a similarity of photoinduced effects between a-Ch and a-Si : H, we attempt to apply this charged defect model to all a-Chs and a-Si : H. However, as the fundamental nature of defects in a-Chs seems to be different from that in a-Si : H, charged defects cannot be invoked to interpret overall features of photoinduced phenomena in a-Si : H.

Finally, we should discuss why a part of commonality exists in all disordered materials. The basic underlying features of a-semiconductors and insulators can be listed as: (i) deformable lattice, which means that there exists a strong carrier–phonon interaction and (ii) localized nature of photoinduced carriers. These two factors lead to "self trapping of carriers" either singly or in pairs. Then STHs and STEs may play an important role in photoinduced changes in a-Si : H and a-Ch, as well as in v-SiO_2. The above scenario is shown schematically in Fig. 10.19.

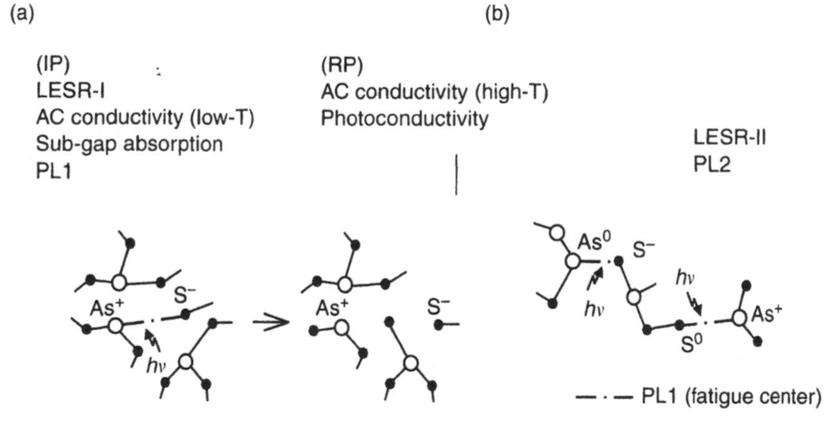

Figure 10.18 Photoinduced defects and related phenomena in a-As–S system (Shimakawa and Elliott, 1995).

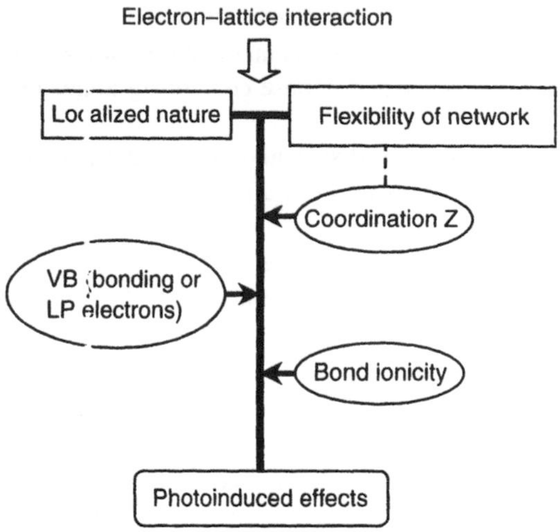

Figure 10.19 Scenario leading to various photoinduced effects in disordered solids.

The small PVE in a-Si : H can be due to less flexibility of its network, since coordination z ($\approx$4) for a-Si : H is larger than that ($\approx$2.4) in a-Ch (Shimakawa *et al.*, 1999).

Remember that small bandgap materials, for example, a-As_2Te_3 (0.84 eV) and a-Ge : H (1.1 eV) show very small photoinduced change or almost no change in usual illumination conditions (Aoki *et al.*, 1989, 1991, 1999; Shimakawa and Aoki, 1992; Hayashi *et al.*, 1997). Several reasons for the small photoinduced changes in small bandgap materials can be considered as follows: (i) If the ratio σ_p/σ_d of the number of photoinduced free carriers to dark free carriers is an important factor for LIMD creation, the number of LIMDs for small bandgap materials is expected to be smaller than for larger bandgap materials such as a-Si : H (1.8 eV), a-As_2Se_3 (1.8 eV), and a-As_2S_3 (2.4 eV), etc. (ii) The small Stokes shift (0.2 eV) and small half width of photoluminescence (PL), for example, in a-Ge : H, suggests a weak carrier–phonon coupling, which can be attributed to the large extent of localized wavefunctions due to small bandgap. Hence LIMD creation is difficult to occur. (iii) A large dielectric constant ε_∞ ($\approx$16), characteristic of small-bandgap materials, reduces the Coulombic attractive force between electrons and holes, suggesting a smaller likelihood for the formation of STEs. (iv) Owing to the small bandgap, the phonon energy released can be small and hence thermal energy is not enough to cause bond switching for establishing stability. However, as we discussed in the section on quantum efficiency η, the released non-radiative energy does not seem to be essentially very important for creating LIMDs. Thus, why smaller bandgap materials show a less photoinduced change may be an important issue in understanding the photoinduced effects in disordered matters.

References

Anderson, P.W. (1975). *Phys. Rev. Lett.* **34**, 953.

Aoki, T., Kato, S., Nishikawa, Y. and Hirose, M. (1989). *J. Non-Cryst. Solids* **114**, 798.

Aoki, T., Nishikawa, Y., Ii, H. and Hirose, M. (1991). *J. Non-Cryst. Solids* **137–138**, 407.

Aoki, T., Shimada, H., Hirao, N., Yoshida, N., Shimakawa, K. and Elliott, S.R. (1999). *Phys. Rev. B* **59**, 1579.

Biegelsen, D.K. and Street, R.A. (1980). *Phys. Rev. Lett.* **44**, 803.

Branz, H.M. (1999). *Phys. Rev. B* **59**, 5498.

Crandall, R.S., Sadlon, K., Salamon, S.J. and Branz, H.M. (1991). In: Stafforerd (ed.), *Amorphous Silicon Materials and Solar cells*, AIP Conference Proceedings No. 234. AIP, New York, p. 154.

De Neufville, J.P., Moss, S.C. and Ovshinsky, S.R. (1973–74). *J. Non-Cryst. Solids* **13**, 191.

Dersh, H., Stuke, J. and Beichler, J. (1981). *Appl. Phys. Lett.* **38**, 456.

Elliott, S.R. (1986). *J. Non-Cryst. Solids* **81**, 71.

Elliott, S.R. (1991). In: Cahn *et al.* (eds), *Material Science and Technology*, Vol. 9. VCH, Weinheim, p. 375.

Elliott, S.R. and Shimakawa, K. (1990). *Phys. Rev. B* **42**, 9766.

Fedders, P.A., Fu, Y. and Drabold, D.A. (1992). *Phys. Rev. Lett.* **68**, 1888.

Fritzsche, F. (1995a). *Solid State Commun.* **94**, 953.

Fritzsche, F. (1995b). *Phys. Rev. B* **52**, 15854.

Ganjoo, A., Ikeda, Y. and Shimakawa, K. (1999). *Appl. Phys. Lett.* **74**, 2119.

Ganjoo, A., Ikeda, Y. and Shimakawa, K. (2000). *J. Non-Cryst. Solids* **266–269**, 919.

Ganjoo, A., Shimakawa, K., Kamiya, H., Davis, E.A. and Singh, J. (2000). *Phys. Rev. B* **62**, R14601.

Ganjoo, A., Shimakawa, K., Kitano, K. and Davis, E.A. (2002). *J. Non-Cryst. Solids* **299–302**, 917.

Gotoh, T., Nonomura, S., Nishio, M., Nitta, S., Kondo, M. and Matsuda, A. (1998). *Appl. Phys. Lett.* **72**, 2978.

Guillaume, C.B., Mollot, F. and Cernogora, J. (1977). In: Spear, W.E. (ed.), *Proceedings of the 7th International Conference on Amorphous and Liquid Semiconductors.* Centre of Industrial Consultancy and Liaison, Edinburgh, p. 612.

Hajto, J., Janossy, I. and Forgacs, G. (1982). *J. Phys. C: Solid State Phys.* **15**, 6293.

Hautala, J., Ohlsen, W.D. and Taylor, P.C. (1988). *Phys. Rev. B* **38**, 11048.

Hayashi, K., Hikida, Y., Shimakawa, K. and Elliott, S.R. (1997). *Philols. Mag. Lett.* **76**, 233.

Hisakuni, H. and Tanaka, Ke. (1994). *Appl. Phys. Lett.* **56**, 2925.

Imagawa, K., Shimakawa, K. and Kondo, A. (2000). *J. Non-Cryst. Solids* **266–269**, 428.

Iovu, M., Shutov, S., Rebeja, S., Colomeyco, E. and Popescu, M. (2000). *J. Optoelectron. Adv. Mater.* **2**, 53.

Jackson, W.B. and Zhang, S.B. (1990). In: Fritzsche, H. (ed.), *Transport, Correlation and Structural Defects*. World Scientific, Singapore, p. 63.

Janossy, I., Hajto, J. and Choi, W.K. (1987). *J. Non-Cryst. Solids* **90**, 529.

Kakalios, J., Street, R.A. and Jackson, W.B. (1987). *Phys. Rev. Lett.* **59**, 1037.

Kimura, K., Murayama, K. and Ninomiya, T. (1985). *J. Non-Cryst. Solids* **77** and **78**, 1203.

Kolobov, A.V., Lyubin, V., Yasuda, T., Klebanov, M. and Tanaka, K. (1997). *Phys. Rev. B* **55**, 8788.

Kolobov, A.V., Oyanagi, H., Tanaka, K. and Tanaka, Ke. (1997). *Phys. Rev. B* **55**, 726.

Krecmer, P., Moulin, A.M., Stephenson, R.J., Rayment, T., Welland, M.E. and Elliott, S.R. (1997). *Science* **277**, 1799.

Kuzukawa, Y., Ganjoo, A., Shimakawa, K. and Ikeda, Y. (1999). *Philos. Mag. B.* **79**, 249.
Liu, J.Z. and Taylor, P.C. (1987). *Phys. Rev. Lett.* **59**, 1938.
Lyubin, V.M. and Klebanov, M.L. (1998). *Semiconductors* **32**, 817.
Lyubin, V.M. and Tikhomirov, V.K. (1989). *J. Non-Cryst. Solids* **114**, 133.
Lyubin, V.M. and Tikhomirov, V.K. (1990). *Soviet Phys. Solid St.* **32**, 1069.
Lyubin, V.M. and Tikhomirov, V.K. (1991). *J. Non-Cryst. Solids* **135**, 37.
Morigaki, K. (1988). *Jpn. J. Appl. Phys.* **27**, 163.
Morigaki, K. (1999). *Physics of Amorphous Semiconductors*, Imperial College Press and World Scientific, London.
Mott, N.F. and Davis, E.A. (1979). *Electronic Processes in Non-Crystalline Materials*, 2nd edn, Oxford University Press, Oxford.
Murayama, K. (1987). In: Kastner, M.A., Thomas, G.A. and Ovshinsky, S.R. (eds), *Disordered Semiconductors.* Plenum, New York, p. 185.
Murray, J.D. (1989). *Mathematical Biology.* Springer, Berlin.
Naito, H., Teramine, T., Okuda, M. (1987). *J. Electrophotographic Soc. Jpn.* **2**, 53 (in Japanese).
Nonomura, S., Gotoh, T., Nitta, S., Kondo, M. and Matsuda, A. (2000). *J. Non-Cryst. Solids* **266–269**, 474.
Pfeiffer, G., Paesler, M.A. and Agarwal, S.C. (1991). *J. Non-Cryst. Solids* **130**,111.
Pfleiderer, H., Kusian, W. and Kruhler, W. (1984). *Solid St. Commun.* **49**, 493.
Phillips, J.C. (1981). *J. Non-Cryst. Solids* **43**, 37.
Redfield, D. and Bube, R.H. (1996). *Photoinduced Defects in Semiconductors.* Cambridge University Press, Cambridge.
Schropp, R.E., Boonstra, A.J. and Klapwijk, T.M. (1987). *J. Non-Cryst. Solids* **97** and **98**, 1339.
Senda, M., Yoshida, N. and Shimakawa, K. (1999). *Philos. Mag. Lett.* **79**, 375.
Shimakawa, K. (2000). *J. Non-Cryst. Solids* **266–269**, 223.
Shimakawa, K. and Aoki, T. (1992). *Phys. Rev. B* **46**, 12750.
Shimakawa, K. and Elliott, S.R. (1995). In: Marshall, J.M., Kirov, N. and Vabrek, A (eds), *Electronic, Optoelectronic and Magnetic Thin Films.* John Wiley & Sons Inc, New York, p. 95.
Shimakawa, K., Inami, S. and Elliott, S.R. (1990). *Phys. Rev. B* **42**, 11857.
Shimakawa, K., Inami, S., Kato, T. and Elliott, S.R. (1992). *Phys. Rev. B* **46**, 10062.
Shimakawa, K., Kondo, A., Hayashi, K., Akahori, S., Kato, T. and Elliott, S.R. (1993). *J. Non-Cryst. Solids* **164–166**, 387.
Shimakawa, K., Kolobov, A.V. and Elliott, S.R. (1995). *Adv. Phys.* **44**, 475.
Shimakawa, K., Yoshida, N., Ganjoo, A. and Kuzukawa, Y. (1998). *Philos. Mag. Lett.* **77**, 153.
Shimakawa, K., Nonomura, S. and Shimizu, K. (1999). *Oyo Buturi* **68**, 1122 (in Japanese).
Shimakawa, K., Mehrun-Nessa, Ishida, H. and Ganjoo, A. (2003). *Phil. Mag. B* (in press).
Shimizu, T. 1993). *J. Non-Cryst. Solids* **164–166**, 163.
Shimizu, K. Shibata, T., Tabuchi, T. and Okamoto, H. (1997). *Jpn. J. Appl. Phys.* Part 1 **36**, 29.
Shulger, A.I. and Stefanovich, E. (1990). *Phys. Rev. B* **42**, 9664.
Singh, J. and Shimakawa, K. (2000). *Asian J. Phys.* **9**, 543.
Song, K.S. and Williams, R.T. (1996). *Self-Trapped Exciton*, 2nd edn. Springer, Berlin.
Staebler, D.L. and Wronski, C.R. (1977). *Appl. Phys. Lett.* **28**, 671,
Staebler, D.L. and Wronski, C.R. (1980). *J. Appl. Phys.* **51**, 3262.

Street, R.A. (1991). *Hydrogenated Amorphous Silicon.* Cambridge University Press, Cambridge.

Street, R.A., Kakalios, J., Tsai, C.C. and Hayes, T.M. (1987). *Phys. Rev. B* **35**, 1316.

Stutzmann, M., Jackson, W.B. and Tsai, C.C. (1985). *Phys. Rev. B* **32**, 23.

Tanaka, K. (1980). *J. Non-Cryst. Solids* **35** and **36**, 1023.

Tanaka, Ke. (1976). *Thin Solid Films* **33**, 309.

Tanaka, Ke. (1983). *J. Non-Cryst. Solids* **59** and **60**, 925.

Tanaka, Ke. (1990). *Rev. Solid State Sci.* **4**, 641.

Tanaka, Ke. (1996). *Current Opin. Sol. St. Mater. Sci.* **1**, 567.

Tanaka, Ke. (1998). *Phys. Rev. B* **57**, 5163.

Tanaka, Ke. (2001). In: Nalwa, H.S. (ed.), *Handbook of Advanced Electronic and Photonic Materials.* Academic Press, San Diego, p. 119.

Tanaka, Ke. And Ishida, K. (1998). *J. Non-Cryst. Solids* **227–230**, 673.

Tanaka, Ke, Ishida, K. and Yoshida, N. (1996). *Phys. Rev. B* **54**, 9190.

Tikhomirov, V.K. and Elliott, S.R. (1995). *Phys. Rev. B* **51**, 5538.

Watanabe, T., Kawazoe, H. and Yamane, Y. (1988). *Phys. Rev. B* **38**, 5677.

Yamasaki, S. and Isoya, J. (1993). *J. Non-Cryst. Solids* **164–166**, 169.

Yamasaki, S., Kaneiwa, M., Kuroda, S., Ohkushi, H. and Tanaka, K. (1987). *Phys. Rev. B* **35**, 6471.

Zhadanov, V.G., Kolomiets, B.T., Lyubin, V.M. and Malinnovskii, V.K. (1979). *Phys. Stat. Sol.* (a) **52**, 621.

Zhao, Y., Zhang, D., Kong, G., Pan, G. and Liao, X. (1995). *Phys. Rev. Lett.* **74**, 558.

11 Applications

One of the most successful applications of amorphous semiconductors (a-semiconductors) is in the field of electrophotography (or xerography) first demonstrated by Carlson and Kornei in 1938. The principles underlying the modern xerographic process are the same as those developed by Carlson and Kornei. Thin films of Se (alloyed with a small amount of As) and Si : H are used for this purpose. Organic photoreceptors are used at present in low-cost xerographic machines. As homogeneous thin films of large area can be prepared relatively easily from amorphous materials, further applications in fabricating devices like solar cells, thin-film transistors (TFT), X-ray image sensors and phase-change memory devices have been developed since. In this chapter, some of such devices fabricated from amorphous semiconductor films are reviewed. However, the xerographic devices are not reviewed here, because they have already been described in several other books (e.g. Elliott, 1990).

11.1 Photovoltaic devices

The photovoltaic (PV) effect was first discovered in 1839 by a French physicist Edmond Bequerel (e.g. Fahrenbuch and Bube, 1983). Fifty years later Charles Fritts, an American inventor, was the first to prepare solar cells, which were made of thin wafers of selenium, covered by a thin layer of gold. These cells had a very low efficiency; only about 1%. The first solar cells prepared from crystalline silicon (c-Si) were reported in 1954. These cells were developed at Bell Laboratories, USA, and had an efficiency of around 6%. In the following two decades, using more advanced technology, c-Si solar cells were prepared with efficiency up to 15%. The first significant use of solar cells was made in 1958 to provide power for the satellite Vanguard I. These solar cells were very reliable but very expensive, which is of course, of little concern in any satellite endeavor. Then onward solar cells were mainly used for satellites and other space projects, because of their high cost. In the early 1970s the oil crisis in the Middle East led to serious considerations of using PV cells as an alternative source of energy. As a result, a significant research effort was devoted to the development of solar cells for both: (1) improving their efficiencies and (2) reducing their costs. This scientific endeavor resulted in the price drop from \$200 per peak watt in 1960 to less than \$20 in the late

1970s, which made solar cells economical for many more applications, particularly where electricity from the "mains" power supply was unavailable or too expensive, for example, powering navigational lights at sea. However, due to the kind of technology involved in producing c-Si at present, it is not possible to reduce the cost of crystalline cells so much that they can be useful for domestic applications.

Therefore researchers tried to study the possibility of using the hydrogenated amorphous silicon (a-Si : H) for fabrication of solar cells. The process of producing a-Si : H is relatively more economical, because it can be deposited in the form of thin films rather easily. The first a-Si : H solar cells were fabricated by Carlson and Wronski (1976) at RCA Laboratories, USA, which had an initial efficiency of 2.4% for a cell of size 5×10^{-3} cm^2, and 1.1% for larger cells of area 3.5 cm^2. In 1979, Kuwano and his coworkers from Sanyo electric Co., Japan, fabricated a-Si : H solar cells on glass substrates. Although RCA began the development of a-Si : H solar cells, Sanyo was the first to market the device. They realized that even quite low efficiency cells could supply enough power to operate small calculators and watches. Another advantage of a-Si : H solar cells is that they have larger efficiency under the fluorescent light compared with sunlight, therefore the solar powered calculators work quite well in an office environment.

As the absorption coefficient of a-Si : H is relatively high in the appropriate (visible) spectral range, only a very thin film of the material (<1 μm) is required to gain the same absorption as a thicker layer of c-Si. This is the reason why in the 1980s a-Si : H was viewed as the "only" suitable thin film PV material. Solar cells can also be fabricated using a-Si alloyed with hydrogen, germanium and carbon to form semiconductors of different band gaps in the range of 1.3–2.0 eV (Dawson *et al.*, 1992). The dependence of the band gap of a-Si : H on the hydrogen concentration has been studied by several groups (e.g. Zhu and Singh, 1993a). The variation in band gaps allows the fabrication of not only single junction but also tandem and triple junction stacked cells to maximize the absorption of the solar spectrum (Yang and Guha, 1992; Yang *et al.*, 1994). However, as also described in Chapter 10, a-Si solar cells degrade in their efficiency after exposure to radiation, known as the Staebler–Wronski effect (Staebler and Wronskii, 1977). Despite the problem of degradation, a-Si technology has advanced significantly in the last two decades and achieved an initial efficiency of more than 13% (Ashida, 1994) for single-junction a-Si : H solar cells.

Furthermore a-Si : H thin films can be deposited on different types of substrates. This enables one to use a-Si : H solar cells for a wide range of applications, for example, they can be integrated into roof tiles or deposited on flexible plastic films (Yoshida *et al.*, 1996).

11.1.1 Cell structures

In order to take the advantage of some of the excellent properties of undoped a-Si : H material, p–i–n and n–i–p heterojunction cell structures are used for fabricating solar cells rather than the classic n–p structures used for c-Si solar cells. A p–i–n cell structure consists of p (a-SiC : H)–i (a-Si : H)–n (a-Si : H) layers and n–i–p structure of n (a-Si : H)–i (a-Si : H)–p (a-SiC : H) layers. The commonly used a-Si : H cell configurations are single junction (p–i–n and n–i–p) and tandem (dual- and triple-junctions) structures as well as Schottky barrier cells, metal–insulator–semiconductor (MIS) cells and several other different types of multi-junction cells.

Solar cells of a-Si : H have been fabricated on substrates such as glass, metal foils and plastics. The highest stable conversion efficiencies (13%) have been obtained with triple-junction structures (Guha *et al.*, 1999). With a-Si/c-Si hybrid structures such as the Sanyo HIT cell (Tanaka *et al.*, 1992), even higher efficiencies (18.3%) have been attained. Since these devices do not have thin film structures, they will not be considered here. We will now review some of the more important a-Si : H based commercial solar cell structures in detail.

11.1.1.1 Single-junction structures

Most of the low-wattage a-Si PV cells used in consumer applications are fabricated using single-junction p–i–n structures on glass substrates (Carlson and Catalano, 1989). The structure is shown in Fig. 11.1. These single-junction devices are deposited on glass substrates that are coated with ~600 nm of a textured tin oxide that acts as the front electrical contact. The tin oxide layer is grown with a sub-micron surface texture that is determined by the grain size of the crystallites. On the tin oxide layer, a 10 nm boron doped a-Si carbon alloy (p-layer) is first deposited, followed by an undoped a-Si : H (i-layer) about 250 nm thick. Finally, about 30 nm thick phosphorous doped a-Si (n-layer) is deposited to form a p–i–n structure. For the back electrode, about 400 nm thick aluminum layer is deposited by sputtering on the n-layer. As stated above, single-junction a-Si : H devices can also be made using a n–i–p structure on metal substrates, but then the stabilized efficiency of commercial modules is achieved only about 4–6%.

11.1.1.2 Tandem structures

Due to a higher stabilized conversion efficiency, several manufacturers have introduced multi-junction a-Si : H PV modules. As the multi-junction cells have higher

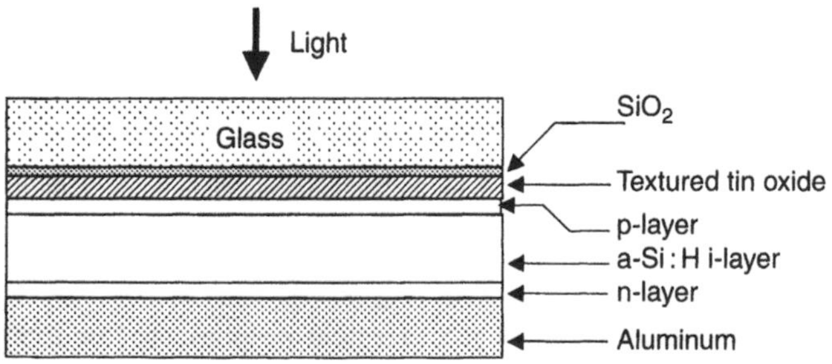

Figure 11.1 A schematic illustration of a single-junction solar cell structure fabricated on a glass substrate.

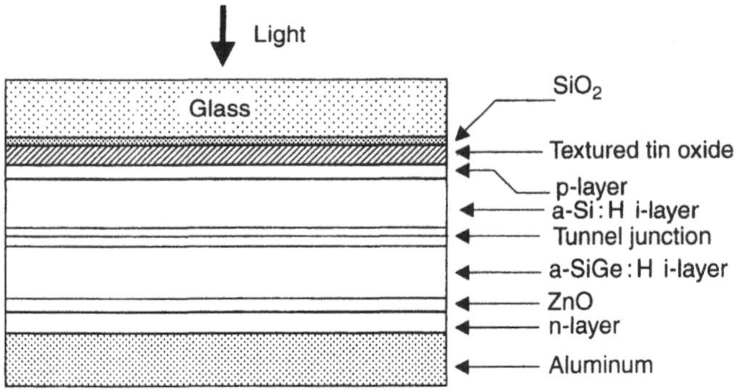

Figure 11.2 A schematic illustration of a device structure used for commercial tandem modules.

theoretical conversion efficiency than single-junction cells, they can utilize more of the solar spectrum at a higher net output voltage. The theoretical efficiency for an ideal tandem device is about 36% (Bennet and Olsen, 1978). Another factor for favoring multi-junction devices of a-Si : H is the improved stabilized performance associated with the use of a thinner i-layer (Hanak and Korsun, 1982; Štulík and Singh, 1998b).

The most commonly used device structure for commercial tandem modules is shown in Fig. 11.2 (Carlson *et al.*, 1996). The front junction is similar to that of a single-junction device as described above with an i-layer of about 1.8 eV optical gap. Another p-layer is deposited after the first n-layer to give a tandem structure. This n–p junction is often referred to as a "tunnel" junction but actually it acts as a recombination junction in a tandem of two p–i–n cells connected in series. The second or back junction is formed by depositing an undoped a-SiGe : H (i-layer) alloy on the second p-layer and then depositing a second a-Si : H n-layer (see Fig. 11.2).

11.1.1.3 Triple-junction structures

Recently, Guha *et al.* (1999) have obtained the highest initial and stabilized conversion efficiencies (15.2% initial and 13.0% stabilized) for a-Si solar cells using a triple-junction structure. The best triple-junction cells have been fabricated on stainless steel foil substrates that were coated with layers of textured silver and ZnO. The structure is shown in Fig. 11.3. A phosphorous-doped a-Si : H n-layer (about 20 nm thick) is deposited on the ZnO, and then a-SiGe : H i-layer with a graded Ge concentration is deposited on the n-layer. The triple-junction contains two "tunnel" junctions each consisting ~10 nm of boron-doped μc-Si : H and ~10 nm a-Si : H.

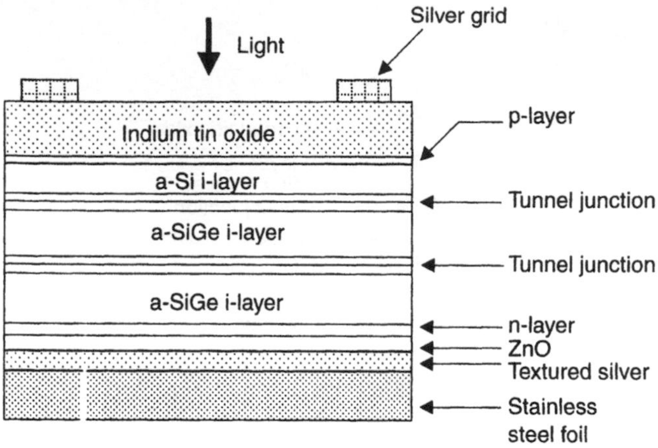

Figure '1.3 A schematic illustration of a triple-junction solar cell structure fabricated on a stainless steel substrate.

11.1.2 Principle of operation

We will bri :fly discuss the mechanism that determines and limits the energy conversion in ι solar cell. The energy band diagrams of a p (a-SiC : H)–i (a-Si : H)–n (a-Si : H) solar cell with a 320-nm thick i-layer of 1.76 eV band gap and $10^{16}\,cm^{-3}$ density of neutral dangling bonds and charged defects are shown in Fig. 11.4(a) at thermal equilibrium (no illumination zero bias) and in Fig. 11.4(b) at the "maximum power point" on its light I–V characteristic under AM 1.5 illumination. The built-in potential, which is determined by the separation between the Fermi level E_{FP} and E_{FN} is 1.3 V. The band bending across the i-layer in Fig. 11.4(a) clearly indicates a large electric field. This type of built-in potential produces a high value of V_{OC}, resulting in built-in electric fields greater than $10^4\,V\,cm^{-1}$ within i-layers less than 1 μm thick. These high electric fields help collecting even the low mobility photogenerated carriers in the i-layer.

The energy state of neutral dangling bonds D^0, located near the middle of the gap, acts as recombination centers. In addition, there is an additional recombination path in the p/i interface region adjacent to the p-contact, which tends to have defect densities that are larger than in the bulk. Even though the interface width extends over a very thin region, it has a large effect on both the carrier recombination and the electric field distribution. As the interface is close to the p-layer, it affects the fill factor and the open-circuit voltage as well (Lee *et al.*, 1998). This point can be illustrated by measuring the dark I–V characteristics of two p–i–n solar cells as shown in Fig. 11.5. The first cell is a standard cell and the second cell has its interface region improved by using the high hydrogen dilution (Wronski *et al.*, 1997; Lee *et al.*, 1998). It can be seen from Fig. 11.5 that

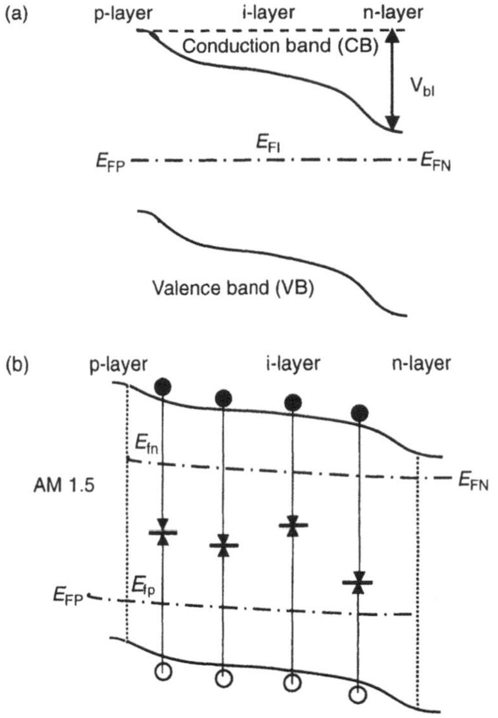

Figure 11.4 Energy band diagrams of a 320-nm thick p(a-SiC : H)–i,(a-Si : H)–n(a-Si : H) cell (a) in dark under zero bias and (b) under AM 1.5 illumination and load. The built-in-potential, V_{bi}, is indicated in (a). The different recombination paths for carriers are shown in (b).

the current due to the generation and recombination from defect states near the middle of the gap is reduced by an order of magnitude by improving the p/i interface. This results in the current being no longer dominated by the p/i interface, but limited by the bulk, which also results in a corresponding increase in V_{OC} by 30 meV.

The operation of a-Si : H solar cells clearly depends strongly on the light-induced defects. Defects that are introduced after a prolonged exposure to light reduce the carrier lifetime, leading to a change in quantum efficiencies as a function of the wavelength. This results in a small decrease in J_{SC} in high quality solar cells, particularly when the i-layer thickness is less than ~300 nm. Defects generated in the bulk of i-layers do not have a large effect on the fill factor. They are the major contributors only to the loss in the cell's conversion efficiency. The light-induced defects in the p/i interface region have the most pronounced effect of lowering the open-circuit voltage. However with good p/i interfaces, high open-circuit voltages are obtained, which also do not degrade in the sunlight.

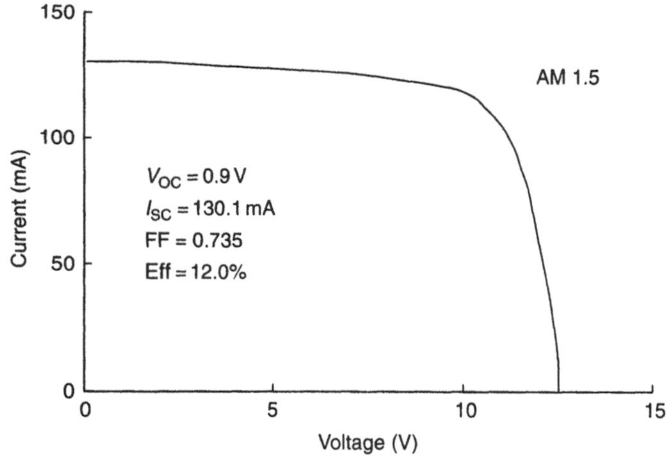

Figure 11.5 I–V characteristics of an integrated type single-junction a-Si : H module under AM 1.5 illumination. The 12% efficient module consists of 14 cells in series and has an area of 1 m^2 (Tsuda *et al.*, 1993).

11.1.3 Optimization and designing of solar cells

As discussed above, the p/i interface region plays an important role in the cell performance, so with better p-type contacts and improved p/i interface region one can achieve improvement in V_{OC}. Using a lower band gap material for the intrinsic layer results in an increase in the absorption that increases the short-circuit current. A single-junction a-Si : H solar cell with a band gap of 1.7 eV is found to have the highest efficiency. With such smaller band gaps, an open-circuit voltage of 0.9 V can be obtained and at the same time a 1 μm thick i-layer absorbs a fraction of AM 1.5 sunlight to generate a short-circuit current of ~18 mA cm^{-2}.

With the development of optical enhancement based on textured substrates and reflectors in the 1980s (Yablonovitch and Cody, 1982; Deckman *et al.*, 1984) a major breakthrough in fabricating thin high efficiency solar cells was achieved. The texture causes a large angle scattering resulting in multiple internal reflections, which allow the weakly absorbed light to undergo many passes through the i-layer resulting in large optical absorption enhancements.

With continuing improvements in materials, various cell components and taking advantage of the optical enhancements, a conversion efficiency in excess of 13% (Tsuda *et al.*, 1993; Ashida, 1994) has been obtained from single-junction a-Si : H solar cells. The illuminated I–V characteristics of such a 100 cm^2 single-junction module having 14 cells connected in series, each with FF in excess of 0.73 and V_{OC} of 0.9 V, is shown in Fig. 11.5. An improvement in the stability of a-Si : H high-efficiency solar cells has also been achieved by developing improved and stable a-Si : H films. Hydrogen dilution of the source gas during a-Si : H growth has led to a significant improvement in the stability by minimizing the light-induced changes in a-Si : H solar cells.

On the theory side also a considerable advancement has been made in designing both single-junction and tandem structure a-Si : H solar cells (Fan and Palm, 1983). In order to enhance the PV performance of a multilayer structure solar cell, one has to maximize the absorbance of solar energy in its i-layer and minimize it in the other layers. This can be achieved by applying the admittance analysis method (Macleod, 1986; Zhu and Singh, 1993b,c; Štulík and Singh, 1996, 1997). At the same time, the recombination of the photogenerated charge carriers should be reduced in the i-layer, which requires the optimization of the collection efficiency of a solar cell. Hubin and Shah (1995) and then Štulík and Singh (1998a) have derived the collection efficiency of a-Si : H solar cells as a function of the defect density (density of dangling bonds) in the i-layer. The collection efficiency χ is defined as (Štulík and Singh, 1998a):

$$\chi = \frac{\int_0^L (G(x) - R(x))\,\mathrm{d}x}{\int_0^L G(x)\,\mathrm{d}x}, \tag{11.1}$$

where $G(x)$ and $R(x)$ are the rates of generation and recombination as a function of the position x within the i-layer and L is the thickness of the i-layer. In writing Eq. (11.1), it is assumed that the collection efficiency, χ, of a solar cell is practically equal to the collection efficiency of its i-layer. For a constant generation rate, $G(x) = G_0$, an approximate form of the collection efficiency is obtained as (Hubin and Shah, 1995; Štulík and Singh, 1998a):

$$\chi = \frac{1}{L}\frac{l_n l_p}{l_n \exp(L/L_c) - l_p \exp(-L/L_c)}\left[\exp\left(\frac{L}{L_c}\right) - \exp\left(-\frac{L}{L_c}\right)\right], \tag{11.2}$$

where $l_n = \mu_n \tau_n |E_0|$ and $l_p = \mu_p \tau_p |E_0|$ are the drift lengths of free electrons and holes, respectively, and $L_c = 2(l_n l_p/(l_n - l_p))$; μ_n, τ_n and μ_p, τ_p are band mobilities, capture times for free electrons and holes, respectively. For calculating l_n and l_p one requires the corresponding product of μ and τ, which is inversely proportional to the defect density, N_d, as: $\mu\tau = 1/(\sigma N_d)$, where σ is proportional to the capture cross-section and its numerical values are known (Street, 1991). Thus, the collection efficiency is obtained as a function of the defect density.

The short-circuit current in a single-junction solar cell is obtained as (Štulík and Singh, 1998b):

$$J_{SC} = q\chi \int A_{\text{i-layer}}(\lambda) F(\lambda) \frac{\lambda}{hc}\,\mathrm{d}\lambda, \tag{11.3}$$

where q is the electronic charge, $F(\lambda)(\mathrm{W\,m^{-2}\,nm^{-1}})$ incident flux of solar radiation, c speed of light and $A_{\text{i-layer}}(\lambda)$ absorbance in the i-layer obtained through the admittance analysis method (Štulík and Singh, 1997). Thus, according to

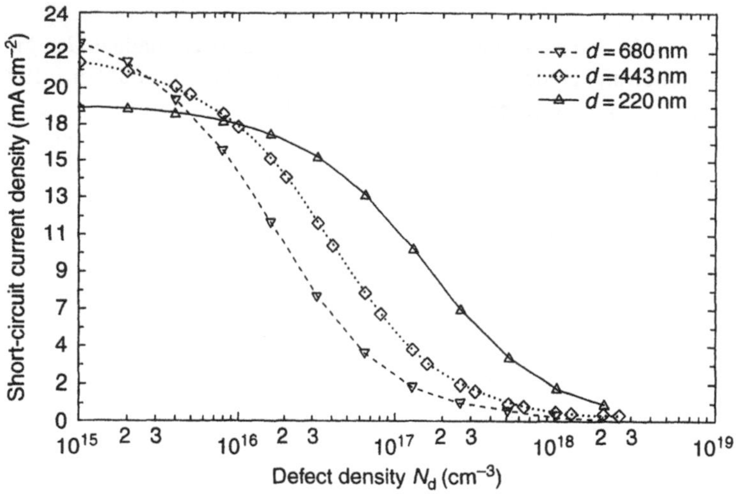

Figure 11.6 The short-circuit current density, J_{SC}, plotted as a function of the defect density, N_d, for a-Si : H solar cells with the thickness of their i-layer as 680, 443 and 220 nm (Štulík and Singh, 1998c).

Eqs (11.2) and (11.3), the short-circuit current depends on the generation and recombination rates and defect density in the i-layer of a solar cell. Although both absorbance and recombination are enhanced in the i-layer when its thickness is increased, for optimal performance one needs to maximize the absorbance and minimize recombination in the i-layer. This is used as the condition for finding the optimal thickness of the i-layer. Accordingly, using Eqs (11.1)–(11.3), one can determine the optimal thickness of the i-layer by optimizing the short-circuit current, and then the optimal thickness becomes a function of the defect density. The short-circuit current, J_{SC}, thus calculated has been plotted in Fig. 11.6 as a function of the defect density, N_d for three thicknesses, d, of the i-layer. It is clear from Fig. 11.6 that the cell of $d = 680$ nm has the largest J_{SC} initially at a defect density of 10^{15} cm^{-3} in comparison with the other two cells of thinner i-layers. However as the defect density increases to $\approx 10^{17}$ cm^{-3}, J_{SC} of the cell with $d = 680$ nm decreases to a much smaller value in comparison with that of cells with thinner i-layers. That means a solar cell with a thicker i-layer is affected more by the increase in the defect density than that with a thinner i-layer. The changes in the defect density can occur due to photo-induced effects (degradation), and the theory allows one to choose the optimal thickness such that there is minimal reduction in the short-circuit current due to increase in the defect density (Štulík and Singh, 1998c). Thus, although the problem of degradation in a-Si : H solar cells cannot possibly be eliminated, the theory provides a way to minimize its influence on the PV performance of a cell.

11.2 Thin-film transistor

Amorphous Si TFTs were first developed in the form of field effect transistors (FETs) (Powell, 1984; Spear and LeComber, 1984). A basic structure of a-Si : H FET is shown in Fig. 11.7. Two most important factors for TFTs are high ON/OFF current ratio and small gate voltage V_G which are achieved in TFTs fabricated from a-Si : H, because they have a high resistance in the OFF state and their properties are almost compatible with modern c-Si ICs (LeComber, 1989). A typical example of source–drain current and voltage characteristics is shown in Fig. 11.8 (Snell *et al.*, 1981). These characteristics are suitable as switching transistors for liquid crystal display (LCD).

Logic circuits using a-Si : H FETs were first proposed by Matsumura and Hayama (1980) and Snell *et al.* (1981). The most important factor limiting the use of a-Si : H FETs in logic circuits can be their frequency response (LeComber,

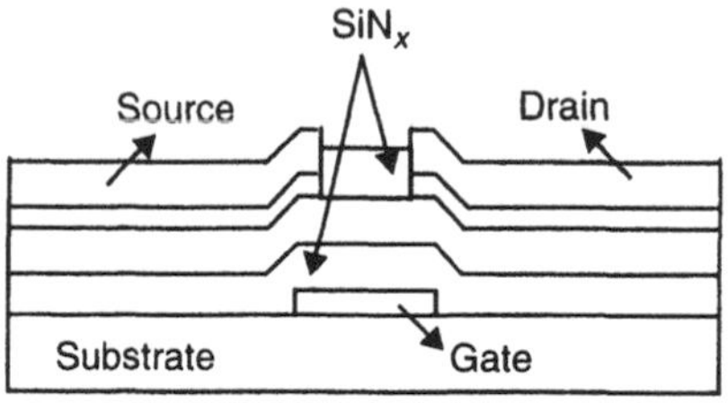

Figure 11.7 A schematic illustration of the basic a-Si : H TFT structure.

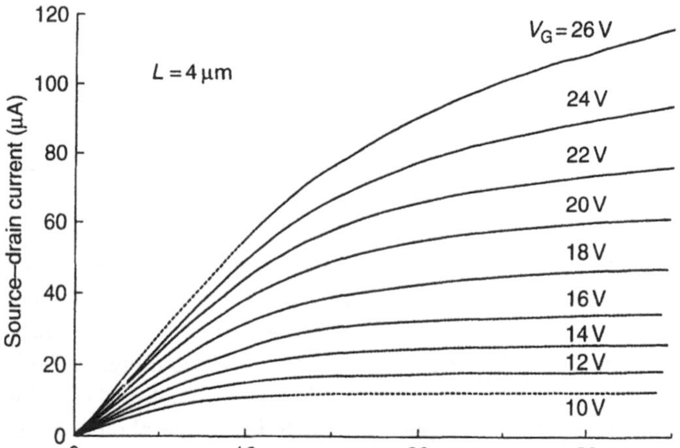

Figure 11.8 Typical source–drain current and voltage characteristics of an a-Si : H TFT (Snell *et al.*, 1981).

1989). This limit is determined by the time required for the ON current to charge circuit capacitances and/or by the carrier transit time between source and drain. The necessary condition to improve FETs is therefore the reduction in the separation between source and drain, which also increases the ON current of devices. There are several ways to achieve this, that is, modification of FET structures (Tuan, 1986). The combination of the switching a-Si TFTs with crystalline or polycrystalline driver integrated circuits is useful for producing TFT–LCD panels. Now a-Si : H TFTs are realized as flat display panels of size 55 × 65 cm (Sameshima, 1998).

As a-Si : H is a good photoconductor, off-state photo-leakage current for the TFT under back light illumination causes a reduction in ON/OFF current ratio (Choi *et al.*, 2000). This type of improvement is also a very important factor for practical device performances. Finally, a-Si : H TFTs are insensitive to radiation damage, for example, for γ-ray (LeComber, 1989). As the conventional crystalline devices can be easily destroyed under high-energy radiation, the use of a-Si : H TFTs is recommended in high-level radiation environments.

Once the large area TFT active matrix array (AMA) as used in flat panel displays becomes available, other sensors, for example, X-ray detectors, will also be developed. These will be introduced in the next section.

11.3 X-ray image detectors

After the availability of a-Si TFT flat panels with small pixel sizes at the component level, it was recently discovered that an X-ray photoconductor using amorphous Selenium (a-Se) can be used to convert X-ray images directly to a charge distribution stored on the pixels of the flat panel (Kasap and Rowlands, 1999). In such a system, X-ray photons are directly converted into charges that are collected and detected. This system has proven to be simple and has many advantages in digital radiography (Rowlands and Kasap, 1997; Rowlands *et al.*, 1997). This is different from an intermediate conversion to photons and then again to charges as it occurs in some other flat panel sensors (Antonuk *et al.*, 1992).

For fabricating an X-ray photoconductor, a-Se is coated onto an AMA and an electrode (A) is deposited on the top of a-Se to apply biasing potential and hence the electric field. An a-Se photoconductor along with its various components is shown in Fig. 11.9. The photogenerated carriers (electrons and holes) in a-Se by the absorption of X-ray photons travel along the electric field lines and are collected at their respective positive biased electrode A and the storage capacitor C_{ij}. A charge signal is thus produced that can be read by self-scanning. Each pixel electrode carries an amount of charge ΔQ_{ij} that is proportional to the incident X-ray radiation by virtue of the X-ray photoconductivity of the photoconductor over that pixel. So, when the ith row in a gate line is activated (see Fig. 11.9(a)), all TFTs in that row are turned "on" and N data lines (from $j = 1$ to N) then read the charges on the pixel electrode in the ith row. This data is then multiplexed into a serial form, digested and then fed into a computer for imaging. The scanning control then activates the next row $i + 1$ and so on until the whole matrix is read.

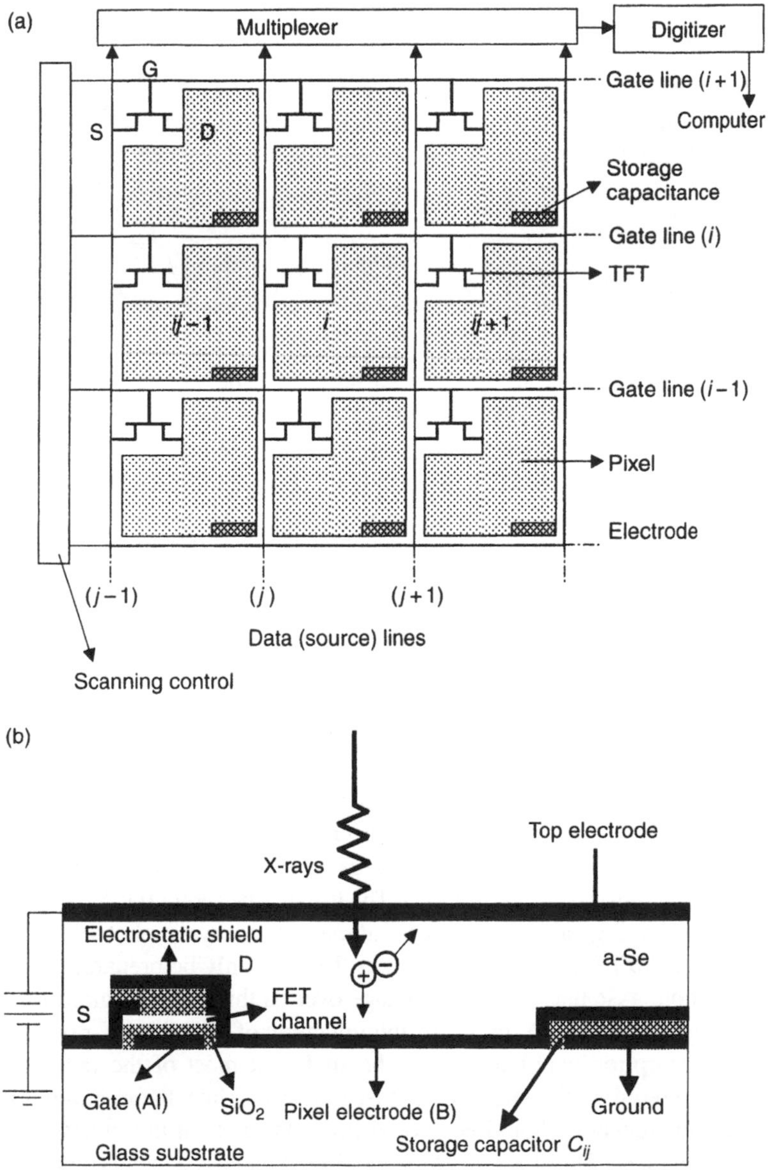

Figure 11.9 (a) Schematic illustration of a TFT AMA used in X-ray image sensor detectors with self-scanned read-out and (b) cross-section of a pixel shown in (a).

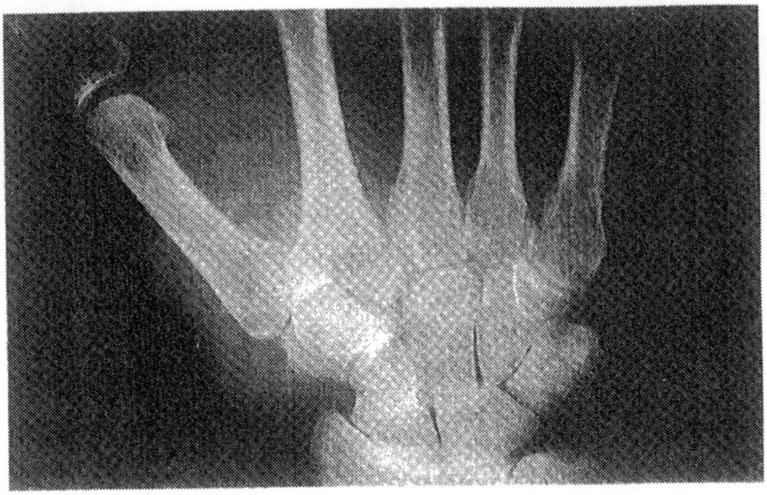

Figure 11.10 An X-ray image of a phantom hand obtained from a flat panel X-ray image detector using a-Se photoconductor. (Courtesy of Sterling Diagnostic Imaging, used with permission.)

This type of detector with an a-Se photoconductor produces excellent results. An X-ray image of a hand using such a detector, developed by Sterling Diagnostic, is shown in Fig. 11.10. The resolution is determined by the pixel size which at present is typically ~150 μm but is expected to become smaller in future detectors.

When X-ray photons interact with the atoms of a material, it leads to the emission of an energetic photon from the inner core, such as K-shell into the conduction band. This is the photoelectric effect shown by the sharp vertical edges in the absorption coefficient in Fig. 11.11. Here, the absorption coefficient of a-Se is shown along with that of some other photoconductors. The use of a-Se as the photoconducting layer has many advantages over other photoconductors. The main advantage is that it can be prepared in large areas uniformly without damaging the underlying AMA substrate (Kasap, 1991). The use of other potential X-ray photoconductors is technologically limited, for example, single crystal photoconductors of GaAs, CdTe, CdZnTe, ThBr and others, as they can only be prepared in small areas. In addition, a-Se has another advantage over its thickness required for use in mammography and chest X-rays. The minimization of the initial X-ray dosage requires the absorption coefficient, α, to be such that most of the radiation is absorbed with a required thickness, L, that is, $1/\alpha < L$. Thus, the required thickness of the photoreceptor should be several times larger than the inverse of the absorption coefficient. This means that the required thickness depends on the photon energy and hence on the particular imaging application and the location of K and L shells. For mammography at 20 keV, the required thickness is about 100 μm,

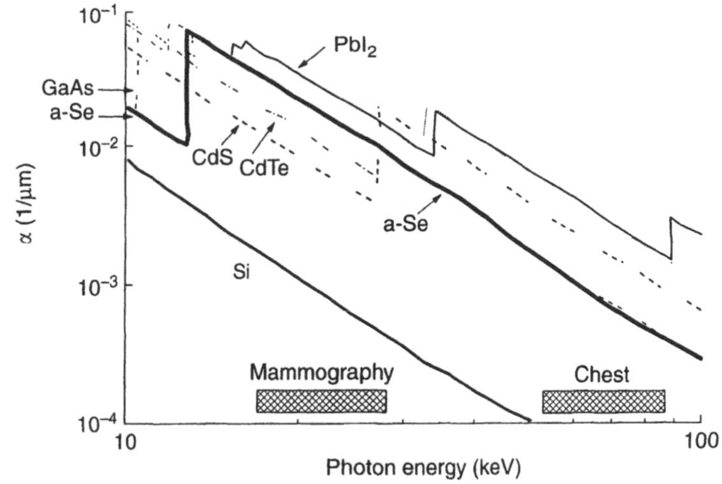

Figure 11.11 Absorption coefficient is plotted as a function of the photon energy for various photoconductors (Kasap and Rowlands, 1999).

and for chest radiology, it is about 2,000 μm with the photon energy of 60 keV for a-Se. The corresponding thickness of a CdTe detector is 160 and 180 μm, respectively. However, it is easier to deposit thicker a-Se layers than thinner single crystals of CdTe.

As a scintillator for converting X-ray quanta into visible photons, cesium iodide (CsI) doped with Tl has also been used on AMA of a-Si : H TFTs (Hoheisei *et al.*, 1998). Any candidate for X-ray photoconductors should also have an excellent X-ray photoconductivity as found in a-Se. This means that the energy required to create a single free e–h pair must be as low as possible.

11.4 Optical memories

With the information technology entering our daily lives, there is a need to develop memory disks to store large amounts of diverse information. For this we need computer memory disks (portable) with large storage capacities and good reliability. At the same time they must be cost effective. The magneto-optical (MO) and phase change optical memory disks, for example, digital versatile disks (DVD) fall in this category. The MO uses the magneto-optical recording system, whereas the phase change memories adopt the phase change recording system. Phase change memories are based on thin films typically incorporating alloys of chalcogenide materials. The main principle of the phase change optical memories is the change in the refractive index of a material on phase change from an amorphous to crystalline state. The use of amorphous chalcogenides as viable materials for optical phase

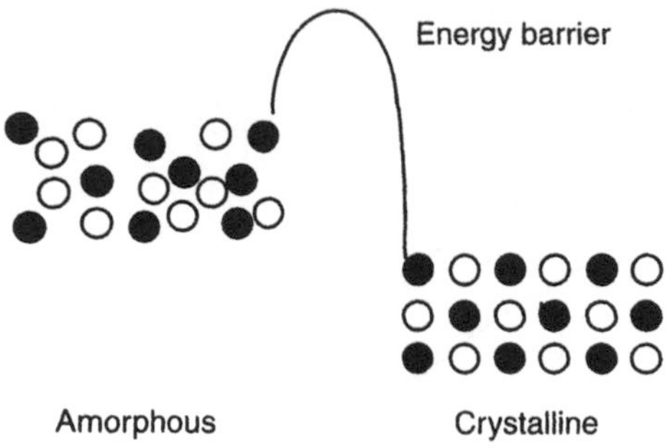

Figure 11.12 A schematic illustration of phase change process for optical memories. Energy barrier must be overcome in going from one phase to another.

change recording was proposed nearly three decades ago (US Patent 3,271,591, 1966; Ovshinsky, 1968; Cohen *et al.*, 1969; Ovshinsky and Fritzsche, 1973).

The basic concept in phase change memories starts with the use of a material that can exist in two separate stable structural states, that is, amorphous and crystalline. An energy barrier must be overcome before the structural state can be changed, thereby providing the stability of the two structures. This phase change process is shown schematically in Fig. 11.12. The physical properties, for example, electrical and optical, of the two structural states are different. Energy to overcome the barrier can be supplied to the material in various ways, including exposure to intense laser beams and application of a current pulse. Laser exposure (optical excitations) is used for recording (overcoming the barrier) and erasing (returning to the initial state), in the case of an optical memory. If the energy applied exceeds the threshold value, the material will be excited to a high atomic mobility state, in which it becomes possible to rapidly rearrange bond lengths and angles by slight movement of the individual atoms.

In the phase change memory technology, the information is stored using a structural phase change in certain thin film alloys containing typically one or more elements from column VI of the periodic table, such as selenium or tellurium. These alloys are stable in both their crystalline and amorphous states. These two different structural states have different optical and electrical properties and regions of the alloy material can be switched back and forth between the two states by the application of pulses (electrical or optical) with sufficient energy to overcome the energy barrier separating the two states.

In a material such as germanium–antimony–tellurium (Ge–Sb–Te), which is commercially used for the optical storage (US Patent 3,530,441, issued Sept. 22, 1970), compositions can be selected in which these minute changes in bonding position of the atoms can cause profound changes in the optical properties of the material, including its optical absorptivity and reflectivity. The importance of the composition lies in the selection of a material composition that can form a crystalline structure without phase segregation. Selection of an appropriate

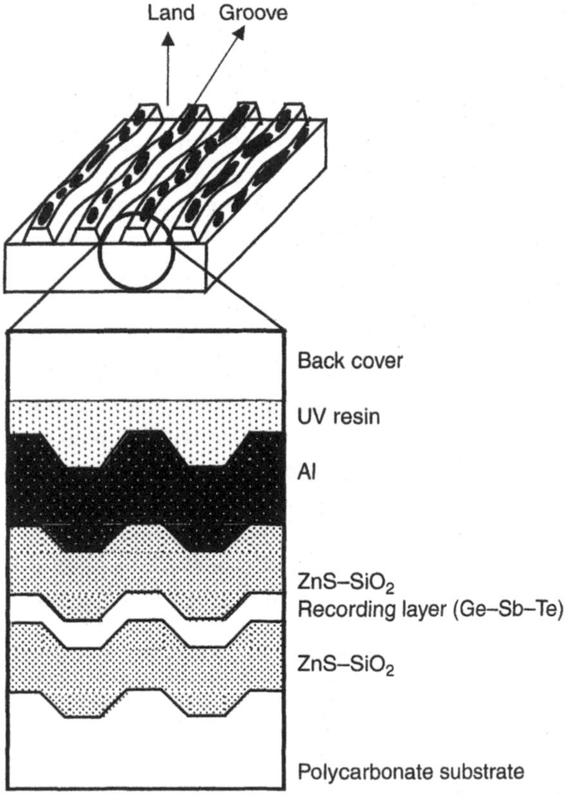

Figure 11.13 Schematic illustration of the structure of a phase change rewritable optical disk (Nagata *et al.*, 1997).

composition and inducing a high mobility state during laser exposure are the two underlying principles in the direct overwrite phase change erasable optical recording media. Ge–Sb–Te ternary alloys are capable of rapid transition between the two states.

The structure of phase change rewritable optical memory disks is shown schematically in Fig. 11.13. Lands and grooves (see Fig. 11.13), embossed in a plastic disk substrate, are over coated with a four-layer thin film stack comprised of a dielectric film (typically $ZnS–SiO_2$), the phase change recording material (Ge–Sb–Te), another dielectric film (typically $ZnS–SiO_2$) and a layer of a metal (Al) to serve as a back reflector. Finally a protective plastic overcoat layer is applied. As the disk rotates in the disk drive, it passes under a semiconductor laser that is used either to write and erase data spots using high laser power or to read data by measuring the reflectivity of previously recorded spots at low laser power. With the phase change recording system, data reading and writing are executed by changing the alloy from its crystalline phase to amorphous phase. By irradiating a strong laser on the writing side coated with a chalcogenide alloy, which is initially in the crystalline phase, the material is heated to a high temperature and changes to an amorphous phase for recording. For erasing, the material is heated just above the glass transition temperature, T_g, to change it back to the crystalline state. The data spots (shown as black objects in the figure) are regions of material that have been made amorphous and which, therefore, have lower reflectivity than the surrounding areas of crystalline material. Current CD-RW phase change rewritable disks use an infrared laser for this purpose, and the data spots area of about 1 μm in size. The new DVD-RAM rewritable disks use a shorter wavelength red laser to write smaller spots to achieve much higher data storage density. The phase change recording memories are currently finding various uses, from storing images to full length movies. The demands and desires for further improvements and uses are increasing with the increase in capacity and reliability in data transfer. The currently used memory devices use a red laser for writing, reading and erasing data. The capacity of the media can be increased by having more shorter wavelength lasers. Also, one of the drawbacks of the phase change memories is the "jitter." During heating of the material to write data, the temperature of the spot is raised so as to melt it. At the same time the disc is moving at a great speed. This results in the melted material to overflow its own position and the next spot

Table 11.1 Current status of the storage, compatibility and capacity of various commercially produced disks

Disk type	*Storage type*	*Read compatibility*	*Capacity*
CD	Read-only (audio)		650 MB (74 min)
CD-ROM	Read-only	CD, Photo CD, CD-R, CD-RW (after 1997)	650 MB
DVD-ROM	Read-only	CD, CD-ROM, DVD, Photo CD, CD-R, PD, CD-RW, DVD-RAM	4.7 GB
DVD	Read-only (video)		4.7 GB (2 h)
Photo CD	Write-once (images)	(No separate drive)	650 MB (100 images)
CD-R	Write-once	CD, CD-ROM, Photo CD	650 MB
CD-RW	Phase change rewritable	CD, CD-ROM, CD-R, Photo CD	650 MB
DVD-RAM	Phase change rewritable	CD, CD-ROM, DVD, Photo CD, CD-R, PD, CD-RW, DVD-ROM	2.7 GB
DVD + RW	Phase change rewritable	CD, CD-ROM, DVD, Photo CD, CD-R, PD, CD-RW, DVD-ROM	3 GB

can be written after the overflow. This gives rise to the wastage of space. The jitter has been reduced considerably over the last few years, but still this problem needs further improvement.

The current status of the storage, compatibility with other readable drives and capacity of various commercially produced and used disks is shown in Table 11.1.

References

Antonuk, L.E., Boudry, J., Wang, W., McShan, D., Morton, E.J., Yorkston, J. and Street, R.A. (1992). *Med. Phys.* **19**, 1455.

Ashida, Y. (1994). *Solar Energy Mater. Solar Cells* **34**, 291.

Bennett, A. and Olsen, L.C. (1978). *Proceedings of the 13th IEEE Photovoltaics Specialists Conference*, p. 868.

Carlson, D.E. and Catalano, A. (1989). *Optoelectronics* **4**, 185.

Carlson, D.E. and Wronski, C.R. (1976). *Appl. Phys. Lett.* **28**, 671.

Carlson, D.E., *et al.* (1996). *Proceedings of the 25th IEEE Photovoltaic Specialist Conference*, p. 1023.

Choi, Y.J., Lim, B.C., Woo, I.K., Ryu, J.I. and Jang, J. (2000). *J. Non-Cryst. Solids* **269–299**, 1299.

Cohen, M.H., Fritzsche, M. and Ovshinsky, S.R. (1969). *Phys. Rev. Lett.* **22**, 1065.

Dawson, R.M., *et al.* (1992). *Proceedings of the 11th European Solar Energy Conference*, p. 680.

Deckman, H., Wronski, C.R. and Yablonovitch, E. (1984). *Proceedings of the 17th IEEE Photovoltaic Specialist Conference*, p. 955.

Elliot, S.R. (1990). *Physics of Amorphous Materials*, 2nd edn, Longman Scientific & Technical, London.

Fahrenbruch, A.L. and Bube, R.H. (1983). *Fundamentals of Solar Cells*. Academic Press, New York and http://www.pvpower.com/pvhistory.html

Fan, J.C.C. and Palm, B.J. (1983). *Solar Cells* **10**, 81.

Guha, S., *et al.* (1999). *AIP Conf. Proc.* **462**, 88.

Hanak, J.J. and Korsun, V. (1982). *Proceedings of the 16th IEEE Photovoltaics Specialists Conference*, p. 1381.

Hoheisei, M., Arques, M., Chabbal, J., Chaussat, C., Ducourant, T., Hahm, G., Horbascheck, H., Shulz, R. and Spahn, M. (1998). *J. Non-Cryst. Solids* **227–230**, 1300. i.

Hubin, J. and Shah, A.V. (1995). *Philos. Mag. B* **72**, 589.

Kasap, S.O. (1991). *The Handbook of Imaging Materials*. Marcel Dekker, New York, Chapter 8.

Kasap, S.O. and Rowlands, J.A. (1999). In: Marshall, J.M., Kirov, N., Vabrek, A. and Maud, J.M. (eds), *Thin Film Materials and Devices-Developments in Science and Technology*. World Scientific, Singapore, p. 13.

LeComber, P.G. (1989). *J. Non-Cryst. Solids* **115**, 1.

Lee, Y., *et al.* (1998). *Proceedings of the 2nd World Conference On Photovoltaic Solar Energy Conversion*, p. 940.

McLeod, H.A. (1986). *Thin Film Optical Filters*, 2nd edn, Adam and Hilger, Bristol.

Matsumura, M. and Hayama, H. (1980). *Proc. IEEE* **68**, 1349.

Nagata, K., Furukawa, S., Nishiuchi, K., Yamada, N. and Akahira, N. (1997). *Extended Abstracts of Symposium on High Density Phase Change Optical Memories in Multi-media Era*. Japan Society of Applied Physics, p. 11.

Ovshinsky, S.R. (1968). *Phys. Rev. Lett.* **21**, 1450.

Ovshinsky, S.R. and Fritzsche, H. (1973). *IEEE Trans.* ED, **NED-20**, 91.

Powell, M.J. (1984). *MRS Symp. Proc.* **33**, 259.

Rowlands, J.A. and Kasap, S.O. (1997). *Phys. Today* **50**, 24.

Rowlands, J.A. and Zhao, W. and Blevis, I. (1997). *Radio Graphics* **17**, 753.

Sameshima, T. (1998). *J. Non-Cryst. Solids* **227–230**, 1196.

Snell, A.J., Mackenzie, K.D., Spear, W.E., LeComber, P.G. and Hughes, A.J. (1981). *Appl. Phys. A* **26**, 83.

Spear, W.E. and LeComber, P.G. (1984). *Semiconductors and Semimetals* **21D**, 89.

Staebler, D.L. and Wronski, C.R. (1977). *Appl. Phys. Lett.* **31**, 291.

Street, R.A. (1991). *Hydrogenated Amorphous Silicon*. Cambridge University Press, Cambridge.

Štulík, P. and Singh, J. (1996). *Solar Energy Mater. Solar Cells* **40**, 239.

Štulík, P. and Singh, J. (1997). *Solar Energy Mater. Solar Cells* **46**, 271.

Štulík, P. and Singh, J. (1998a). *J. Non-Cryst. Solids* **242**, 115.

Štulík, P. and Singh, J. (1998b). *J. Non-Cryst. Solids* **231**, 120.

Štulík, P. and Singh, J. (1998c). *J. Non-Cryst. Solids* **226**, 299.

Tanaka, *et al.* (1992). *Jpn. J. Appl. Phys.* **31**, 3518.

Tsuda, S., *et al.* (1993). *J. Non-Cryst. Solids*, **164–166**, 679.

Tuan, T.C. (1986). *MRS Symp. Proc.* **70**, 651.

Wronski, C.R., Lu, Z., Jiao, L. and Lee, Y. (1997). *Proceedings of the 13th IEEE Photovoltaics Specialists Conference*, p. 587.

Yablovovitch, E. and Cody, G.D. (1982). *IEEE Trans. Electron Dev.* **29**, 300.

Yang, J. and Guha, S. (1992). *Appl. Phys. Lett.* **61**, 2917.

Yang, J., *et al.* (1994). *Proceedings of the 24th IEEE Photovoltaics Specialists Conference*, p. 380.

Yoshida, Y., Fujikake, S., Kato, S., Tanda, M., Tabuchi, K., Takano, A., Ichikawa, Y. and Sakai, H. (1996). *Technical Digest of 9th International Photovoltaic Science and Engineering Conference*, Miyazaki, Japan, pp. 291–294.

Zhu, F. and Singh, J. (1993a). *J. Appl. Phys.* **73**, 4709.

Zhu, F. and Singh, J. (1993b). *Solar Energy Mater. Solar Cells* **31**, 119.

Zhu, F. and Singh, J. (1993c). *J. Non-Cryst. Solids* **152**, 75.

Index

9 780415 287708

Milton Keynes UK
Ingram Content Group UK Ltd.
UKHW031533071024
449327UK00005B/88

9 780415 287708